NO
EARTHLY
POLE

'Not here! the white North hath thy bones, and thou
Heroic Sailor Soul!
Art passing on thy happier voyage now
Towards no earthly pole.'

Epitaph to Sir John Franklin from his nephew by marriage,
Alfred, Lord Tennyson

This book is dedicated to the memory of Dan Weinstein (1934–2017).
A true, loyal friend whose tolerance, humanity, compassion, altruism
and understanding radiated to all who knew him. His prestige was an
honour to his family and to his country.
'Just do it!'

NO EARTHLY POLE

THE SEARCH FOR THE TRUTH ABOUT THE FRANKLIN EXPEDITION 1845

Ernest C. Coleman

AMBERLEY

First published 2020

Amberley Publishing
The Hill, Stroud
Gloucestershire, GL5 4EP

www.amberley-books.com

British Library Cataloguing in Publication Data.
A catalogue record for this book is available from the British Library.

ISBN 978 1 3981 0211 8 (hardback)
ISBN 978 1 3981 0212 5 (ebook)

Typesetting by Aura Technology and Software Services, India.
Printed in the UK.

Contents

Acknowledgements

One of the chief joys of embarking on an enterprise such as I undertook between 1990 and 1995 proved to be the great number of people I met who were keen to lend a hand, or give support, when needed. There are far too many to name them all, but I would particularly like to thank the following, on many of whom depended the entire impetus of the project.

The Royal Navy, especially those senior officers who, masking their doubts, allowed me to proceed. To Rear Admiral J. A. L. Myres FRICS CBE, Hydrographer of the Navy for his support and encouragement, also to Chief Yeoman Brian Wing who looked after things in my absence to such a degree that hardly anyone knew that I had been away.

To John and Inga Howse for their many kindnesses and to other members of the 'Calgary Light Horse' including Bob Shaw, David Parker, Rick Valentini, Lt Cdr Bill Larkin RN, Dr Sandy Morrison, and Ron Staughton. To Lyall and Margaret Hawkins and the staff of the Arctic Islands Lodge, Cambridge Bay and to the wonderfully hospitable Sarah Allen and family both for accommodation and a place of retreat.

To Willy and Mrs Laserich and the staff of Adlair Aviation Ltd who frequently came to my rescue and, with special thanks to two of Adlair's pilots, Mike Carr-Harris and D.J. 'Smitty' Smith. To Bob O'Connor of Aero Arctic for the use of one of his helicopters and the services of 'Buck' Rogers to fly it.

To Lt I. D. ('Jim') Crack RN and Cameron Treleaven for being prepared to place themselves in a remote corner of the world with only me for company.

To Lincolnshire County Council, the Trustees of the Medlock Charitable Trust, and West Lindsey District Council for their substantial help with funding, and to Bateman's Brewery Ltd for their financial 'safety net'.

To the Royal Canadian Mounted Police, especially Superintendent Brian Watts, Chief Inspector Gene McClean, Staff-Sergeant (Pilot) Bob Martin, and Constable Terry, and Mrs Joan, Zeniuk.

Acknowledgements

To the Canadian Armed Forces, in particular Major Chris Grant and Lieutenant(N) Colin Perriman.

To the Directors, Curators, and Staff of the Glenbow Museum, Calgary. Barbara Tomlinson, the Polar Curator at the National Maritime Museum. Robert Headland, William Mills, the splendid and patient Shirley Sawtell and other Staff of the Scott Polar Research Institute, Cambridge. And to Dr Charles Arnold, the esteemed Margaret Bertulli and the Staff of the Prince of Wales Northern Heritage Centre, Yellowknife.

To Dr Peter Wadhams and to Pia Casarini Wadhams for their support, experience and freely shared knowledge. To the indomitable Dan and Jonolyn Weinstein for their kindness, generosity, shelter, transport, leadership, and good old American 'get up and go'. And to Max Wenden for sheer, rock solid, dependability.

To Anne Paterson for my personal pennant and to Berghaus Ltd for the superb outdoor clothing. To The Photolab, Gainsborough, for their help in rescuing old photo slides. For their help and encouragement, Mrs E. Diamond, and Ralph Lloyd-Jones. For their support and inspiration Rawden Crozier, N. C. McClintock CBE, and other members of the McClintock family.

To the late A. G. E. Jones for allowing me to tap into his wealth of polar knowledge and for introducing me to the fact that there is more than one way of looking at history. To Max Kameemalik for taking pity on a stray 'Kabloona' and guiding him over the ice.

And finally, to my wife for taking the absences, disappearances, and adverse effects on the bank balance without too much comment.

Foreword

It is important that the reader understands that this work is not, and was never intended to be, an academic dissertation or thesis. It is not offered for 'peer review', as I am totally unaware of anyone who would consider me to be a 'peer' or would even want to. My reason for writing is to speak out in opposition to the clustering together of some academics and experts who have closed their minds to the wider reasoning and experiences needed to understand the travails of mid-nineteenth century maritime exploration in the Arctic. It is intended to be a quiet 'Enough is enough!' that exposes the condemnations of those who have learned their history from textbooks written by their fellow scholars who, in turn, were tutored from yet more books – and so on, ad infinitum. The result of such a system is the regurgitating of the same unwarranted contempt they have just ingested, scorn that disfigures the subject. The academics involved have made the disparagement of the Royal Navy's achievement of the North-West Passage obligatory so that it has become an article of faith in which no dispute is permitted and, if such a challenge may surface, to discount it out of hand.

The 'subject' in this case is the endeavour, determination, labours, and pains of men – plain, ordinary, men – to benefit the rest of humanity by the establishment of a water passage from the Atlantic to the Pacific. That they did not succeed was not the fault of poor leadership, incompetence, foolhardiness, vanity, or impracticality. Their misfortune was nothing more or less than the unpredictable conditions of environmental adversity.

The first part of this work is my account of the visits I made to the Canadian Arctic in search of surviving evidence from Franklin's 1845 expedition, and a few subsequent events. It was written in the first place as a record for the Royal Geographical Society and the Scott Polar Research Institute in the hope that, should anyone choose to continue the search, they might find it of interest. I hope that it will provide some form of assessment of my ability to enter into a discussion of such matters. The later part is an examination of the blizzard of misinformation that carried with it the disparagement and tarnishing of a great enterprise and the sullying of the memory of those who died whilst striving to complete it.

E. C. Coleman FRGS

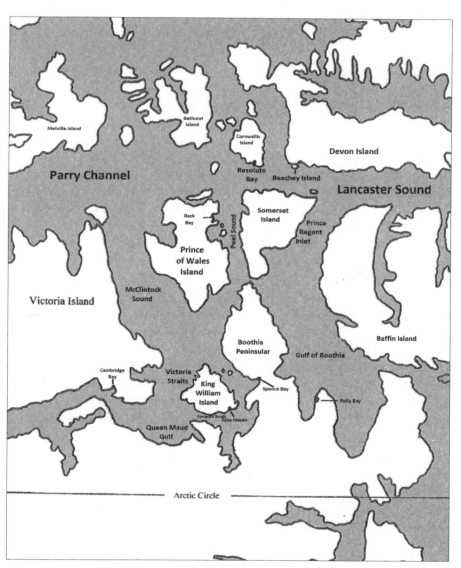

Map of the Canadian central Arctic.

1

An Urge to Head North

A tap on the shoulder made me turn away from the aircraft window. I was handed a note from the pilot, Staff Sergeant Bob Martin of the Royal Canadian Mounted Police. Unfolding the piece of paper I read the words 'We are just crossing the Arctic Circle.'

The last time I had been so far north had been as a seventeen-year-old 'Junior' onboard an aircraft carrier in the Davis Straits. Now, more than a quarter of a century later, I was back as a Lieutenant, Royal Navy, trying to unravel a 140-year-old mystery.

I probably first heard the name Franklin at school where, along with Flinders, Banks, and Bass (all Lincolnshire explorers), it had been the name given to of one of the school's 'houses'. Decades later, I had a book published setting out the many centuries of the Royal Navy's involvement in the area. There, amongst the distinguished admirals, explorers, Victoria Cross winners, submariners and naval airmen, glinted the story of Captain Sir John Franklin who had set off in 1845 with 129 men and the ships HMS *Erebus* and HMS *Terror* in an attempt to find the fabled North-West Passage. After bidding farewell to a whaling ship in the north of Baffin's Bay, neither men nor ships were seen by their countrymen again. They had simply vanished into the fog and ice of the Arctic.

After eleven years of searching, a note was found in a cairn on the northwest corner of King William Island. It held the only evidence of the missing expedition's fate. After circumnavigating Cornwallis Island, Franklin's ships had wintered at Beechey Island 1845-46. They had then – almost certainly via Peel Sound – reached the northwest corner of King William Island where the ships were 'beset' in the ice that pours down the McClintock Channel from the Polar Sea. Two winters were then endured as the trapped ships slowly drifted south through the northern end of Victoria Strait. The note revealed that Sir John Franklin had died on 11 June 1847, and that Captain F. R. M. Crozier had assumed command.

With the expedition's supplies running low and almost certainly with scurvy breaking out amongst the ship's companies, Crozier decided to 'desert' the ships and lead the 104 survivors south to safety down the Back River, a waterway whose estuary breached the North American coastline three hundred miles to the south. From there they were faced with a journey of a thousand miles over the Canadian 'Barren-lands' to the nearest Hudson's Bay Company post. Fate, however, decreed that none were to survive, and there is no evidence that any of the men reached beyond a desolate beach on the lip of the North American continent later to be named 'Starvation Cove'. Even the ships had disappeared without trace.

It was a desperately sad story that had burned itself into the Victorian Age. The Franklin searcher, Captain Leopold McClintock, considered the Franklin Expedition as 'Perhaps the noblest episode in England's naval history', whilst a popular series of cigarette cards issued by John Player in 1915 recorded that 'Amongst the illustrious British explorers of the nineteenth century, the central figure will always be that of the gallant Sir John Franklin.' The legend was only to fade against the bright, white tragedy of Captain Scott and his companions at the other end of the Earth.

The succeeding years shed little light on the mystery of the missing expedition. A large amount of stores had been found near Victory Point on King William Island where the ship's companies had landed prior to the departure southwards. Bones and a few graves marked their journey and a sledge-mounted boat indicated their means of passage over the land and sea-ice. Beyond that, little was known, and many questions remained unanswered. Why, for example, had Crozier chosen to head south when, to the north, there was the strong possibility of supplies at Fury Beach in Prince Regent Inlet and, only a little further to the north, lay Barrow Strait and Lancaster Sound – the obvious entry into the Arctic for any search expeditions? Both Fury Beach and Lancaster Sound were considerably less than half the distance to the nearest Hudson's Bay Company post. Why had no further messages been found? If Franklin had found Peel Sound free of ice, it was reasonable to assume that he would not have wanted to stop to build marker cairns containing messages along his route but no message had been found on Beechey Island and, apart from the note telling of Franklin's death, no other message had been found on King William Island.

For me, however, the key mystery lay in the burial place of Franklin himself. Popular mythology claimed that he had been buried in the ice of Victoria Strait but, of course, there was not the slightest hint of proof. In fact, there was strong evidence that pointed away from a burial in the ice. There were few recorded examples of such interments prior to the Franklin Expedition, and one had even resulted in the grisly sight of a corpse floating alongside the ship when the ice melted. The McClintock Expedition of 1858 had been forced to entomb one of their number in the ice due to their distance from land, and Commander Albert Markham of the 1875-6 Nares Expedition was forced to resort to the ice to bury one of his sledge team for the same reason. When Markham wrote his biography of Franklin, he accorded his subject the same chilly fate.

I, on the other hand, felt that Franklin's body would have been taken the short distance to the shore of King William Island and buried there. It was quite possible that the grave – if it existed – had been missed. The number of people who had visited the area had been very small indeed, even the native Inuit were said to consider it to be a poor hunting ground and generally kept well away. The last Royal Naval officer known to have visited the site was Captain Francis Leopold McClintock in 1859. There had been a number of small American and Canadian expeditions and one large Canadian Army expedition had swept over the area, but no sign of a formal grave had been found. I believed, however, that the chance still remained of Franklin's grave being discovered.

But how does a bulkily built 47-year-old, with no experience whatsoever of surviving in the Arctic, get such an expedition off the ground? I could at least point out that I hoped to go in the short-lived Arctic summer. There would be no point in going at all if a thick blanket of snow obscured the very signs I hoped would be left to mark a burial site. I tried the Royal Marines 'Arctic and Mounted Warfare Cadre' but was met by a mixture of hilarity and outrage culminating in the advice that I should drop the idea and get back to lecturing school children on the benefits of life in the Service. I then tried the Canadian High Commission where I found a totally different attitude. Both the military and the Royal Canadian Mounted Police (RCMP) on the staff of the Commission were encouraging in their response to my enquiry. They were somewhat curious regarding the reason behind my attempt to get into the Arctic, but did nothing to impede my progress and, in the end, were to prove a vital prop to my ambitions.

By now the time had come to acquaint my superiors with my ideas. My immediate boss, a Commander, told me he was happy with the idea if the Captain (the Director of Naval Recruiting) was. The Captain, who had recently commanded the ice-patrol ship, HMS *Endurance*, said he was happy if the Commander was happy. So, in the end, we were all happy.

Whilst all this was going on I had been sounding out a number of my colleagues around the country to see if any of them were interested in going with me. So far, I had a list of volunteers that numbered just one.

When the time for decision came, after thinking long and hard I chose the single name on the list, Lieutenant I. D. ('Jim') Crack. A good choice as it turned out for, with his submariner and engineering background, his practical approach to problems proved to be of constant value to the expedition.

After a hard struggle to raise funds for the enterprise the 'Franklin Search Expedition', weighed down with tents, supplies, and Arctic kit, took off in the middle of June 1990 on a routine Royal Air Force flight to Calgary, Canada.

On arrival we were met – much to our surprise – by a Canadian Army corporal driving a van and offering us a lift to our accommodation at a nearby Army base. A Canadian Army Major, on the staff of the London High Commission, had been pulling the right strings on our behalf. That night in the officer's mess, fortune continued to smile upon us when we were adopted by a group of expatriates from Britain and Australia, on whom I bestowed the title 'The Calgary Light Horse'. As might be expected, this was all too good to last. After lunch the following day, a Canadian Army officer approached me with a written message. It read 'Expedition cancelled' followed by an instruction to telephone the British High Commission in Ottawa. Totally baffled by this turn of events, I called the number to find myself talking to a British Army officer (senior in rank to me) who seemed deeply annoyed that an administrative document that I had completed months before had not arrived on his desk. He then stunned me by demanding that Jim and I return home by the next flight. My only immediate defence was to tell him that the expedition had been covered by the British national press (a short note in the *Daily Telegraph*) and it would look bad if he was to cause its cancellation at this stage. This information was met by a heavy groan and a demand that we remain where we were until we heard from him again, but, whatever happened, we need not count on going up north – that was definitely 'off'.

When our friends from the Calgary Light Horse found me fuming into my beer that night they were outraged at the turn of events and offered to get the Canadian press in on the act. I could ill-afford the risks that path offered and declined, only to find that word of my predicament had reached the ear of a Canadian Naval Captain. Called to the telephone on the bar, I was treated to a full broadside of contempt for people who did not realise that 'Canada was a free country and nobody, but nobody, could stop you from going where you want, when you want – and what are you going to do about it?' I decided that the time had come for a bout of selective amnesia and, instead of waiting at Calgary for the next military transport aircraft to leave for Yellowknife, I would book Jim and me on the next civilian flight northwards.

We arrived unannounced in the capital of the Northern Territories but managed to obtain accommodation in an Armed Forces building known as 'Plywood Mansion'. I had hoped to be able to hitch a ride to King William Island in the Canadian Army twin-engined Otter aircraft I knew to be based at Yellowknife, but I found its planned programme was to take it nowhere near where I wanted to go. At this I decided to take up the invitation of an RCMP Inspector on the staff at the London Canadian High Commission to call at the local RCMP Headquarters. It was not long before I found myself explaining my hopes to Staff Sergeant (Pilot) Bob Martin who, having gained permission from the Superintendent, readily agreed to give us a lift up to the Inuit settlements. He turned out to be a quite spectacular find who was to play a vital role in much of my search for clues to the Franklin mystery.

The following morning Bob picked us up and drove us and our kit out to the airport where we joined a small group of official passengers. Our pilot, having loaded the aircraft (again a Twin Otter with balloon-like 'Tundra' tyres and with the ribbon of the RCMP Long Service Medal on its tail plane) towed the machine out of its hanger. Safe in the hands of 'Cap'n Bob's Airlines' we flew north along the route of the Yellowknife River before, in response to a request from me, we veered westward just enough to catch a glimpse of the Coppermine River – the path chosen by Franklin on his first attempt to find a North-West Passage.

Beneath us the 'Barren Lands' unrolled in a grey-green emptiness sprinkled with hundreds, possibly thousands, of lakes glinting beneath a slate-coloured sky. Then, with the last tree many miles to our rear, stretches of snow began to line the northern slopes of low hills and ice scattered the light from the lakes.

About an hour after we had crossed the Arctic Circle, Bob banked the aircraft to reveal the rare spectacle of the Wilberforce Falls as they poured the waters of the Hood River (named after Midshipman Hood, murdered whilst on one of Franklin's overland expeditions) down a two hundred-foot deep chasm. The river had been Franklin's return route after exploring part of the rim of North America. His desperate journey down the river had earned him the title of the 'man who ate his boots'.

After a flight over the frozen Bathurst Inlet and Queen Maude Gulf, we landed at the largest of the Inuit settlements, Cambridge Bay, on the south-east corner of Victoria Island and 540 miles north of Yellowknife. Bob refuelled the aircraft within sight of the iced-in bay that had sheltered HMS *Enterprise* under the command of Captain Collinson during his 1850 search for Franklin.

We then headed east over Victoria Strait and towards my first sight of King William Island. When over the Royal Geographical Society Islands,

Bob turned the aircraft northeast in order to give me (by now sitting alongside him in the co-pilot's seat) a chance to see the coastline from the Graham Gore peninsula northwards.

There it was below me. The very path taken by Captain Crozier and the two ship's companies as they tried to reach the far-off safety of the nearest Hudson's Bay Trading post; the southeast corner of Erebus Bay where Lieutenant Hobson of the McClintock Expedition found a sledge bearing a ship's boat; Point Le Vesconte, Point Franklin and the great slash of Collinson's Inlet. On to Back Bay where Hobson had found the written note and where an American 7th Cavalry officer, Lieutenant Frederick Schwatka, found the bones of Lieutenant Irving in 1879. Then over Victory Point – James Ross's 'Furthest' of 1830 – and on northwards to Cape Felix, the northern tip of King William Island. It was a few miles northeast from Cape Felix that the *Erebus* and *Terror* had found themselves locked in the ice of the Victoria Strait during the winter of 1846.

From Cape Felix, Bob swung the aircraft southwards and flew over the waterlogged interior of the island. Meltwater lakes and ponds of every shape and size, each linked by rivers and streams, flashed in the harsh northern sunlight. The grey-brown streaks of 'eskers' (raised, inland beaches) scored their way across the landscape, their surfaces frequently cracked by frosts into strange polygonal shapes.

As we came into land at Gjoa Haven, I could clearly see the small bay that had given sanctuary to Roald Amundsen during the winters of 1904/5. He had called it 'the greatest little harbour in the world' as his tiny vessel, *Gjoa*, sheltered during the first successful achievement of a North-West Passage. The *Gjoa* (originally pronounced 'Yer', but now modified to 'Joe') had given its name to the modest Inuit settlement that now clustered around the eastern side of the bay. Again, we did not stay long before we were off once more, this time eastwards to Pelly Bay, a tiny Inuit hamlet on the western shores of the Gulf of Boothia.

When we landed among our destination's black, snow-capped hills it seemed that the entire population (approximately 300) had come out to meet us. They smilingly showed us around their wind-blasted settlement, which for many years had been run by a priest who had been the only white resident. His stone-built church remained, but now only as a museum. For me, a special interest in the site arose from the fact that in 1854, Doctor John Rae, on an overland expedition on behalf of the Hudson's Bay Company, came across Inuit at Pelly Bay and later, Repulse Bay, who had in their possession silver tableware and other items that could only have come from the Franklin Expedition. They even had the star of the Guelphic Order that Franklin had worn around his neck for his official photograph prior to the expedition's departure from England. John Rae also heard tales of men abandoning ships and attempting to make their way overland to the south.

From Pelly Bay we returned across the base of the Boothia Peninsula and landed at another Inuit settlement, Spence Bay. By now I had so bombarded Bob with facts about the Franklin story that he became very keen to get even further involved. I had explained that one of my theories had been that Franklin could have been buried at the one spot in the Arctic which (it was erroneously believed) could be located be anyone with a compass – the magnetic North Pole. Although that 'spot' (actually, quite a large area) was now situated on Bathurst Island, during Franklin's time it was believed to be at the same site it was located at by James Ross in 1830. As Ross's Pole had been situated on the west coast of the Boothia Peninsula, almost opposite the northern tip of King William Island, it seemed reasonable to take a look if it was possible. However, when I arrived at Yellowknife I had learned that my idea was not original and had been put forward by a Royal Navy Rear Admiral some decades earlier. It had been rejected out of hand by every Franklin expert. Nevertheless, as Bob could legitimately take his aircraft northward, he left most of his passengers behind to amble through the settlement by the light of the midnight sun whilst he took Jim, Gillian (with a hard 'G', a female RCMP Constable), and myself in the direction of Pasley Bay. Below me the succession of dark, forbidding, troughs and ridges that were revealed by the cold slanting light quickly put paid to any notions that I could have reached Cape Queen Adelaide (the site of Ross's location of the magnetic North Pole) overland from Spence Bay.

When, according to the aircraft's compass, we arrived over the site, we found beneath us a vast plain of light brown shingle with no evidence that anyone had decided to mark the site of a grave. Bob made the aircraft revolve on a wingtip on the exact spot where Ross claimed to have built a cairn marking his discovery, but there was nothing to be seen. A final slow trawl up and down the area revealed nothing except that (as was to be confirmed by later experience) the only way to search was on foot.

Before long we had reached the ice-bound inlet that was Pasley Bay. In 1941, the Royal Canadian Mounted Police vessel *St Roche* sheltered in the bay during her epic voyage through the North-West Passage from west to east. During her stay one of her crew died and was buried on the bay's southern shore. As we flew along the length of the bay, Bob spotted a tall cairn just off the southern shore. 'Do you want to go down and take a look at it?' Of course I agreed and soon found myself gripping the sides of my seat as Bob brought the aircraft down on an impossibly bumpy stretch of shingle. Having warned everyone to hold tight, Bob applied the brakes as soon as the wheels touched ground. The aircraft rattled and shook to a halt on the rock-strewn esker.

Once we had left the aircraft (still in our Yellowknife street clothes), we could see that we were about two miles from the cairn. Closer at hand, however, was another reason for Bob to have landed the aircraft. A cairn and plaque marked the burial place of the RCMP constable who had died whilst wintering onboard the *St Roche*. Bob tidied the site until it was as smart as a war grave and added his name to a list of visitors that was kept in a container close to the plaque. Then, carrying a rifle over his shoulder, our pilot led us towards the mystery cairn.

A walk of about half an hour that was brisk enough to keep the chill of the early morning at bay (it was about three-quarters of an hour past midnight) brought us up against a fast flowing meltwater stream that ran alongside a ridge topped by the cairn. Bob and Jim went towards the ice to see if they could get around the stream that way, whilst Gillian and I went in the opposite direction. We had been going only for a short time when Gillian suddenly pointed to a section of the stream and followed up by breaking into a sprint in the same direction. As she reached the near bank she leapt and, with one foot, landed on a small rock that hardly breached the surface of the stream. Without a moment's hesitation she continued her flight across the icy water with yet another leap to reach the far bank, landing with both feet firmly on dry land. Then, with tiresome inevitability, she turned to me and invited me to do the same. Common sense should have urged me to ignore the challenge, but I am made of simpler stuff and I soon found myself charging at the bubbling water which I met with a jump such as I had not done since those far-off days on the assault course days after entering the Navy. Having made a split-second contact with the rock in the centre of the stream (which wobbled violently under the impact) another prodigious jump brought me gasping but dry-shod to the other side. Unfortunately, in my desperation not to be outdone by someone who was both younger and fitter, I had not realised the situation I would find myself in on the other side of the stream. The far bank was actually a narrow ledge from which reared up a sixty-foot snow wall. My choice was to do the obvious and try to re-cross the stream without the benefit of a run-up, or to follow Gillian who was already showing the way by punching and kicking handholds and steps into the frozen snow. Lacking the courage of my convictions, I turned and started my way up the snow face. Pushing aside any surges of panic as I climbed higher and higher I found that the example set by Gillian was not too difficult to follow and, apart from a great deal of gasping and wheezing as I rounded the final shoulder of the ridge, I achieved the top with the same glow normally associated with the scaling of the north wall of the Matterhorn.

By the time we reached the cairn, Bob was already standing by it. Jim had been forced to remain on the other side of the stream as his footwear was totally inappropriate for the exertions needed to get across. As it was, the cairn turned out to be relatively modern and was almost certainly erected by the crew of the *St Roche* during their enforced stay locked into the ice of the bay. Under the gaze of a few curious caribou who watched us from a nearby ridge, we now had to get back across the stream. Bob had found a stretch where the waters were shallow enough for him to wade across with his tall flying boots and Gillian's footwear was high enough for her to dash nimbly for the far bank. I, on the other hand, seemed to have no choice but to remove my shoes and socks, roll my trousers up, and simply take the plunge. Bob, however, had other ideas and hoisted me onto his back before carrying me safely across. I was so impressed that I considered nicknaming the RCMP 'Mounties', but someone else had apparently beaten me to it.

Our short visit to Pasley Bay had given me a brief insight into the summer Arctic away from the shelter of the Inuit settlements. Preparation, initiative, and teamwork were all to be vital.

On our return to Spence Bay, Jim and I said goodbye to Bob who left to take his aircraft back to Yellowknife. I had decided to remain at Spence Bay to see if there was any way we could get up the Ross Strait to Cape Queen Adelaide. It still seemed possible that a close look at ground level might reveal some sign left by Commander James Ross to mark his magnetic North Pole. What if the Admiral had been right, and Franklin had been buried on the spot?

After a flight in an elderly aircraft of over 1250 miles since 9am the previous day we were extremely grateful for the accommodation offered us by the local RCMP. The following morning I left Jim to struggle with the Royal Marine cooking stoves we had brought with us (they were proving difficult to light and tested Jim's engineering skills to the limit) whilst I went into the settlement with the RCMP Corporal to see if a guide could be found who could take us northwards. I was to be disappointed. After being introduced to the local Elders I was constantly met by an attitude that bordered, in some cases, on hostility. No-one had a sledge or skidoo to hire and, when I called on the local priest, even he seemed agitated and eager to get rid of me. By the end of the day the RCMP Corporal advised me to cut my losses and take the flight the following morning to Gjoa Haven. It was to be some time before I learned the reason behind the strange hostility I had experienced.

In the nineteenth century the Hudson's Bay Company had been accused of spreading poisoned blankets among the northern Indians and the Inuits resulting in fifty per cent of their population being wiped out.

Certainly, there was a great loss of life, but almost certainly as a result of the common cold and influenza rather than poisoned blankets. Unbeknown to me, a new myth had grown up. This asserted that the 'the Navy' was sending men infected with aids into the Arctic to complete the job started by the Hudson's Bay Company. My reception may be imagined when I was introduced as 'Lieutenant Coleman of the Royal Navy'.

Fortunately, no one from Spence Bay seemed to have had the forethought to use the local radio network to warn the people of Gjoa Haven that we were coming and so we were met by a mixture of polite curiosity and outright amiability. Much of this was helped along by the local RCMP Constable Terry Zeniuk who was to prove yet another excellent example of that superb body of men and women. An Inuit Special Constable assisted Terry, and through him I was able to arrange for a guide to take us to the northwest corner of the island. The distance was about one hundred miles as the crow flies but considerably more with the deviations caused by the massed lakes, ponds, rivers and streams.

Our guide was to be Paul, a strongly built, cheerful Inuk who spoke good English heavily laced with Americanisms picked up from watching too much television. Through him I arranged to have two All-Terrain Carriers (ATCs), one a double-seater for him and myself and a single-seater for Jim, who had considerable motorcycle experience and could quickly come to grips with the eccentricities of the three-wheeled machines. In addition, he provided a steel box trailer for all our kit, spares, tools, rifle, and the fuel needed for a journey of two days – the length of time he claimed that it would take us to reach our destination.

After a wildly expensive night in the Gjoa Haven 'Hilton' – a wooden hut whose owner charged the same whether the guest slept in one of his cramped dormitories or on the restaurant table – we were ready to face the journey north. Before long, however, we soon came to appreciate that Paul, in common with many of his fellow Inuits, operated something known as 'Eskimo time'. With twenty-four hours daylight, the local people operated an 'eat when hungry, sleep when tired' policy. So, instead of turning up at the appointed time, Paul was nowhere to be seen. After a search, I tracked him down to his home where I was told that he was fast asleep and would join us when he was ready. I had already been long enough in the Arctic to realise that nothing would be achieved by making an issue of a cultural difference. I just had to kick my heels.

When Paul did show up, at well past midday, he could hardly control his mirth at the pitifully small amount of kit we had with us. He had at least twice the combined amount that we intended taking with us. Nevertheless, we were soon on our way with me seated behind the guide, the trailer bumping along behind us, and with Jim acting as outrider.

Once we had left the few tracks in and around Gjoa Haven, I soon realised that the tundra, when viewed from the air, is very different when seen at close quarters. At best it is very much like a rough meadow which merely causes the machine to vibrate constantly. For most of the time, however, it consists of beachball-sized hummocks that are difficult to walk on, much less to try and control an ATC over. The vehicles bucked and reared like bad-tempered broncos, an effect that continued for mile after mile. Add to this the frequent crossing of streams of unknown depth that invariably led to the machine being bogged down and having to be hauled out, and rocky 'eskers' (long sand and gravel ridges formed by ancient glaciation) with their wide frost cracks and boulders that can be as big as a car, and you have quite a journey. Finally, and worst of all, add the dreaded 'muskeg' – a freezing grey mud that sucks the vehicles down and requires them to be heaved clear immediately before they become irretrievable.

After just one hour, the ATC Paul and I were riding on broke down. I had no other option but to send the guide back astride the limping machine to get a replacement, the smell of a burnt clutch fouling the otherwise sparkling air beside a frozen lake. At any other time it would have been a delightful spot to pass a couple of quiet hours. The tundra glowed with flowers; golden Arctic poppies competed with purple saxifrage and scarlet lichen. A snowy owl circled before perching at a safe distance to keep an eye on us. Ptarmigans and loons called out into the silence and were echoed by whistling swans. The enforced reverie was ended by the return of our guide. He had come back with a single-seater ATC. I could have sent him back to get a double-seater, but we had lost enough time already, so I chose to continue the journey by lying full-length and face down on the trailer – a bad mistake.

The springless carriage turned an awful journey into a nightmare. Holding on by a thin length of rope, I was thrown into the air with every jolt of the trailer, crashing down hard against its steel sides. My shins and elbows suffered greatly with each unpredictable leap and soon, with my ribs aching from the constant pummelling, I found that going through a stream would send up a shower of icy mud and water to soak my head and shoulders. Further refinement came when the trailer became stuck in a stream. At this I had to roll off the back and heave the combination out whilst the guide's revving of the ATC engine caused the wheels to spin rapidly, drenching me in a bitterly cold spray. The final touch came when, as so often happened, I had just pushed the trailer and ATC clear of a stream only to see the vehicle start to sink into a patch of muskeg. Eventually, after just ten minutes I felt that I could take no more, yet was destined to spend many more hours thudding across the tundra hanging on to the pounding trailer.

Occasional relief to my agonies came when we encountered the smooth, gravelled, surfaces of eskers which proved to be no worse than an average farm track. Other eskers, however, were made of large angular boulders. When these were met, I would gratefully roll off the back of the trailer and resort to hauling, heaving and lifting the ATC and trailer over the obstacles. A further break came when, sometime during the early evening, our guide began to repeatedly leap onto the seat of the ATC and scan the horizon through his binoculars whilst making a clucking sound. Shortly afterwards we came across a wide, strongly flowing river that was clearly impassable. When I asked the guide why he had brought us this route when earlier he had claimed to be able to get us to our destination in about two days, he replied that he had only done the journey before in the winter when there was a thick layer of snow upon the ground and a snow-mobile ('skidoo') could skim over the surface with little or no difficulty.

Pushing any disturbing thoughts to the back of my mind, I decided that, as the river flowed from the northwest, it made sense to continue along its southern bank as that would, at least, take us in the right direction. Unfortunately, it turned out that the river was one of a series that connected a succession of large frozen lakes. Eventually, at about one-thirty the following morning, it was decided to call a halt and pitch camp on the tundra-covered slopes of a low hill that bordered a large ice-bound lake. It had been a long day and my body felt as if it had been run over by a tank with every inch bruised and battered.

We decided to erect just one of our tents in the face of a gathering wind that came at us across the ice. Our guide, in the meantime, put up a tent the size of a bungalow, produced rolled-up rubber foam to place underneath his luxurious double-sized sleeping bag, and very soon had a pot of coffee rattling away on the top of his stove. Jim, on the other hand, shivered as he tried to get our stoves to work, but managed only to get a limp flame over which he produced a lukewarm, lumpy, porridge while I scrounged some hot coffee from Paul.

The night that followed was our first night in a tent under an Arctic sky. And a pretty miserable affair it was too. With nothing but a thin ground sheet and two or three inches of soil between the permafrost and us, it did not take long before the cold began to seep through to chill the bones. In the meantime the tent began to flap and shake alarmingly as a biting cold wind raced across the lake to strike us with a howl that chilled the blood even further. When, after a night of fitful sleep with the sun blazing through the thin canvas, we crawled from the tent, we found that of the two metal spikes that tipped the tent-poles, one had been bent at right-angles whilst the other had completely sheared off.

Taking pity on us, Paul gave us hot coffee before we once again packed the trailer ready for the next part of our journey. Having become aware of our guide's roll of foam rubber, I commandeered it to help cushion my body from the inevitable hammering it was about to receive as we continued our journey. In practice, however, its help was minimal and my shins, arms, and occasionally, head continued to be brought into sharp and painful contact with the steel sides of the springless carriage. After a number of hours, the shaft that coupled the trailer to the ATC snapped under the hammering, and the trailer had to be secured by rope, which introduced an interesting yawing effect to my passage over the malevolent tundra.

The day was spent as before, the relentless pounding being relieved only by the dragging through ice-streams and muskeg and the lifting over cracks on the eskers or piles of frost-shattered boulders. I was in good company in finding difficulty in crossing the terrain. No less an Arctic traveller than Dr Rae, after his searches for the Franklin Expedition, wrote, 'In the months of June and July it is all but impossible to travel across that country.' Describing the same season, the chronicler of the Schwatka expedition recorded:

> Sometimes we would sink to our waist, and then our legs would be dangling in slush and water without finding bottom. The sled would often sink so that the dogs could not pull it out, light as was the load, and when we would gather round to help them, we could gain an occasional foothold, perhaps by kneeling on a hummock, or holding on with one hand while we pulled with the other... Without the assistance of dogs and natives it is altogether probable that we would not have been able to accomplish more than two or three miles at best.

As we approached midnight, we breasted a low hill surrounded by lakes and small meltwater ponds. This, I decided would do for our second camp. With his own tent up, Jim helped me with mine as I was having difficulty due to the fact that my ribs felt as if they had caved in and even the slightest movement caused me agonies. With the task completed, Jim decided to check the fuel. There was soon evidently something about the two remaining full containers that Jim did not like. He called me over and asked me to take a sniff at the contents of the containers. I couldn't smell petrol. We then pointed this out to Paul who, with a beaming smile, told us that – after Jim had checked the fuel before we had left – he had swapped two containers of petrol for two containing fuel for his stove. His smile refused to disappear even when Jim pointed out in a forceful manner that we now had much less fuel than we needed to get us to the northwest corner of the island. There was nothing for us to do but to send our guide back immediately to get more fuel. Jim drained the tank of his ATC to ensure that Paul had enough fuel to get him back to Gjoa Haven and, after a meal and a hot drink, our guide was, once again on his way southwards.

The following day, whilst waiting for the return of our guide, we decided to walk westwards towards a large area of ice which, through binoculars, looked like a coastline. After a comfortable trek of about eight miles, we found ourselves on the shores of a large lake whose icy surface sparkled with deep blues and whites beneath the midday sun. The lake's shores were cloaked with glittering snowfields and enveloped in a deep silence punctuated only by the crunch of our boots and the occasional cry of a bird. On our return journey, we came across the hollowed-out piles of stones which (we later learned) were Inuit 'rock graves' where the dead were placed in the hollow centre, covered with caribou skins and left for wild animals to dispose of. Close by was a line of rocks that had all the appearance of being arranged by the hand of man, but whose purpose was unknown to us.

At six-o-clock that evening, whilst Jim struggled to get the stoves to work, I made radio contact with Terry Zeniuk. I had intended merely to acquaint him with our situation and ask him to prompt our guide into an early return, but his news for us was far more disturbing. Paul, it seemed had returned to inform the RCMP Constable that he had had enough of the overland attempt and did not want to go any further. It was only under pressure from Terry that he had agreed to come back and pick us up. I was dumbfounded. To have got this far – probably less than two days travel to our destination – only to fail as the result of our guide's whim, seemed a cruel end to our attempt to reach the northwest corner of King William Island. I tried to persuade Terry that Jim and I should go on alone, but he urged us to avoid any such ideas as there were no 'search and rescue' facilities available to him should anything happen to us. In the end I could see no answer but to return to Gjoa Haven with our guide.

There was not a happy atmosphere at the camp that night as Jim and I brooded over the outcome of events. The only bright spot was the appearance of a large number of lemmings that darted between our legs as they scavenged for scraps of food.

As the sun circled towards the western horizon, I assembled a large number of rocks on the southern side of the hill and spent most of the evening arranging them in the shape of an 'Admiralty Pattern' anchor. Above the anchor stock, I marked out the letters 'RN' and, on either side of the flukes, the date '1990'. It seemed a rather proud symbol for such a poor show, but I thought it might at least cause interest to anyone passing that way in the years to come. By the time I had finished, Jim had retreated to his tent where he was tormented by lemmings racing about underneath his groundsheet.

We did not stray far from the camp the following day in case Paul arrived whilst we were away. When he had not arrived by early evening, I contacted Terry Zeniuk once again to be told that Paul had left and could expect to be with us later that evening. In fact, it was to be about three the next morning before I heard the far-off sound of his ATC. When the sound grew closer, I realised that there was more than one machine approaching, a fact confirmed when I heard Paul speaking in Inuktatuk to another person. I merely snuggled down in my sleeping bag feeling in no mood to lay out a red carpet to welcome our guide's return.

When I awoke the next morning I found that Paul had been accompanied by another Inuit who had ridden a two-seater ATV. This was splendid news for me as, with our two guides sharing the two-seater, Jim and I would have a machine each to get us back to Gjoa Haven.

The journey back was, for me, almost a pleasure. I could take any amount of pulling and heaving the trailer and various vehicles out of the muskeg and streams just so long as I did not have to ride the trailer again. Then, about thirty miles outside of Gjoa Haven, over the noise of my engine, I heard a loud shout to my rear. Looking over my shoulder, I just caught sight of Jim as he rolled forward on the tundra grass and his machine toppled over sideways.

Bringing my machine up alongside the now standing Jim, I was aghast to see that the rear right wheel of his ATV had parted company with the rest of the vehicle. It appeared that the wheel had been simply held on by a single nut which now lay somewhere behind us. We searched as minutely as we could but could find no sign of the missing part. When we tried to find the tools and spare parts that Jim had checked before leaving, we found that our guide had left them behind at Gjoa Haven. However, a rummage through all the panniers of the machines produced a single tool, one that saved the day. From somewhere, someone produced a pair of vice grips, a sort of adjustable spanner that could be locked 'closed'. With the wheel back in position, the vice grips were clamped onto the protruding axle. This simple tool ensured that the machine carried Jim all the way back to the settlement.

About a mile outside Gjoa Haven, a rudimentary road was picked up and, leaving Jim to look after the stores, I raced ahead to find Terry Zeniuk and to acquaint him with our current situation. I spotted him walking with Joan, his wife, towards the RCMP Station ('Detachment'). They stared in alarm at the strange shouting and waving individual bearing down on them at high speed aboard an ATC. Even when I had drawn up alongside it took them a moment or two to realise who I was. What I had not appreciated was that, after just 48 hours or so, not only did I cut a desperately unkempt and wild looking figure, but that I had been burnt a mahogany colour by exposure to wind and sun.

After commiserating with me over my predicament, Terry yet again came up trumps. I was very concerned about the expedition funding and could really not afford to book Jim and myself into the local 'hotel', so it seemed that we would have to pitch our tent on the outskirts of the settlement. Terry, however, offered us the two spare bunks in the RCMP Detachment's cell and after using the prisoner's shower, he invited us to his accommodation for a most welcome dinner.

After a night in the cells and a breakfast of caribou sausages at the hotel, Jim went to check our equipment whilst I went to the RCMP Detachment to have a word with Terry about what we were to do next. There was no plane out of Gjoa Haven for seven days and I was desperate to waste as little time as possible whilst I was in the Arctic. I bent over a map of the region on Terry's desk, contemplating the possibility of Jim and I going overland westwards from the settlement, when one of the lenses fell out of my reading glasses. Life, it seemed, was determined to extract every minor irritant out of my current situation.

The loss of the lens led me to search for somewhere in Gjoa Haven that had a small screwdriver to repair my glasses. Eventually, it turned out that the settlement's Nursing Station was the most likely place to have such a tool. Whilst I was waiting for the job to be done, I got into conversation with the station's porter, an Inuk named Max. When I mentioned that I wanted to go westward to Booth Point but expected difficulty with the transportation, he suggested that he could take us over the ice of Simpson Strait – if we were prepared to take the risk. The ice, apparently, was, in parts rotten. Cracks were beginning to spread in all directions and the thickness of the ice could not be guaranteed to stand up to a snowmobile hauling a sledge. But if he was prepared to take us, I was not prepared to turn the offer down. Yet another idea struck me. If he could take us to Booth Point (on the south coast of King William Island) could he take us to Todd Island, a few miles to the south of Booth Point? In 1869, the American, Captain C. F. Hall (later to be murdered on an Arctic expedition) reported that he had found the remains of five of Franklin's men on the islet. A Doctor Gibson rediscovered the remains of four of the men in 1931.

After an amiable discussion over costs the matter was settled. Max and his son would take us down Simpson strait (named after a Dr Simpson who shot two of his companions during a later expedition before being shot himself), and hired us his tent, the large, local, variety that Paul had used. He also threw in as part of the deal caribou skins to sleep on, his stove, and his wife's best kettle.

Thus prepared for our journey, Jim and I joined Max on the settlement's southern shore. A short paddle in a canoe brought us over to the ice where Max and his son, Andy, were preparing their skidoo and sledge with its plywood boxed-in space for stores and passengers. With our kit secured, Jim and I climbed onboard whilst Andy started up the skidoo. Max mounted the rear of the sledge where he could steer by throwing his weight from one side to the other.

We set off with about thirty miles of ice to cross before we reached Todd Island. I found the sensation of gliding over the ice a strange and exhilarating combination of contact between the elements that is lost in the turning of a wheel. The friction of the runners beneath us was melting the top of the ice leaving us actually aquaplaning over the glistening surface with a most satisfying hissing sound. Ahead of us I could see Andy crouched with one knee on the seat of the skidoo. I asked Max why his son did not sit astride the machine and was somewhat alarmed to be told that should the ice give way beneath him, the boy could leap off the machine thus giving him a chance of avoiding a very cold ducking. As for the sledge passengers trapped inside the contraption they could expect to follow the skidoo to the bottom as the sledge steerer – Max – rolled off the back. I smiled weakly and pretended to be unconcerned whilst deciding to tell Jim about this unconventional lifeboat drill (crew away first) on arrival at our destination.

It was not long before we began to see the size of the problem. Andy was having to steer around huge holes in the ice and circle great floes that had broken away and were forced up by the pressure of the ice around them. After about three hours, we stopped for a warm drink from flasks we had topped up at the hotel. Ahead of us, we could see the dark smudge that marked Todd Island whilst, to the south, Max pointed out on the thin line that marked the rim of North America, not merely Richardson Point – named after Franklin's companion on his overland journeys – but also the direction of Starvation Cove, the most southerly point known to have been arrived at by Franklin's people as they tried to reach sanctuary down the Back River. At last we were getting close to the Franklin mystery, albeit further to the south than we had intended.

Another hour on the sledge brought us up against the practically barren pile of rocks that made up Todd Island. Once ashore, Max helped us erect our tent and showed us how to get fresh water from the sea ice as the island had no fresh water of its own (the ice was brought ashore and piled on rocks to let the salt leech out over a few hours). Then he and Andy set off back to Gjoa Haven with the promise to pick us up three days later and take us over to Booth Point.

I soon found that the caribou skins loaned to us by Max were a huge asset, not merely in acting as ground sheets (which his tent did not have) but also as some protection against the cold. We had set up camp on the easternmost point of the island, which meant that we had ice on three sides (an arrangement, I was later informed, that was perfect for encouraging polar bear attacks) and the temperature at night dropped well below freezing. An attempt to melt ice in the tent overnight met with failure.

Our first day scouring the island revealed little apart from confirming that fact that the place was little more than a pile of rocks that rose gradually to a short plateau in the centre. Plenty of caribou and seal bones littered the surface, and ancient stone tent rings showed that the place had seen human habitation at one time or another. Several eider ducks had built their downy nests amidst groups of yellow flowers I thought of as 'Todd Island dandelions' (actually Arnica).

On the second day, Jim and I were making our way westward along the southern coast of the island when I suddenly grabbed his arm, bringing him to an abrupt halt. In front of me I had seen what appeared to be a human skull rising out of the ground. On closer examination it was in fact the top part of a skull sitting on the surface. Alongside lay other bones that may, or may not, have been human. The fissures of the skull were clearly visible indicating to my extremely limited anatomical knowledge that it had belonged to a young man and, as there were more ratings than officers on the Franklin expedition, I named him 'Jack'. Jim suggested that his surname should be 'Todd' – so 'Jack Todd' he became. Of course, he could have been an Inuit, but neither of us were expert enough to determine the racial characteristics of the remains and, until we learned anything different, he was to be the fifth of Captain Hall's Todd Island discoveries.

I had a common-sense agreement with the Canadian authorities that if I should find something of interest that was clearly a random artefact unrelated to its surroundings, I could remove it from the site whilst making sure that the site of the find was carefully recorded. If, however, the link to the site was less than random, I should do no more than mark and record the position leaving everything as it was for later inspection by experts. Accordingly, I marked the site on our map, drew a rough sketch-map of the southern shore of the island and built two small cairns as visual markers. Unfortunately, due to the nature of the find, there was something else I had to do. The RCMP had warned me, as they wished our little expedition well, that the one piece of evidence that would be most unwelcome would be a human skull. Such a find would, inevitably, bring upon our own heads the full weight of the grinding administration needed when human remains are found. However, there was nothing for it but to inform Terry Zeniuk by radio of our find. His first question was rather unexpected. Over the atmospheric howls and whistles of the receiver, I heard Constable Zeniuk ask 'Are there any signs of violence?'.

Once Terry had got the full picture, he asked us to stay on the island for an extra day. He would send out an Inuit Special Constable to view the site, whilst Max was to be told of the delay. Twenty-four years later, a party of journalists from *Outpost*, a Canadian travel magazine visited Todd Island with a local historian, Louie Kamookak as their guide. He took the party to a spot where he claimed he had discovered a human skull. It was actually the top part of a skull which – from a photograph – bore exactly the same random marks inflicted by exposure to the elements as the one I had come across. It was clearly the same 'Jack Todd'. Oddly enough, however, the large fragment seemed to have been moved to a different site. It now rested on fine sand rather than the rock chippings on which I had found it, and there was no sign of the associated bones alongside it.

We spent the extra day minutely scouring the island for any further signs of occupation but apart from a few ancient tent rings, there was nothing to be seen. Around ten o'clock that night the sound of a skidoo could be heard across the ice. At first we thought it was the Special Constable, but it turned out to be Max towing a low sledge on which he had secured a boat. He had decided that the state of the ice was now so 'rotten' that there seemed every likelihood that it could give way beneath us. With Jim and I sat in the boat he felt we would stand a better chance should the ice crack and crumble beneath us. In addition, he handed me a sharp knife. If the skidoo went through the ice I was to use the knife to sever the towrope and thus prevent us from following the heavy machine to the bottom of Simpson Strait.

As we drank a final cup of tea before setting out, Max expressed surprise that the Special Constable had not been out to see us; he had certainly set out earlier that day taking his wife along for company. We were later to find out that the Constable and his wife had gone through the ice and had been forced to make a chilly journey back to Gjoa Haven on foot.

Fortunately, they both arrived safely.

It was almost exactly on midnight when we set off from Todd Island towards Booth Point. A thin layer of cloud shaded the low sun, but the light was easily enough for us to see the problems that had caused Max to bring the boat. Ahead of us the ice was riddled with areas of open water.

Some were simply ponds of meltwater on top of solid ice, but others were the open mouths of huge holes that could have swallowed a London bus. From our position in the boat we could not tell the difference until we were actually on top, or alongside, the open water. There were many heart-stopping moments as we saw Max leave the solid ice and skim across a small lake of shallow meltwater or feel the boat jerk to one side as he suddenly steered the skidoo around a hazard leaving us to gaze into its blue-grey depths as we slid along its crumbling edge.

At one stage, about a mile from the shore of King William Island, Max was halted by a four-foot-wide lead that ran across our path. After a moment's thought, Max unshackled the towrope and headed off on his skidoo to find a part of the lead that was narrow enough for him to cross with the machine. He seemed to continue forever but, after about a mile and a half, we saw him turn around and head back towards us, this time on the opposite side of the crack. I threw the towrope across to him and climbed back onto the boat as he re-secured the rope to his machine. Telling us to 'hang on', he put the clutch of the skidoo into neutral, opened the throttle to its maximum, and then kicked the machine into first gear. We were catapulted forwards and shot across the gap as if airborne.

With the shore safely reached, we unloaded the boat and bade farewell to Max who was eager to get back to Gjoa Haven. To the retreating sound of his skidoo, Jim and I set about pitching our camp on the bleak, rock-strewn shore we had found ourselves. It was not long before we found ourselves wishing that we had joined Max heading for civilisation, the rotten ice notwithstanding. During our attempt to cross the island we had occasionally been pestered by Arctic mosquitoes, obnoxious creatures that were designed to penetrate the thick coatings of Arctic animals, and for whom humans provided an easy target. For the most part, the wretched things could be kept at bay by insect repellents, but the mosquitoes that waited for us at Booth Point believed in the security of numbers. They swarmed about us as we put up our tent, the air vibrating to their menacing whine, and ignored our flailing arms as we tried to drive them off. We tried putting on our balaclava helmets, but they crawled over our heads and tried to enter our mouths and noses. They marched along the walls of the tent in legions and forced us to close the tent-flap whilst we wiped the canvas with gloved hands until we had exterminated every single one.

At last, exhausted by the day, we opened out our sleeping bags on top of the caribou skins, allowed ourselves a warming chocolate drink, and turned in to the muffled whine of a million frustrated mosquitoes as their shadows were freckled onto the tent by the weak rays of a rising sun.

When we awoke, we found that the sun itself had vanished behind a thick layer of clouds and that the temperature had dropped to such a level that our water inside the tent had frozen solid, and the ground outside was littered with dead snow buntings. Sad though it was to see the frozen corpses of the birds, the fact that the cold had seemed to have driven off the mosquitoes almost made the tragedy worthwhile.

The grey skies over Booth Point made an already dreary place grim to the point of depression. The place had been named (from across the Simpson Strait) by Captain George Back RN as he reached the end of his journey down the Great Fish River in 1833. Thirty-six years later, the American, Captain Hall, had come across the bones of one of the Franklin Expedition amongst the rocks before finding a few miles to the west the remains believed to be those of Lieutenant Le Vesconte, the Second Lieutenant of HMS *Erebus* (the bones were brought back to England and buried beneath the Painted Hall at Greenwich before being removed to the Franklin memorial in the college chapel). In 1981, a Canadian, Owen Beattie, found some human bones at Booth Point, which led him down the contentious path of supporting native claims of cannibalism amongst Franklin's men.

Whilst roaming over the rock-strewn terrain, I spotted something that was clearly man-made. At first I thought it was made from bone, but later learned that it was caribou antler. About seven inches long and an inch and a half wide, flat on one side and slightly rounded on the other, it contained a series of countersunk holes such as might be used for screws. It had two circular depressions, one larger than the other, carved out of the flat side and one end had been chamfered. All the 'screw holes' and the circular carvings were contained in a wide, regular, groove neatly gouged out along the length of the piece. Whatever it had been intended to be, it had not been completed; one end remained as a rough untreated mass. At first it seemed to be a type of scraper, but that would not account for the regularity of the centre groove or for the 'screw holes'. I felt that it had the dimensions appropriate to a musket lock-plate and could be a local attempt to reproduce such an item to replace a damaged one. Any attempt to obtain the heat required to repair a metal plate would have been hopeless, and perhaps someone had attempted to carve one out of caribou antler. Could it have been done by a native skilled in such carving? If so, what level of contact did the two parties have? Like the skull, however, any such suggestions would be better coming from experts, and the artefact was later deposited with the Canadian authorities.

The following day (our final complete day), I intended that we should make our way westwards along the coast towards Cape James Ross, but we were thwarted by a deep, fast-flowing river. Instead we followed the river up to its source as it flowed out of the huge, freshwater Koka Lake. On that bitterly cold grey day, neither bird nor animal made its presence known. The only sound came from shards of ice giving off an attractive tinkling sound as a biting wind blew them over the surface of the lake. After some miles of walking along the lake's southern shore we came across the bizarre sight of a huge ridge made up of almost symmetrical two-foot square blocks of ice piled up at the lake's edge. Whether the ice had expanded and forced itself into this configuration, or whether the wind had forced it up onto the land, I could only guess.

As we were now entering into late evening, we decided to head south by climbing one of the few large hills on the island. Arriving at the top exhausted and, on my part, aching (I was still suffering from the effects of my battering on the trailer), we could see Booth Point and, beyond it, Todd Island. Further still, across the Simpson Strait, we could clearly see Richardson Point. To its right, visible only as a thin line on the horizon, was the area of Starvation Cove where the last of the Franklin party remains were discovered by Lieutenant Schwatka in 1879.

On our way down the southern slopes of the hill we came across another mystery. An isolated boulder, about the size of a large car, had several sets of caribou antlers placed on its top. On the south side of the rock, two large areas of soil, each almost twenty feet across, had been savagely ground up. The rents in the earth were so deep that in places, sub-surface rocks had been forced up to lie shattered and broken on the top. The damage seemed recent, certainly no more than a day or two since. Yet there were no tyre marks to suggest the use of earth-moving equipment necessary to produce such an effect. Indeed, apart from the placing of the caribou antlers, there was nothing to suggest that any human had passed that way before. It was as if two bolts of lightning had struck the earth. I was never able to find a satisfactory answer to the mystery.

After a warming meal at our tent, I took a final lone walk by the light of the midnight sun as it hovered in the west before continuing its journey around the northern sky. I had just crossed the narrow sand-strip that connected Booth Point to the rest of the island when I turned to look out across Simpson Strait. Beyond the silver and grey expanse rose a majestic range of mountains with grey and brown lower slopes topped by snow-capped summits whilst numerous wide glaciers poured down to meet the ice of the strait. Then two things happened simultaneously. Firstly, I realised what I was looking at and, secondly, the 'mountains' began to dissolve before my eyes. Soon, only the occasional mushroom-shaped blob of brown remained and, before long, the magnificent sight had been reduced to flickering line on the horizon. I had been privileged to witness an Arctic mirage; not, by any means, a rare occurrence, but startling in its clarity and definition.

After a final night under canvas we emerged into a grey morning with the Simpson Straits blurred by fog. Breakfast had not long been finished when we heard the sound of an engine approaching. It was Max returning to collect us. This time, however, the sound of his engine did not come from the ice but came instead from our rear. Through my binoculars I caught sight of movement at the end of the sandy spit connecting Booth Point to the island. There was not just one but two ATVs coming towards us. Some minutes later, a broadly smiling Max rattled up followed a rugged looking middle-aged man who proved to be a gold prospector. He had volunteered to come along with Max to help in case of any difficulties in returning Jim and me overland, the ice of the straits being now too rotten to risk passage. They had brought with them a sturdy sledge, which was not only to take our kit back but also two naval officers.

With our tent and packed rucksacks securely lashed to the low-slung sledge, I sat astride the conveyance and wedged my feet against the up-curved leading edge of the sledge runners. Behind me, Jim locked himself onto the rear of the sledge. It was with some trepidation that I shouted to Max that we were ready, as I was deeply concerned that I was about to undergo yet another bone-crunching journey across the tundra. But my fears proved to be groundless, even on the roughest rocks the sledge skimmed over the surface with little more than a modest shaking and shot across the tundra with a smoothness that almost matched the ice, with eider ducks, ptarmigans, and plovers scattering as we raced through their territories. The occasional river we had to cross proved a challenge, but all were successfully negotiated with only a minor loss of kit.

We arrived at Gjoa Haven in the early evening. The sledge, which had served us so well on our return journey, collapsed into matchwood during the final few yards. That night was spent in luxurious indolence at the Gjoa Haven 'Hilton' followed by a breakfast of caribou sausages before saying farewell to Max (who gave me an 'ulu', a musk ox horn-handled seal skinning knife, as a parting gift), and boarding Bob Martin's RCMP Twin Otter for the flight back to Yellowknife.

During our stay at Yellowknife we were accommodated by Constable Terry Zeniuk and his wife Joan (who had earlier returned from Gjoa Haven) whilst I reported our expedition to the Prince of Wales Northern Heritage Centre. The Director, Dr Charles ('Chuck') Arnold, identified my Booth Point find as caribou antler and promised to keep me informed should its true purpose be revealed by experts. He then passed me over to the Centre's archaeologist, Margaret Bertulli, to describe the site of the skull I had found on Todd Island. Margaret proved to be of great assistance and encouragement in the future, never failing to help where she could despite an unwavering belief in the proprieties of Arctic archaeology (anyone amiably ambling in with something they had picked up in the barren lands or beyond could expect a high decibel lecture on archaeology and the law which would have them shuffling out as broken men – Margaret never took prisoners).

All that remained was to say our farewells to the friends we had made during our time in Yellowknife (including Bob Martin who proved to be the finest 'Margarita' cocktail maker in the Northern Territories) before boarding a Canadian Air Force transport aircraft which took us south via Whitehorse in the Yukon. After an all-too-short period of royal treatment by the members of the 'Calgary Light Horse', the Royal Air Force flew us back into the English heat wave of 1990.

So, what – if anything – had been learned? Apart from the expansion of our experience very little could actually be seen to have come about as a result of the expedition. The skull on Todd Island might turn out to be of interest, but it would not have been the first to be found, nor would it be likely to open any startling new avenues of investigation. The artefact I had found at Booth Point could be revealing, but only if it retained enough evidence of its origins. What did turn out to be of interest, however, was the re-kindling of awareness in the Franklin story among a wide and varied section of the population – especially in Lincolnshire, Franklin's county (and my own).

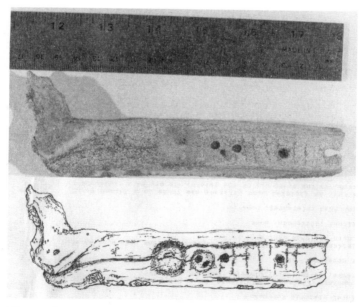

The artefact found by the author on Booth Point, King William Island. The authorities have retained the author's suggestion that it could be an attempt to replicate a musket cheek plate.

2

Raising Steam

The return from my first attempt to reach the northwest corner of King William Island and the possible burial place of Sir John Franklin may have seen the end of the foot-slogging across the tundra for the foreseeable future, but it was soon to be replaced by many hours of treading the boards. Over the following two years I gave over more than eighty talks on the expedition to groups ranging from primary school children (including my own primary school) to retired businessmen, from ship and aircraft enthusiasts to Women's Institutes, and from Sea Cadet Units to Rotary Clubs. Wherever I spoke, I was always amazed at the interest and knowledge of the audience. Many were intrigued by the mystery of the disappearance of so many men with so little trace, others in the technical aspects of voyaging in northern waters, still others in the environment of such a remote and rarely visited part of the world. The questions I had to field were always guaranteed to keep me on my toes. They ranged from the broadly obvious ones such as 'Did they simply take the wrong route?' to the controversial 'What caused their deaths?' Some specialists would ask questions like 'Could the ships have got through the ice if their bows had been more raked and weighted in order to act as an ice-breaker?' or 'Would a better design of screw have helped?' Not all questions, however, were quite so testing. At one primary school silence reigned when the teacher asked if there were any questions. After some prompting a single hand rose slowly into the air followed by the query 'How much did your uniform cost?'

One question, however, rarely failed to be raised. 'When are you going back to complete the search?' In answering I had to be very circumspect. As a serving naval officer I could not make any commitment without the approval of my superiors, and I felt that they had been extremely generous in allowing me time off for my 1990 visit. Even if I did manage to obtain permission for a second attempt there would always be the tiresome difficulty of raising money, and the economic climate at the time was less than encouraging.

Oddly enough, it was not my answering questions that prompted my taking a closer look at a return visit, but my asking them. I had learned of an Arctic Seminar being organised by the National Maritime Museum and had obtained a ticket. The event was held in one of the white and gilded rooms of the Queen's House. It was interesting to note that there appeared to be no empty places; the subject, even at this level, could manage to attract a full house. The speakers were, I felt, uneven in their quality and I repeatedly found myself on my feet asking questions or raising points that I felt had been missed. Usually I received an academic brush-off or a 'Well, that's just your opinion' type of answer. For the most part I allowed my misgivings to pass without further comment, but there came a time when I simply could not keep the lid on. One speaker had been asked from the floor, 'What do you think caused the deaths of the men who headed south towards Back's Great Fish River?' She replied that there was no doubt that the tragedy had been caused through the inexperience of the leader of the escaping party, Franklin's second-in-command, Captain F. R. M. Crozier RN. Instantly I found myself on my feet demanding to know how someone who had been on a number of Arctic and polar expeditions, who had been Ross's second-in-command in the Antarctic, had a Cape named in his honour on the southern continent, and had been considered as a possible leader of the expedition in question, could be considered 'inexperienced'. Silence reigned for what seemed an eternity before the organiser of the event stepped in to announce that it seemed to be a good time to have a lunch break. (I instantly felt that my interjection had been somewhat oafish. It was certainly ill timed as the speaker in question held a very important post at the National Maritime Museum, and it cost me a good lunch to wheedle my way back into their good books).

Despite my own misgivings, a number of people came up to me to express agreement with my points of view on the variety of subjects that I had raised. As they did so I became aware of an immaculately dressed elderly man making his way over to me with the help of a walking stick. As I turned towards him he said, 'Are you the gentleman who defended Crozier?' When I admitted that I was, he stunned me by saying 'Perhaps you would allow me to introduce myself. My name is Rawden Crozier.' Whilst I recovered from the shock, he went on to explain that he was a direct descendant of the very same Francis Rawden Moira Crozier I had so recently been ranting about. He was also a former Lieutenant, Royal Navy. Fortunately, I was relieved to learn that he had come over to express agreement with my opinion of his forebear. During a subsequent chat during the lunch break he suggested that I might like to get in touch with a lady who was currently writing the biography of Crozier, Mrs Betty Diamond. I promised so to do.

During the afternoon of the same day, the seminar assembled in the rooms containing the Museum's Arctic display where I found myself talking to the lady who had organised the event and was outlining my recent visit to King William Island (and whinging on about the fact that some of her speakers appeared never to have got further north than Watford) when I became aware of the tiny figure of another lady who was listening to the conversation. Under any circumstances it would have been difficult to ignore the sparkling Italian eyes and radiant smile of the woman with whom I soon found myself in conversation. She turned out to be Maria Pia Casarini-Wadhams, the wife of Doctor Peter Wadhams, the Director of the Scott Polar Research Institute. She also proved to have encyclopaedic knowledge of polar exploration in her own right and was currently researching a book on Jane, Lady Franklin. Her enthusiasm for all things concerned with the subject was infectious and that, combined with my meeting with Mr Crozier, planted the seeds of an idea; perhaps another attempt to go north could be feasible?

The germination of those seeds came about after I had written to Mrs Diamond. She had responded with an invitation to lunch at the Royal Geographical Society and, as I entered the Society's august portals, I could not help but reflect upon those that had gone this way and had become household names, explorers who had roamed from pole to pole, climbers who had ascended the highest mountains and travellers who had left their footprints in desert sands or had hacked their way through unknown jungles. Almost all of them (the exception, in my opinion, being Shackleton), in company with their forebears and their successors, formed a constellation of achievement that was both forbidding and inspiring.

I could never get over the first impression I had of Mrs Diamond – that she was exactly the type of woman Agatha Christie would have written about as a heroine in one of her detective novels. Everything about her declared her to be splendidly English in presence, in attitude, and in inclination. The wife of a retired Royal Naval captain, she was clearly used to directing junior officers along the right course and very quickly had me feeling comfortable in my strange surroundings. She first of all introduced me to another of her guests, Mr Donald Bray, who was a great grand-nephew of Able Seaman John Hartnell whose body had been disinterred by the Canadian academic, Dr Owen Beatty, during his visit to Beechey Island in 1986. After a visit to the map-room to see original charts of King William Island, we retired to an ante-room where I showed

Mrs Diamond photographs of my own expedition, and gave her a piece of white quartz I had picked up on the south of the island.

During lunch, Mr Bray showed me a copy of a report he had received from an American banker, Steven Trafton (later published in *The Mariner's Mirror*) covering Trafton's visit to the northwest corner of King William Island in 1989, during which time he and his team located a number of cairns he believed to be from the time of the Franklin expedition. There was no mention of any trace of Franklin himself or any graves despite a detailed search having been carried out.

Before I left, Mrs Diamond asked me whether I would care to be proposed for a Fellowship of the Royal Geographical Society. As I had never considered such an honour being available I was delighted with the opportunity (although I felt that there was every chance of being turned down when compared to those who had actually achieved something with their efforts). Even more extraordinary was Mrs Diamond's suggestion that she would ask A. G .E. Jones to second her proposal. Mr Jones had been the secretary and close associate of the doyen of Franklin experts – Dr R. J. Cyriax – and was well-known for his forthright views, which often refused to sit comfortably with the accepted versions of events and personalities. During subsequent correspondence with Mr Jones, he was to help my efforts greatly, particularly on the question of the location of Victory Point. I was elected a Fellow of the Royal Geographical Society on 11 November 1991.

After considerable research into burials in ice, I discovered that Captain Albert Markham RN (a cousin of Clemence Markham, the chief sponsor of Captain R. F. Scott's expedition to the Antarctic) wrote in his *Life of Sir John Franklin* of 'that last sad and solemn scene on the ice floe ... round a grave that has been dug out of solid ice'. As a Commander, Markham had taken part in the Nares Expedition of 1875-76 and had made a brave attempt to reach the North Pole. On the return journey, a Royal Marine had died of scurvy and was buried either in the deep snow covering the land they had reached three days before, or in the sea-ice over which their route to their ship lay. His book on Sir John Franklin was to become the standard work on the subject for many years. In it he set the pattern for many succeeding works on Franklin. Whilst I could do nothing but stand in admiration of Markham's achievements, I reserved the right to disagree with his conclusions regarding Franklin's burial, conclusions that I believed were coloured by Markham's own experiences.

By now I felt I had a case that I could lay before my immediate boss, a Commander. Without blinking an eye he agreed to my ideas and approved my plans with the single proviso being that he should be kept informed of progress at all times. It was now time to get moving.

The search for funds could not begin soon enough. I struck gold at my first shot. A Lincolnshire manufacturer, Ian Medlock, had in the past shown interest in supporting Lincolnshire-based projects and might be worth getting in touch with. I wrote to Mr Medlock with a copy of the report from my previous expedition. My delight can be imagined when a couple of weeks later I received a cheque through the post for a staggering £1000. With me (and Joy, my long-suffering wife) putting up a similar amount, I looked as if I was on course to reach the figure I had given myself – more in hope than expectation – of £5000.

Now it was the turn of the Royal Navy to lend a hand. When I had got in touch with the Navy's expedition organisation prior to my previous visit, the Lieutenant-Commander manning the appropriate desk simply said, 'Don't worry, old boy, we'll get you out there one way or another.' In addition to arranging a flight with the Royal Air Force he had even been able to find £150 from his funds to help us on our way. Unfortunately, he had now moved on and a different officer sat at the desk. His reply to my letter asking for similar help told me bluntly that the minimum numbers for an expedition was to be four and, as I had decided that, this time, it would be a solo effort, he could not help. This was devastating news as the cost of a flight out to Calgary would have played havoc with my funding plans. I telephoned him to see if he would reconsider his decision if I could explain that all I wanted was his authority behind an application for a RAF flight. But he insisted on sticking to the rulebook. I asked him if he could give me the telephone number of the RAF section who dealt with flights. At first he refused, saying that it would do me no good by going down that path, but eventually relented and gave me the number I wanted. I immediately rang the number and got through to a cheery Squadron Leader who, in answer to my enquiry as to whether or not there was any problem about arranging a flight for a single-member expedition replied, 'No problem at all, I'm totally in favour of such efforts – after all, I've just flown a chap half-way around the world to referee a football match, so you should be no problem.' My heart leapt at this encouraging sign. Could I book such a flight through him? 'Sorry, I'm afraid it's got to come through your expedition people.' Game, set and match. It now looked as if I would have to take the cost of an airfare into my financial accounting.

There was one last hope. A newly appointed Director of Naval Recruiting (a 'four-stripe' Captain) was coming up to my office to pay a call and introduce himself. On the day of his visit, I went through the normal marks of respect and courtesies prior to a forenoon of escorting the Director around his Lincoln estate and introducing him to the rest of the local team. With that successfully out of the way, the Director, the Commander and I went out to lunch at a restaurant in the shade of Lincoln's cathedral. I was pleased to find that I did not have to bring up the subject of my proposed Arctic visit myself as the Commander had already briefed the Director, and I was soon able to put the Director in the picture regarding my ambitions. When I told him of the difficulties I had encountered with the Navy's own expedition department, he seemed surprised at their inflexibility and offered to raise the matter on my behalf. He did try to put my case but came up against the same rigid application of the rulebook. However, whilst giving me the bad news, he also broadly suggested that any naval officer worth his salt should be able come up with some sort of answer to the problem. He further felt that, if I could come up with a workable solution, it would be likely to gain both the Commander's, and his, approval.

It was to take something of a leap of imagination to crack the puzzle. One of the difficulties that naval recruiting was experiencing lay in convincing the British ethnic population of the value of a career in the Royal Navy. Despite several well-meant enterprises, the cultural backgrounds of many suitably qualified young men and women of Caribbean, Asian and African descent seemed to prevent them from looking to a future with the Royal Navy. By sheer coincidence, I knew of a naval officer who had problems with such recruitment that made ours look tame by comparison and yet he was making often spectacular breakthroughs.

I had met Lieutenant Colin J. S. Perriman of the Royal Canadian Navy during my last visit to Yellowknife. At that time he had been second-in-command of the Yellowknife Canadian Forces Recruiting Centre Detachment and, despite being 'naval', gloried in the title 'Military Career Counsellor'. By now, he was in charge of the outfit as 'Detachment Commander'. I felt the best approach would be to talk to him in person and so reached him by telephone. To my very great surprise he recognised my voice immediately and it was not long before we were chatting like old shipmates. I explained to him my problems and ended with a possible solution: could he invite me out on an official visit to see how he dealt with the cultural problems arising from the recruitment of Inuits (formerly 'Eskimos') and Native Canadians (formerly 'Indians')? With a commendable lack of hesitation, he agreed and even offered accommodation (along with the loan of camping equipment in order to avoid my having to carry such items out with me). Within days, a letter landed on my desk asking that such a visit be arranged. I forwarded the letter to the Commander who gave his approval. Before long a 'duty' flight had been arranged with the RAF, only the final date remained to be fixed.

My next step was to ensure that I had permission from the Canadian authorities to visit the northwest corner of King William Island. Not unreasonably, the Canadians did not see it as continuous open season across their northern territories. Both for reasons of diplomacy (an American vessel had recently entered Canada's northern waters denying Canadian sovereignty and sparking a diplomatic row), and for questions concerning Inuit land claims, official authorisation was necessary. The Canadians were very sensitive about the prospect of damage being done to any archaeological remains by amateurs. There was also the question of safety; rescues were possible but were very difficult to mount and tremendously expensive.

I decided to write to Miss Margaret Bertulli MA, the Arctic Archaeologist at the Prince of Wales Northern Heritage Centre. She had been of assistance during my last visit and had shown interest in our modest 'finds'. I explained in my letter what my intentions were and asked her advice. Her reply was moderately encouraging; quite properly archaeological permits could only be given to qualified archaeologists. She could, however, gain permission for me to simply visit the area to carry out a search on the clear understanding that I would not attempt any archaeological work, nor remove any artefacts I might find (except in cases where there was clearly no relevance to the surrounding area, such as a small item that might have been dropped during passage. Even then, the site had to be marked, photographed, and its location recorded). In a subsequent letter, she was to inform me that 'solo expeditions are not recommended.' Nevertheless, I now had my authorisation.

By now I had permission from my superiors to make the visit, a flight out to Calgary, Alberta, accommodation and equipment at Yellowknife, and a permit to visit the target area. A letter to John Howse of the 'Calgary Light Horse' garnered a kind offer of accommodation whilst in Calgary, and a letter to Constable Terry Zeniuk of the Royal Canadian Mounted Police (who had kept an eye on us during the first visit to Gjoa Haven) provided reserve accommodation at Yellowknife. I also got in touch with Staff Sergeant (Pilot) Bob Martin of the RCMP who had provided such spectacular support during my first visit. He had been studying for his Inspector's examination (subsequently passed) and was currently in the US learning to fly executive jet aircraft. He hoped, however, to be in Yellowknife as I passed through.

There remained just one problem. The money was simply not coming in. I had written to all the organisations and firms who had helped me last time, but none could spare any money. I spread the net further. Some small (but nonetheless most welcome) donations had come in, Mrs Diamond had contributed, as had the members of the Boston Royal

Naval Association. I had a letter with a cheque from a 'Miss Crozier' of Northern Ireland and, after I had talked about the subject on BBC Radio Lincolnshire, a cheque arrived from a Yorkshireman passing through the county on his way home. A substantial cheque arrived as a result of a Lieutenant Commander friend steering me towards a particular naval fund when he heard about my difficulties.

As the time was getting close to when I could expect my flight date I decided to contact Adlair Aviation Ltd, an aircraft charter company based at Cambridge Bay that had been recommended by Bob Martin, in order to get some idea of the costs involved in flying to northwest King William Island. I spoke to Rene Laserich, a son of the owner, and explained what I hoped to do. He was keen on the idea and wished to be encouraging. The main problem, as he saw it, was the fact that the only aircraft he had available which could land in the area at the time of year when I expected to go was a twin-engined Otter. The price of such an aircraft for a round trip was in the region of £5,500. The stunned silence he heard at this figure probably prompted him to add that he would do all he could to help, but at the moment, the best suggestion he could come up with was that I should pay for a one-way flight in order to get me on to the island. With this achieved, there was the possibility of a flight coming over the area whilst on passage to one of the Arctic settlements; it may be possible to drop in and pick me up with the cost shared with the original charterer. There was little I could do at this stage but thank him for his help and suggest that I might be in touch with him later. I then (behind my wife's back) went to see my bank manager to arrange extensive overdraft facilities in case the worst came to the worst.

In the meantime, I felt I should take what opportunities remained to catch up with every item of information on the northwest corner of King William Island that was available at the National Maritime Museum, the Royal Geographical Society library, and at the Scott Polar Research Institute. At the latter establishment I was particularly helped by the Head Librarian, William Mills, and his wonderfully encouraging assistant, Shirley Sawtell. My intention was to pore through every record of every visit to the area from Commander James Ross in 1830 to Trafton in 1989. The visits, as I expected, turned out to be few, and not all had published records. Ross had been followed by the Franklin Expedition in 1847-48. Lieutenant W. R. Hobson RN had reached the area in May 1859 where he was joined by Captain F. L. McClintock RN (this was the last visit to the area by the Royal Navy). Lieutenant Frederick Schwatka of the US Cavalry made his way into the area ('an abomination of desolation') in 1878, after which, the site was left at peace until the 1930s when the Canadian Army's Major Burwash explored the region. He was followed in 1949 by Inspector H. Larsen (who in 1941-42 had captained the RCMP vessel *St Roche* in the first ever west to east voyage through the North-West Passage). The 1960s saw an expedition by over fifty men of the Canadian Army to search the area in conjunction with an attempt to trace Franklin's ships. The Colonel in charge was reported to have said of his time on the island, 'The only thing I need to be happy for the rest of my life is to know that I shall never have to go to King William Island again.' In the early 1980s a group of Canadian dignitaries under the Regional Commissioner flew in for a short visit to the south of the region. Dr Owen Beattie searched the southern end of the area in 1982, with Stephen Trafton locating cairns and tent rings seven years later.

My broad plan was to set up a base at Cape Maria Louisa, about twelve miles from Cape Felix, the northernmost part of the island, with Crozier's Landing a similar distance to the south. From my mid-way position, I could carry out the three main projects I intended to undertake. The first, and most important, was to range as far and as wide as possible in the search for a grave, or graves, connected with the Franklin Expedition. An interesting book that had been published earlier in the year, David C. Woodman's *Unravelling the Franklin Mystery: Inuit Testimony*, suggested that Franklin might be buried on 'two small islets' noted by Inspector Larsen a mile north of Cape Felix. Larsen had failed to reach the islets himself due to a combination of bad weather and fog.

The book also contained a copy of the 'Peter Bayne' map. I had long had an interest in this map for when Bayne sketched the map under instruction by local Inuit – the map itself showing the site of a burial of an 'important man' – he was accompanied by another seaman named Patrick Coleman. When Bayne and Coleman returned to tell their leader, the American Charles Francis Hall, he became abusive. Bayne retreated from the confrontation, but Coleman returned the tirade in full measure and was shot by Hall for his trouble, dying a few days later (Hall was later poisoned by his men). The map was to be passed from hand to hand, re-drawn from memory, and offered for sale to the Canadian Government in the 1930s. The whole thing might be nonsense but, with a name like Coleman being involved, it might be tempting fate to ignore it.

My third project was to visit Crozier's landing. There was unlikely to be much new to be learned from this site and it was more of a pilgrimage than anything else. It was from this spot that 105 men of the Royal Navy had set off in hope of survival, but none were to see their homes again. Whilst passing down the coast, I also intended to have a closer look at the site of Victory Point, the furthest point reached by Commander James Ross in 1830. To help me I would take a copy of the sketch made by Ross of the actual spot.

In addition, I had promised Mrs Diamond to keep my eyes open for and evidence of surface coal that might have been used as fuel by Franklin's people; and the Hydrographer of the Navy, Rear Admiral J. A. L. Myres, had asked me to let him know of any mapping errors that I might come across in order that he might bring them to the attention of his Canadian counterpart (a tall order as my surveying expertise extended little beyond that picked up in my Boy Scout days).

On the day I received my air tickets from the Joint Services Booking Agency, the total cash I had collected came to just half the return fare from Cambridge Bay to Cape Maria Louisa. What was I to do? I could hardly be blamed for pulling out through lack of finance. On the other hand I could legitimately get as far as Yellowknife on my ethnic recruiting project. I decided to take my expedition motto from the anonymous lines I had picked up from a book of First World War verse:

When the day looks sort of gloomy
And your chances sort of slim,
When the situation's puzzling
And the prospects mighty grim,
When perplexities keep pressing
Till all hope is nearly gone,
Just bristle up and grit your teeth
And keep on keeping on.

I decided to 'keep on keeping on'.

3

Stumbling North

My flight instructions stated that I was to fly from RAF Brize Norton to RAF Gutersloh, Germany, where I would be given overnight accommodation before picking up my flight to Calgary. At Brize Norton, my ticket and Flight Order were checked and, with everything in order, I boarded the aircraft for Gutersloh. After a short flight and brief airfield formalities, I found myself at the entrance to the arrival area, surrounded by my baggage, and without a clue what to do next. Was there any transport to the Officer's Mess? 'No Sir, nobody was expected for the Mess on your flight.' Would any transport be available? 'No, there's nothing available.'

Fortunately, the Mess was apparently only about half a mile from the arrival area, so I commandeered a luggage trolley and set off under my own steam through light drizzle. I arrived at the Mess reception desk in a very damp condition and was informed that they had never heard of me. Eventually the Mess Manager (a Warrant Officer) arrived and offered me the only accommodation he had available, a cabin without heating or hot water. It did, however, have a broken window. Of course I accepted the offer along with the towels he managed to dig out for me. I was also to need German currency as I would have to pay in cash for any meals I had whilst in the Mess. With this organised, I was happy to pass the time reading magazines in an anteroom until supper, after which I thought I might usefully spend some of my Deutchmarks in the bar. But the place was packed with a variety of men wearing flying suits who, apart from darting the occasional curious glance at the bulky, bearded figure propping up the bar, decided to ignore him. I had not been there long when one of the green-overalled types staggered into the room bellowing 'The Wing Commander's gone belly up! Let the party begin!' before tripping over a low table and diving headfirst into an armchair. At this I hoisted a white flag and sloped off.

After a chilly night with the rain dripping through the broken window, and an even chillier attempt at a shower, I packed my bags and went for breakfast. From the reception desk, I learned the time of my flight, and the fact that there was still no transport available for me, or my baggage, to get to the airport, so I loaded up the trolley (I had hidden it behind a bush just in case) and set off to the flight reception desk feeling like a tramp. But such a feeling was as naught compared to what I felt upon my arrival.

The Corporal continued to appear dumbfounded as he looked at my ticket and checked his computer screen. 'I can't understand it, Sir,' he repeated. 'Your ticket is correct, and you are listed for the flight, but the aircraft is carrying ammunition and cannot carry passengers.'

The fact that I appeared to have been booked to fly from Germany to Canada in a lumbering, propeller-driven Hercules transport aircraft had been depressing news in itself. Now I found that I could not even get on board. Eventually, a Senior NCO arrived to take a look at the situation. There was a blurring of fingers over the computer keyboard and more head-scratching, which was to grow even more agitated when a young Army officer turned up clutching a large rucksack and also claiming a seat on the same flight. The problem, having by now become far too difficult, could only be answered by everyone ignoring it. Accordingly, the Army officer and I, having handed in our baggage, made our way across the tarmac to the waiting aircraft. The crew seemed unconcerned that they had suddenly acquired two passengers, and the two members of the RAF Police who were carried to guard the cargo appeared to welcome the chance to chat to someone different.

The vast blimp-like body of the aircraft was filled to capacity with large silver containers, only the tail ramp itself providing an area where our baggage could be stowed. A narrow passage down the port side of the aircraft was lined with canvas seating facing inboard. To combat the noise of a thousand joints screaming in unison, earplugs had to be worn. Above this the pilot could be heard giving us details of the flight: 'Do not let the fact that you are surrounded by twenty tons of paraffin detract from the pleasure of your flight.' He went on to further add to my catalogue of frustrations by announcing that our first port of call was to be RAF Lyneham, in Wiltshire; just the news I needed after having left England only the day before.

After we had landed, we were told to leave the aircraft and proceed to the flight lounge. There an RAF Police sergeant told us that there was likely to be a two-hour delay. Outside, beyond a wire fence, a large crowd watched an air display which included the RAF flight display team, the Red Arrows, whose antics helped to pass the time away (despite the fact that I could regularly see them training from my garden, and often drove through their smoke trails on my way into the office).

At last the waiting was over and we boarded a small bus which took us the short distance back to the aircraft. Once again we threaded our way down the fuselage and were strapped into our seats with our ears plugged. The engines built up to a climax: and then died down. The Loadmaster appeared from the cockpit door and told us that there was a minor problem and we would have to disembark once again. After another hour watching the air display from the flight lounge we were taken back out to the aircraft and repeated the procedures for take-off. Yet again the engines had to be shut down. As we clambered out of the side door, the pilot came round to us and suggested that we stay with the aircraft as the problem was (this time) really minor, and it was not worth going back to the flight lounge.

With the flight crew back on board, and ourselves strapped in, the Hercules finally lumbered into the air. At last I was on my way, even being able to ignore the pilot who informed us that, as we were late in departing, our fuelling stop at Keflavic in Iceland would have to be missed, but he 'thought' he had enough fuel to get him direct to our new first stop, Gander in Newfoundland.

Thirteen hours in the back of a Hercules cannot be recommended. The horizon is generally limited to the sheer wall of a very large crate about two feet from the eyes. By turning and kneeling uncomfortably on the bar of the canvas seat it was possible to look out of the tiny windows at the tops of clouds creeping past. The tedium was broken by the regular distribution of sausage rolls and confectionary bars whilst the Loadmaster brewed excellent cups of tea. Otherwise it meant reading or attempting to sleep.

I awoke from one of these naps to take a look out of the window. There, through gaps in the clouds, I could see icebergs floating southwards past the coast of Newfoundland. Before long we were at Gander where the RAF Detachment were bemused to find passengers on the aircraft. However, they shrugged their shoulders, put us down as 'crew', and obtained accommodation for my new soldier friend and me in a local hotel (the unfortunate RAF Police having to remain on the aircraft to guard the contents of the crates). Gander itself had little to offer the short-term visitor and seemed to consist of little more than a collection of low, angular, buildings painted in pastel or primary colours. Apart from the staff of the hotel and the occasional passage of a small truck, the place seemed to be deserted. In the event, I was pleased that there was little to distract me from my chief aim of getting over any jet lag as quickly as possible.

The following morning, after a breakfast of coffee, eggs 'over easy' and cinnamon toast, I re-joined the aircraft, which headed for Winnipeg as a fuelling stop. Whilst the aircraft was being re-fuelled, I called in at a Canadian Airforce detachment to see if they could tell me what the chances were of a flight from the Airforce base at Edmonton (north of Calgary) to Yellowknife. There was a flight due to leave within a day or two, but it was fully booked up. I then telephoned Colin Perriman who told me that a military aircraft was actually coming into Calgary about four days after my arrival, which would be able to give me a lift up to Yellowknife.

For hundreds of miles before touching down at Calgary the earth beneath resembles a monstrous tiled floor, each square 'tile' of exactly the same size with complete perfection marred only by the occasional small lake or modest rise in the ground breaking into the pattern. Then Calgary itself rises from the floor, its heart a cluster of elegant towers that glinted silvery and copper in the early evening sun. Surrounding this spectacular centre, the suburbs spread, scored with arrow-straight streets and avenues. From the air the whole city gained a dramatic backdrop from the purple peaks of the Rocky Mountains.

The aircraft landed and taxied close to the main airport buildings. With the engines stopped the Loadmaster opened the side door and I dropped to the ground followed almost immediately by my baggage. As I pulled the bags clear of the aircraft I was approached by a very military looking type who slammed his foot down and wagged a wide salute at me. 'Lieutenant Coleman, Sir? I'll just get your bags on the transport and get you through customs. You have a party waiting for you in the Arrival Lounge.'

This, I felt was much more like it. It was probably nothing more than the innate courtesy of the soldier (who I think was a Warrant Officer) but whatever it was, it felt like a breath of fresh air and, having said goodbye to my fellow passengers, I took a seat in the car and was soon heading towards the Arrival Area. There, much to my delight, I was greeted by David Parker, a friend I had last seen over lunch at the George Hotel in Stamford, Lincolnshire. He was one of the 'Calgary Light Horse' who had looked after Jim Crack and me during our first arrival in Calgary. His role in life was to persuade film makers that Alberta was a good place to practise their art and since we had last met he had taken the actor-director Clint Eastwood up in a helicopter and persuaded him that the Rocky Mountains would make an excellent site for a film about Wyoming.

Along with his wife and son, David took me into Calgary to the home of another 'Light Horse' member, John Howse, where I was to stay for the next few days. I was delighted to find a barbecue in full swing. Apart from John and his wife, Inga, there was Bill Larkins (former Lieutenant Commander RN), Bob Shaw (former Lieutenant RN), Campbell (former professional footballer, now energy consultant), and Rick Valentini (dentist). Everyone wanted to know my plans, and when it became clear that I was to have a few days in Calgary before heading north a scheme was immediately put into action for a night to be spent at Rick's log cabin in the foothills of the Rocky Mountains.

In less than forty-eight hours after my arrival in Alberta, I found myself being driven along a grassed-over track that led through woodland to Rick's authentic western-style log cabin. From its western facing veranda ('deck') the ground fell away to a valley before rising again to tree-covered hills forming a pleasing skyline. The same view could be gained from inside the cabin through a large picture window. As the wood-burning barbecue would take some time to reduce its fragrant fuel down to a useable, glowing pile of ashes, it was decided that I should be taken down to the valley to be shown a genuine beaver dam. The journey through woods and across meadows was enlivened by Rick falling full length into a deep stream with only his straw Stetson hat and a hand clutching a tin of beer showing.

The dam itself was of great interest to me as I had never seen such a thing before. Completely spanning the thirty feet or so of the river's width, it rose up about three feet from the surface of the water. Its base must have been close to six feet in width. I had been assured that there would be little chance of actually seeing a beaver as they are shy creatures. But, much to my delight, a large beaver swam out from under the centre of the dam and made its stately way upstream. I had now come face-to-face with the prime reason for the foundation of the Hudson Bay Company which had, in turn, triggered off the exploration of the Canadian Arctic. Having ticked off the sight of a beaver on my 'to do' list, I put my energy into fishing in magnificent surroundings.

After a night of disgracefully large barbecued steaks, cold beer and a deep sleep, and awakening to the valley filled with mist, I only had one complaint. I had been assured that, throughout the night, I would hear the woods ringing to the sound of coyotes howling at the moon. I didn't. I was somewhat compensated on the way back by the glimpse of a large grey wolf loping through a clearing in a wood, and by the unexpected sight of a field full of buffalo.

My stay in Calgary was completed by a call at the excellent Glenbow Museum to get a 'feel' for the history of the area I was about to visit. John Howse then took me to the airport to meet the aircraft that was to get me up to Yellowknife. It turned out to be an old friend, the same bright yellow military Twin Otter that shared its hanger with Bob Martin's RCMP Otter. The pilots – Captains Ken and Wayne – were cast from the same mould as almost every Canadian Serviceman that I was ever to meet. Relaxed, friendly, and informal, their almost casual attitude providing a cover for a professionalism that was second to none and a competence that was constantly reassuring.

Remaining at a height of about fifteen hundred feet, the aircraft clawed its way north over vast forests sliced through with wide survey lines of such length that they could be seen from outer space. After some six hours of flying over the interminable forest followed by a flight over the Great Slave Lake – slate blue from horizon to horizon – we came in to land at Yellowknife and taxied up to the hanger I remembered well from my last visit. Within half an hour, the two pilots and I were at Wayne's house enjoying a meal provided by his wife. As we waited for Colin Perriman to join us, Wayne showed me his collection of firearms. He asked me what weapon I was taking up north with me as 'polar bear repellent'. When I told him that I had, as yet, to organise such a weapon, he offered to lend me one of his and came up with a .303 bolt action rifle. It was almost identical to the type of rifle I had done my basic training with more than thirty years before. Manufactured in the year of my birth, it had much of the front part of the wooden 'furniture' beneath the barrel cut back. Apart from that modification, everything else was the same as I remembered from rain-soaked days on Portsmouth's Tipnor rifle range.

Eventually, Colin Perriman arrived, short, broad, and with a wide smile he looked a most unlikely naval officer, dressed in baggy tee shirt and baseball cap. There was a slight problem with accommodation as Colin was in the throes of house moving and so I contacted my first reserve, Constable Terry Zeniuk. Terry was more than obliging. He was somewhat concerned that the room he had offered me was only available for two nights as he and his wife had a lady guest and her new baby coming to stay. This did not matter as Colin would have sorted himself out by then, and I was further delighted to find out that the lady visitor was none other than Gillian who had shamed me into leaping an ice-stream and climbing a vertical (alright, nearly vertical) snow-wall at Pasley Bay on the Boothia Peninsula two years before. She was now a mother with a little girl named Ariel.

On the following morning, Colin wasted no time in trying to get my quest off the ground. As the General commanding the northern area was not in town, it was arranged that I should meet his Chief of Staff, an Airforce Colonel who looked immaculate in blue flying overalls. He proved to be friendly and with a positive attitude towards my scheme until Colin gently raised the question of a flight onto King William Island. Then his smile faded as he recounted the problems he had with his limited budget and how he 'did not feel comfortable' with the idea of providing a flight solely for my use. Had he been able to help, he assured me, he would have done, but he could not see how when subjected to his funding difficulties. There was one tiny ray of hope. 440 Squadron (with whom I had flown from Calgary) might just have some training-time left which could be usefully deployed in the same direction as I was heading. It was a slim chance, but one worth pursuing, so Colin arranged a meeting in the officer's mess with the Squadron's commanding officer.

The Major was a workmanlike individual who had earned a first-rate reputation flying jet fighters. He now he seemed happy to plod around the sky at the statelier pace provided by a Twin Otter. When I put my case to him, he thought long and hard and eventually came up with a possibility. The Squadron could do with some further training in landing float planes and if I was prepared to pay for their accommodation at Cambridge Bay, there was no reason why the landings could not take place on the lakes in the region of Cape Marie Louisa on King William Island. I readily agreed. Just at that moment Captain Ken entered the mess with a large black cloud in tow. He had just been in touch with Cambridge Bay to find out about the conditions on King William Island and had learned that the proposed lakes were still frozen over. Any chance of landing on floats would have to wait. The best that could now be offered was a flight up to Cambridge Bay when the Squadron visited the area on a reconnaissance flight prior to the visit of a VIP to the settlement. The only snag was that the flight was not due to take place for another eighteen days. Could I justify waiting around that long, especially as I had expected to be on King William Island within a week of leaving England? Having thought it through, I decided that it might look just as bad on my return to have abandoned the enterprise having got so near. I would not be completely at a loose end as I could use the time studying at the Prince of Wales Northern Heritage Centre. I sent a fax to the Commander telling him of the situation and began to plan how best to use my enforced stay in Yellowknife.

Colin took me to the military stores department to pick up my Arctic camping kit. I collected a magnificent double-thickness sleeping bag, a stove, a rucksack, and rations for eight days. The rations came in 25lb boxes, each box containing eight separate meals. I took one box of luncheons and one of suppers, local advice suggesting that the breakfast rations were not worth the effort. As each meal was extremely substantial, I did not feel any undue alarm at being on the island for fourteen days on eight days rations through the belief that, with the strong possibility of bad weather restricting movement, there was bound to be days when only a minimum of food would be required. Equally, I had to think very carefully of the weight I intended to move around the Arctic, as small aircraft are sensitive to being overloaded.

Weight was a consideration as to the tent I intended to use on the island. The Canadian military had offered me a substantial tent intended not only for the Arctic but also for four men. Although, no doubt, an excellent item of kit to have, its sheer weight made me look elsewhere. Both my ideas on the rations and on the tent were, at the best, misguided, and, at the worst, simply naive. Colin had a tent of his own which he offered to let me use should I require it. It seemed to be ideal as it was very light and easy to erect. The fact that it was made of material that was probably thinner than the shirt I was wearing escaped my attention.

When we went to collect the items from the military stores, Colin, who believed that I was under-supplied with rations, had added another 25lb box of lunches; an act for which I was soon to feel extremely grateful.

The time had now come to announce my presence in town to the Prince of Wales Northern Heritage Centre. I reported to the reception desk and asked for Margaret Bertulli, the Centre's Arctic Archaeologist. When she arrived to meet me, I found she had not changed at all in the two years since we had last met. She would still go straight for the throat should anyone threaten her attempts to treat the entire Canadian Arctic as one large archaeological dig. Well-meaning travellers who had come in proudly bearing an Inuit bone or part of a sledge they had picked up tended to leave with a heavy limp after Margaret had explained to them that such items were best left in situ where the experts could study them. Fortunately, I had already agreed to her stipulation that I should not attempt any archaeological work so she had not placed any barriers across my path.

There was little that Margaret could do for me at this stage. She was shortly to be leaving to undertake archaeological work at Dealy and Beechey Islands and her time was being used in preparation for her visit. However, she did introduce me to the ladies who operate the Centre's library and archives and who were to prove a great help and encouragement to my immediate efforts.

The one contact that I desperately wanted to make had so far eluded me, so it was with great delight that after ringing Bob Martin's number without success for many days, I heard the telephone being picked up. Apparently, after arriving back in Yellowknife the day before, he had heard that I was in town and was going through a process of elimination in an attempt to find me. I was soon round his house sampling his ice-cold Margaritas and bringing him up to speed with my progress so far. After talking late into the night we arranged to meet at his house the following day for brunch. Sadly – or so it seemed – he was leaving Yellowknife after only two more days to take up a new appointment in Ottawa.

The previous night, whilst we were reminiscing about the last time I had been in Yellowknife, Bob had given me a copy of David C. Woodman's book *Unravelling the Franklin Mystery*. Before I turned in, I opened the book and started to go over the problem of the 'Bayne Map' again. The map did not make sense as it was. For a start it had a promontory marked as 'Victory Point', an area which had been already been searched several times without result. Even more relevant, perhaps, was the fact that the map did not fit that stretch of coastline at all. There was only one deep bay and that was Wall Bay, some fifteen miles to the north of Victory Point. And there was only one way of making the map fit the Wall Bay area, and that was by turning it upside down. Upside down? Could it be possible that, during the time since the map was first sketched to its purchase by the Canadian Government in 1930, it had been redrawn? Perhaps redrawn from memory, a faulty memory? Perhaps passed on by word of mouth? It might all be nonsense but as we allowed our ample meal to go down, I talked it over with Bob. He agreed that if I was to try and find the site on the map, Wall Bay looked a better bet than Victory Point.

Bob ran through the names of the people who, he felt, might have been able to help me. The list was not very long, and it soon became clear that just about everyone on it had already been tried. It was now only a matter of hours before he would be leaving Yellowknife, and he had set his heart on getting me a lift onto King William Island. There was just one last hope. Sat at his desk, Bob trawled through the names of aircraft charter companies in the area to see if there were any names he could recognise, and 'contacts' that he could use on my behalf. The first couple of telephone calls provided sympathetic replies, but no aircraft in the region. I paced up and down beside his desk as he rang Aero Arctic Helicopter Services and spoke to the Company's President, R. W. T. ('Bob') O'Connor. 'Bob, I've got a small problem and wonder if you can help out.' He went on to explain who I was and what I was attempting to do. I continued to pace up and down as I heard my pilot friend say 'Coffee sounds great, we'll be there in about ten minutes. He put the telephone down, looked up at me and said with a broad smile, 'You might just owe me big.' I took this strange abuse of the English language to indicate that there was just the possibility of success in the air.

We walked up the path towards the entrance to a large red building marked AERO ARCTIC. Immediately inside the door was a small, dark, lobby area and a reception counter. Maps could be seen pinned to walls giving the place an atmosphere similar to a World War One flying base not far behind the front lines. A tall, well-built and casually dressed man appeared at the counter and introduced himself as Bob O'Connor.

Another figure came into view, shorter and wearing a baseball cap. This was Bob's brother, Frank. They listened whilst I – with a dry mouth – outlined what I hoped to do. Bob O'Connor asked a few highly pertinent questions about Franklin and my reasons for searching for him. Satisfied that I was serious in my aims, he thought for a moment before saying 'I haven't anything actually in the Cambridge Bay area, but I'll have a helicopter just to the south in about a week's time. We could probably get that to Cambridge Bay without too much difficulty, and then onto King William Island. Let's have a word with the pilot.'

Within moments I was shaking the hand of 'Buck' Rogers. A former Canadian Air Force pilot who had served in the Gulf War, Buck spoke with a deep, slow voice as he asked me to explain exactly where it was I wanted to go. He was very concerned about the weight of my kit as enough fuel would have to be carried to get us from Cambridge Bay to Cape Maria Louisa. The two-hundred-mile journey would stretch him to the limit of his range. Extra fuel would therefore have to be carried in cans which would be enough, it was hoped, to get him to the US radar station on the south of King William Island, or to a similar base on Jenny Lind Island in the south of the Victoria Straits. At either of these places, it would be possible to organise a couple of barrels of fuel to get him back to Cambridge Bay. Whatever the problems, Buck was game to give it a go. All I had to do was to be at Cambridge Bay by 10 or 11 July; the actual day would depend upon his work to the south of Cambridge Bay.

Bob O'Connor then asked me how long I expected to be on the island. I explained that I thought fourteen to sixteen days should be enough to get my work finished. He then told me that although he did not have a helicopter in the area, he would have a small aircraft that should be able to land providing that there was at least 1200 feet of level ground to get down on. Bob Martin offered to let me have some reels of fluorescent tape to mark out a landing strip.

Within the space of half an hour I had gone from being resigned to a long wait with an uncertain end to an organised system of transport to get me on and off King William Island. My clumsy attempts to express my gratitude were waved aside with the view that Franklin was of enough interest in himself to trigger the kind of help that I was being offered. Before I left, Bob O'Connor handed me a hand-held Global Positioning System (GPS). With my gratitude expressed yet again, Bob Martin and I went off to celebrate with a thoroughly English late lunch of fish and chips in 'old' Yellowknife (1930s). I was not to see him again on this trip as he had to leave for Ottawa and his new appointment. As with my previous visit to Canada, he had played a vital part in my efforts to get to where I was going. His enterprise and initiative, his goodwill and support lived up to every expectation of the Royal Canadian Mounted Police and I was deeply indebted to him.

Colin Perriman was concerned that I should try out my rifle on the local range and arrange a firing through a Warrant Officer in the Canadian Airforce and a member of the Yellowknife Gun Club. The trial shoot proved to be an interesting experience as the rifle turned out to have a couple of defects that needed to be sorted out (which the Warrant Officer offered to undertake). In the meantime, Colin offered me an opportunity to fire his Winchester rifle. This weapon had a telescopic sight and, as I brought the sight up to my eye and squeezed the trigger, the resulting recoil slammed the sight back onto the bridge of my nose and blood poured from the resulting cut. For several days I had to walk around looking as if I had got into a brawl in one of the bars patronised by miners or Indians that Colin insisted I visit ('Don't ask for a glass to drink from. Laugh if anyone tells you a joke').

I also spent much of my remaining days at Yellowknife in the research facilities at the Northern Heritage Centre. In the library, the ladies produced books and maps that were relevant to my visit, and Margaret Bertulli took me into the museum's storage area to see the wealth of artefacts that had been collected from around the Arctic.

Before leaving Yellowknife, I arranged with the RCMP to collect one of their radios from the detachment at Cambridge Bay. Through Colin, I booked a flight to Cambridge Bay. Almost incredibly, the cost proved to be almost the same as a civil flight from the UK to Calgary, a sharp reminder of the cost of living in the far north of Canada. The runway at Cambridge Bay seemed quite incapable of taking a modern jet passenger aircraft. It looked far too short and its surface of crushed rubble seemed designed to inflict the kind of 'foreign object damage' so feared by aircraft maintainers. But as I watched silver metal hoods slide down to cover the engine exhausts and produce a braking effect, I realised that to the pilots on this regular run there was nothing startling about the huge plume of dust that billowed out in our wake. At the red painted box of a terminal building I (along with a group of tourists visiting the Cambridge Bay to hunt and fish) was met by transport from the Arctic Islands Lodge. With our baggage on board, we were soon off to the settlement itself, about a mile from the airport.

I signed in at the hotel and was asked by the young woman at the reception desk what I was doing in that part of the world. When I had briefly explained my mission, she insisted that I met her father, Lyall Hawkins, who had always had an abiding interest in the Franklin story. As the owner of the hotel, he had designated a room on the first floor of the hotel as 'The North-West Passage Room' and he took me up to see his collection of Arctic prints that lined the room's walls. We had a long talk in the hotel's restaurant where we were joined by others including his wife, Margaret. It soon became apparent that the story of Franklin had lost none of its potency in the very region of its tragic conclusion. The current notion that was foremost in the minds of much of the population of Cambridge Bay was the possibility of finding Franklin's ships. With my turning up in the search for the man himself, the local Franklin lobby received a boost to their frequent gatherings in the hotel's dining room.

The following day (after awakening to a grey sky and flurries of snowflakes) I learned that Buck Rogers would be 24 hours late in reaching Cambridge Bay. With a day to spare I decided that I would take a closer look at Cambridge Bay.

One of the main interests I had in the settlement was the fact that in the summer of 1852 Captain Richard Collinson RN brought HMS *Enterprise* eastward from the Baring Strait through Coronation Gulf to spend the following winter frozen in at Cambridge Bay. This extraordinary feat of navigation was to win praise from Roald Amundsen, who found it extremely difficult to find his way through in the opposite direction with a much smaller, steam-powered vessel. Unfortunately, Collinson, despite sledging journeys up the Victoria Strait, narrowly missed King William Island. Had he steered just a few miles further to the east he might have uncovered the tragedy long before any light was eventually shed on it.

As I walked along the unmade roads of the settlement, I could see that, for the most part, the bay was still locked with ice. A meltwater stream, however, fed one arm of the bay, and the higher temperature of this water had forced the ice back for about half a mile down the length of the inlet. From where I stood, looking across this stretch of open water, I could see the silver-grey bones of an almost entirely submerged wooden vessel. This was the *Maude*, a ship designed and owned by Amundsen himself. She had ended up in the hands of the Hudson Bay Company and had fallen into disrepair, eventually sinking at her moorings. I had trodden the decks of the sainted Nelson's HMS *Victory* as her Officer of the Day, and now hoped for the chance to clamber over the decaying remains of a vessel once sailed by the great Amundsen himself (whom I had never quite forgiven for reaching the South Pole before Captain Scott).

To reach the *Maude* I had to walk for about a mile and a half around the bay. Eventually, I reached the settlement's cemetery with its row of new white crosses erected at a recent mass burial (the deceased have to be stored during the winter until the ground can be dug deep enough to prevent the graves from being disturbed by Arctic foxes). At that moment a vehicle drew up alongside me and the driver asked me, with a hint of a German accent, if I would like a lift. When I told him I was headed off to the *Maude*, I was surprised to learn that was exactly where he was going. I climbed into the back of the truck and joined two baseball-hatted passengers. The driver introduced himself as Willy Laserich – the owner of the charter company I had contacted from England to ask about the cost of a flight to Cape Maria Louisa. To call Willy Laserich 'famous' in arctic terms is to underplay his impact on the area. He had earned huge respect for his medical evacuation work, trips numbered in thousands and which included having six babies born in his aircraft whilst still in the air. Despite over 50 years without any form of accident, he had been charged 250 times with disregarding flying regulations and only once had he been found guilty; the judge gave him a $250 fine, to be paid 'at his convenience sometime over the rest of his life'. When he went to pay the fine, Willy found that someone else in the Arctic community had already paid it. The other passengers were Pete and 'Smitty', two pilots who were hoping to do some flying during the Arctic summer. As we rounded the tip of the inlet, I explained what I was doing in the area.

The *Maude* was three-quarters submerged about thirty yards offshore and I expected to get no closer than the edge of the beach. Willy, however, could do better than that. On a small spit of land just to the north of the wreck an Inuit had erected a tent for his summer fishing camp. Willy 'knocked' on the tent flap and the occupant poked his head out. He was then asked if he could take us out to the *Maude* in his small boat, a request to which he agreed (Willy, it seemed, knew everyone and everyone knew Willy.)

The ship had gone down by the port quarter and rested on the bottom with her stern and most of her port side underwater. We boarded at the point where the port side emerged from the water and made our way along the ten feet of ship's side towards the small forecastle. From this dry platform with its rusting winch I could see into the iced-up hold of the vessel with its solid timber 'knees' and thick cross-members. The *Maude* had been designed on a similar principle to Nansen's *Fram* so that if she was trapped in the ice the pressure on her sides would simply force her up to sit on the ice itself. By doing this the ship would avoid being 'nipped' and having her sides caved in by the enormous pressure. Now only four feet at the reinforced bows remained above the water with strip metal 'chains' reaching up from her starboard side to support shrouds and masts that no longer existed.

That night, as I chatted to Lyall Hawkins and the others at the Arctic Island Lodge, he leaned towards me and quietly told me that the hotel would be happy to cover my costs for the two days under their roof. Yet again, I had cause to be grateful to the Canadian attitude that readily offers help where help is seen to be needed.

The next day, having packed my bags and dressed ready for the business I was about to embark upon, I called at the local RCMP Detachment to pick up my loan radio. The Constable asked me to complete a 'Wilderness Report' as he checked and tested the radio. I was gratified to hear the instrument crackle loudly as he swung through its frequencies whilst remarking, 'These things go on forever.'

Back at the hotel, I found Buck Rogers tucking into a revolting peanut butter and jelly sandwich. When he had completed his elegant repast, we loaded up the hotel's truck with my kit and set off for the airport. There, on the other side of the concrete apron, sat Aero Arctic's tiny two-bladed Bell 206B helicopter, dwarfed by a North West Territory's Air Boeing 737. Whilst Buck checked the weather and confirmed that fuel would be available for him at Jenny Lind Island, I ferried my kit across to his increasingly fragile looking helicopter.

There were a number of tasks that had to be completed before we could get airborne. Buck pumped up and checked the pressure of his rubber floats before removing from the aircraft a number of items that could be left behind. He had to retain his large survival bag and a heavy metal pump that would allow him to fill his tanks up from the fuel barrels at Jenny Lind Island. There were a number of fuel containers that he had to fill in order that he had enough fuel to get him to the supply barrels. Fuelling the aircraft itself was something of a pantomime due to the fact that without wheels, it had to be flown the 25 feet from its parking position to the fuel pumps. Finally, stripped of every unnecessary item in order to reduce weight, we wafted up and headed just north of east towards King William Island and Cape Maria Louisa.

At 150 feet above the tundra I could see small groups of musk ox and caribou threading their way between lakes and meltwater pools. Once we were beyond Admiralty Islands and over the ice of the Victoria Strait, Buck pointed out seals basking among the towering, jagged peaks of blue-white ice that rose like miniature Alps in all directions. Some enormous blocks of ice sat like grand cathedrals whilst others with squared corners tilted over like stately mansions frozen in the act of sliding into a glittering swamp. This was the stretch of ice that had beaten Franklin and his men. How a sailing ship, even one assisted by a puny, primitive steam engine, could have hoped to have forced its way through such a barrier was beyond imagination.

Soon a thin, sandy-coloured line appeared on the horizon. It was King William Island. We headed directly for Cape Maria Louisa with the fuel gauge looking decidedly uneasy. As we crossed the coastline it was easy to believe that the barren land below was flat and easy to cross. The only obvious high ground was a flat-topped hill several miles to the east.

Everything else looked like a sand-coloured desert pockmarked with brown-green tundra, lakes and pools.

After two hours in the air, Buck brought the helicopter down just to the north of a wide meltwater stream that flowed westwards towards the Cape. The ground underfoot was not sand but rubble, which extended north, south and east as far as the eye could see. To the west, beyond a shingle crest, lay the dramatically ridged ice of Victoria Strait. We unloaded my kit and the fuel containers and took turns to pour the fuel into the helicopter's tanks, the other standing downwind in order not to lose any of the precious liquid to a stiffening breeze. With that task successfully completed, there was little else for me to do but shake hands with Buck, thank him, and wish him a safe return journey.

With a final wave from its pilot, the helicopter banked away and headed southwest. It was not long before the sound of its engine faded and I found myself surrounded by silence.

4

The Peter Bayne Map

With the southward departure of Buck Rogers and his helicopter in the direction of Jenny Lind Island, my most urgent task was to get my tent erected. The breeze that had greeted my arrival showed signs of a further stiffening that would do little to help control the fragile shelter. In the end, it was only by means of placing heavy rocks inside that I could feed the flexible rods through their sleeves and get the tent up, as the tent pegs were useless on such a base.

Now that my accommodation was up, and my kit stowed inside, the next most important thing to do was to set up and test my radio. Having run out the aerial, I switched it on, but no dial flicked, no sound emerged. I changed the batteries – but still nothing happened. It was an interesting experiment in personal reaction. There was absolutely nothing I could do about the situation; there was no back-up and no-one to turn to. The trick was to forget all about the radio as a means of communication and use it as a pillow.

Fortunately, my stove (manufactured by a firm named 'Coleman') worked very well and, using water from the adjacent stream, I soon had water boiling ready for a mug of coffee. My thermometer, which I had placed on the ground about 4 feet from the entrance to my tent, registered 2°C above freezing in the early evening air. I had intended to somehow hang my small white ensign and 'sledge flag', with its design of a crown and anchor alongside the badge of the City of Lincoln (my birthplace) on the side of the tent. But, now with a useless radio on my hands, I could find a better use for its aerial as a flagstaff. Soon my flags were fluttering in response to a sharp wind sweeping up from the south.

Having secured the base camp, it now seemed prudent to have a look at the estate – in particular to search for a suitable place to mark out a landing strip for the aircraft that was to pluck me from this place in sixteen days' time. The ground in all directions consisted almost entirely of sand-coloured fist-sized rocks. Here and there, a few yards of brown-green indicated a strip of tundra grass. To the north, and across the river to the south, the distance to the horizons was impossible to gauge as the rock-strewn terrain merged into itself and lost all definition. Occasionally a lone, darker, rock would add a speck of interest to an otherwise bland vista, but it was impossible to use them to gain perspective as their size could not be appreciated.

About two hundred yards to the east of my tent, the ground rose into a low north-south ridge breached by my stream. Through the gap, several miles away on the far horizon, could be seen a substantial hill with snow draped around its shoulders. I crunched my way across to this eastern ridge and walked up the ten feet to its crest. The far side sloped away to a vast area of tundra, interlinked lakes and ponds. Through my binoculars, I could follow the Arctic bog into the distance, right up to the high ground and seemingly beyond. The tundra continued to the south behind the western eskers whilst to the north, an esker ran almost directly eastward as far as the eye could see.

As I turned around and headed back in the direction of my tent, I became aware for the first time of a strip of land projecting into the ice on the far southern horizon. Even with the naked eye I knew what it was and through my binoculars with the ice glistening in its foreground, I found myself looking at Franklin Point. At its base the land curved eastward creating Back Bay. I had seen the area from the air but now I knew that, within days, I would be walking the eskers towards that fated shore.

Keeping close to the stream on my left, I continued to walk westward towards a ridge that formed a bottom edge to the view of the ice in Victoria Strait. As I breasted the rise, I could see that I had about a mile to go before I reached the ice itself. A large patch of tundra occupied much of the space in between. Looking at this view of the shoreline, I thought of the few people, apart from the occasional native Inuit, who had come this way. Commander James Ross and his small sledging party in 1830, and in 1859, Lieutenant Hobson passed through to find the cairn and its enclosed message at Crozier's Landing before returning to his ship, the *Fox*. He was followed from the south by his expedition leader, Captain Leopold McClintock. It might well be that I was the first Englishman and the first member of the Royal Navy to walk this stretch of coast since McClintock's and Hobson's sledge parties made their way back to their vessel one hundred and thirty-three years previously.

Lieutenant Schwatka of the US Cavalry passed through in 1878 and, in the twentieth century, Canadians Major L. T. Burwash and RCMP Inspector H. Larsen had made their way up the coast. In 1981 Dr Beattie's party had reached the river flowing just to my left but had turned back south in the belief that any further progress north 'would be impossible on foot' (*Frozen in Time – The Fate of the Franklin Expedition* by O. Beattie and J. Geiger). His party was followed in 1989 by the American Stephen Trafton, who had based his camp near Cape Felix and had journeyed south to Crozier's Landing for a brief visit.

Skirting around the tundra (although not carefully enough as the muskeg managed to get in an early bite when one of my boots disappeared into the sand-coloured sludge), I made my way down a gentle slope towards the ice. The distance between my tent and the edge of the Victoria Strait was about a mile and a half and, as I approached the shoreline, what had appeared to be a ragged edge to the ice became a jumble of jagged peaks, giant slabs heaved on top of one another, and blocks of ice carved by nature into fantastic shapes. Just to the left of my view, a huge barrier of ice reared up higher than a two-storey building and as wide as Admiralty Arch. Where the warmer water of the stream reached the shore, a shallow curve of clear water had been carved into the retreating ice. Water vapour rose in spirals from the water causing a light mist to hover above the scene, filtering the evening sun as it dipped towards the west. The colour of the ice as it rippled out from the shore took on a surprising amethyst shade that deepened in patches to purple shadow.

To the north, the eskers began to grow markedly rougher with large rocks standing vertically like tombstones. I was soon to become used to the varieties of rock terrain along the coast, both to the north and the south, but nowhere else was I to find so many of these larger rocks huddled together. Perhaps they had been brought to this spot by ice forcing itself up the shore and depositing 'erratics' collected from a distant coast.

I had been surprised by the lack of bird life in the region. During my time on the south of the island there had been a wide variety of birds continually being scattered before our approach. I was therefore pleased to come across a white-plumaged ptarmigan, its peculiarly feathered legs and feet acting as snow-shoes as it warily kept a respectful distance. Despite it being now well into July, the bird had retained its white coat – at least one bird was not convinced that summer had arrived. The next two birds I came across were more interested in each other than in me. Circling around each other on a meltwater pond on the edge of the shore were two King Eider ducks, their bright yellow beaks bobbing in unison, the male distinguished by a white bar on his wings.

I then heard a sound that I had not expected but was to hear on more than one occasion in the days to come. A deep booming sound rolled across the ice followed by a hissing as ice floe rubbed against ice floe.

The mass of ice pouring down the Victoria Strait was moving irresistibly southwards and breaking under the pressure of its own prodigious weight. The ice that reared up in front of me looked rock solid, however, and invited the wanderer to clamber over its ridges. I chose to work my way round and over the ice towards the south side of the river as its miniature delta estuary was laced with muskeg, mud, and water too deep to wade through.

There is something fine about being on ice, especially once the pile- up ice wreckage on the shore had been crossed and the more gently rounded ridges could be traversed with relative ease. The pulse raced, nevertheless, when patches of paper-thin ice gave way beneath my feet causing me to drop an inch or two on to solid ice beneath. What created this thin layer of ice just above the main ice remains a mystery to me. Other excitements came with the gauging of the width of small leads in the ice. Could I jump across safely? Was the snow-covered ice where I intended to land solid? Confidence very quickly asserted itself only to be rapidly dented when I came across a solitary line of footprints, very broad footprints almost a foot across. The only animal I knew of that could have made those tracks was a polar bear. For a while I treated every pressure ridge and block of ice with the greatest caution with my rifle loaded and ready, Colin Perriman's warning that 'It is very embarrassing if the next time you see daylight is from the rear end of a polar bear' constantly in my mind.

No polar bear made its appearance before I left the ice and stepped back onto the shingle beach about half a mile south of the river. Having made my way up towards the eskers forming the eastern skyline I came across a number of small areas of dark green tundra. Whilst crossing one of these, I was made to leap back in alarm when a sharp 'beep!' sound came from the ground beneath my feet. The sound (closely resembling that of a horn on a child's bicycle) squeaked out once again when I put my foot back onto the grassy tundra. It was followed by a faint scratching noise. I took out my knife and gently scraped away the surface soil causing the beeping to become even more indignant and the scratching even more frenetic. With my probing into the ground revealing nothing I turned my attention to a flat rock about a foot across. With the point of my knife I flipped it up and, much to my surprise, exposed a mass of waving claws and bared teeth. There, on its back, defiant to the end, was a small brown-grey lemming in an absolute spitting fury. The animal continued its whirring of legs for a further two seconds and then stopped, looking up at me with ill-disguised annoyance. Then, having decided that I was harmless, it rolled over and waddled off to hide under another rock with just its head peeping out to keep an eye on me. It had never occurred to me that a four-footed animal with enemies that attacked from above had no choice but to roll on its back to defend itself when escape was denied. A design fault.

Behind me, another booming sound rolled across the ice as I made my way up the rocky eskers with the stream now on my left. The ankle-turning tundra on this side (south) of the stream covered a much smaller area, so it was not long before I could see once again the dome shape of my tent with the flags fluttering just beyond. As I had seen nothing resembling a suitable landing ground on the northern side of the river, I now began to roam around the huge expanse of open rock desert before me in search of 1200 feet free from aircraft hazards. Time after time I began pacing along an apparently flat stretch of shingle only to find that it sloped away to one side or the other, or dipped and reared over hummocks that had previously blended in with their surroundings. Sometimes a large crack, sufficient to rip the wheels off a small aircraft, would slice across the intended runway, or large rocks – too big for an individual to move – would render the ground useless. In the end I gave up. There was no point in marking out the longest stretch I had found as it was far too short, and there was no sense in risking an aircraft and pilot by giving a false guide. Any aircraft landing would have to be at the pilot's discretion rather than my inexperienced guesswork.

By now I had been roaming around for some time and it was getting very close to midnight. The air was proving to be quite chilly as the sun reached its lowest point (several degrees above the horizon) and so I made my way back towards my tent, clearly in sight to the north as a dark dot against the now greying eskers.

When I reached the stream flowing by my camp, it appeared in full spate, the waters being fed by the ponds and lakes of melted snow just to the east. In addition to the sounds of the water as it tumbled over large rocks projecting from its surface, there also came a persistent clattering and clinking as the rocks themselves were rattled along by the water's force. As a result of the volume of water, the stream proved difficult to cross without getting a boot full of icy water, and I only succeeded by throwing rocks from the bank to form a series of stepping-stones.

I had not spent long at my tent before I had the stove 'flashed up' and my metal coffeepot filled with water from the river. The Canadian rations I had brought were quite excellent in that each meal came ready prepared in a sealed metal foil bag. All that was required was for the bag to be placed into boiling water for about five minutes. It was then taken out, the top torn off and the hot bag held in an associated cardboard sleeve whilst a spoon was used to get at the contents. Not only was there the very minimum of washing up, but the water could then be used to fill my flask and make a hot drink. The ration packs also contained biscuits, chewing gum, boiled sweets, soft drink powder, toilet paper, toothpicks, and a form inviting the user's comments on the contents.

With a warm meal inside me, I curled up in my sleeping bag with the sun burning through the tent and the gurgle and rattle of the river for a lullaby. In the distance I could hear the weird, haunting cry of loons (Great Northern Divers) as they paddled around the lakes beyond the eastern esker. When McClintock wrote his account of his 1859 expedition, he included a recipe for 'Loom Soup' – he always referred to the birds as 'looms'. The recipe is as follows (vegetarians and vegans might care to look away at this point):

> Take 8 looms, skin and take off the two white lumps near the tail; clean and split them into pieces; wash them well, also the livers. Put them into a large saucepan, and cover them well with water, and boil for four or five hours. An hour before serving up, put in ½ lb. of bacon cut up small; season with pepper and salt, 2 tablespoonfuls of Harvey source (blood?), a little Cayenne pepper, half a wineglass of lemon juice, a teaspoonful of ground allspice, and a few cloves; thicken with 4 tablespoonsful of flour mixed in cold water, then stir gradually into the soup. Add ½ pint of wine, after which let it boil for a few minutes. The result will be 4 quarts of rich soup.

It was no wonder that he ended his naval career as an Admiral with a knighthood.

I awoke the following morning at about six-o-clock to a glorious day. There was not a cloud in the sky nor a breath of wind to raise any life in my flags. It seemed to be a perfect day for me to test my 'inverted' Peter Bayne map theory. I breakfasted on chicken soup and coffee with the intention of having a substantial meal on my return. I estimated that I would be away no more than four to five hours as Wall Bay, my destination, was little more than six or seven miles to the north.

It also seemed a good time to try out my GPS. If I was to find anything, accurate positioning would be invaluable for retracing steps over such a bland terrain. I took the instrument out of its case and switched it on. There was a flickering as its tiny screen came to life. Unfortunately, the words that came up were 'LOW POWER – REPLACE BATTERIES'. Still, there was no problem, as Bob O'Connor had given me a spare set of batteries. But alas, they proved to be the wrong size. As my compass tended to point in any direction it felt like, often finding great attraction in my rifle, any geographical co-ordinates would have to be by God and by guess.

There remained just one task to be completed before I could leave. With no runway on which to use the fluorescent tape provided by Bob Martin, I used it to spell out the words 'BATTS U/S' and 'I OK' in letters tall enough to be seen from an aircraft. I also used the tape to make a large arrow showing the direction in which I was headed. There was no reason for anyone to come looking for me and I had deliberately avoided any commitment to a regular radio check-in. Nevertheless, I felt it could not hurt to let any passing aircraft (should such an unlikely event happen) know what my situation was.

At last, with my rifle over my shoulder, camera round my neck, and with two boiled sweets and a packet of chewing gum in my pocket, I set off northwards, my navigation plan consisting of little more than ensuring that the ice of Victoria Strait remained on my left.

Walking over the eskers proved to require little effort with the slight risk of turning an ankle on the loose rubble. I had walked for about twenty minutes when I came across a deep, dry riverbed. At some time in the past this must have been a very powerful river, but the lakes that supplied it must have shifted their position for only the stream by my camp now made its way towards the ice. I continued for another ten minutes before I lost sight of my tent and began to understand the problems of judging distances in a land with nothing to act as a reference point. A horizon that seemed half an hour away could still be trudged towards two hours later.

It was not long before I breasted a ridge and saw the curve of Wall Bay come into view. Through my binoculars I could see that the shoreline of the bay appeared to be a shingle beach with a large area of green-brown tundra stretching eastward. Patches of dark water indenting the ice suggested that rivers or streams were flowing through the tundra to arrive at the shore.

By now the type of rock I was walking on had changed from the fist-sized stones to which I had become used. Now it was small, flat, frost-shattered slabs that clinked beneath the feet like broken crockery. As I crunched my way across this desolation I came across what appeared to me to be the bones of a human hand. I have no claims as an anatomist, so I could only depend upon what I believed the bones of a hand to look like. There was another possibility. Somewhere in the back of my mind I could recall seeing the arrangement of bones in the flipper of a seal. As I recalled, they also looked somewhat hand-like. Against that, I was at least three miles from the shore and the bones in front of me looked extremely 'hand-shaped'. I photographed the macabre object and continued walking.

With the weather continuing fine and with no need to rush, I spent a considerable time wandering up and down the eskers that crossed my path just on the off chance that I might find something of interest. Consequently, it was past midday before I reached the southern edge of Wall Bay. A number of things attracted my interest as I looked north across Wall Bay from my slightly raised vantage point. The ice in the bay looked like a glistening pile of rubble as it had been forced to pile broken floe upon broken floe by the pressure that reached as far back as the Pole itself. In places the ice had even forced its way beyond the shingle beach, and massive fingers of white and blue reached out as if to claw at the land. Elsewhere, large blocks of ice remained stranded beyond the beach, their undersides eroded by the water from streams that washed beneath them.

The land to the east of the bay remained low for about two miles until a north-south esker sealed off the eastern edge. This esker was met at right angles by another that formed the northern rim of the stream-laced tundra filling the low land. Surprisingly, the air was filled with the sweet, sickening, stench of decay that I could only imagine came from the green vegetation that covered much of the low ground.

Most disappointing of all, however, was the evident fact that the bay could not fit the Peter Bayne map in the inverted style that I had hoped for. The esker on which I stood flattened out as it approached the ice and, not even by the greatest stretch of the imagination could it be made to fit the pattern of ridges shown in the map. I was not overly concerned, as the idea had only been a casual thought, and was really only a prompt to a field reconnaissance, itself no more than a test of the type of terrain I would be passing over on longer projects.

I decided to press on into the bay to have a closer look at the beach route I hoped to be following in a few days' time. With the ice piled up on my left, I made my way along a shingle strand intersected at intervals by patches of clinging mud. My path was also crossed by wide, shallow streams whose waters had nibbled away at the ice exposing the clear water beneath. Two much broader streams proved difficult to cross, especially as their banks were lined with stretches of muskeg whose discovery I frequently only made when my entire foot and half my leg would disappear into the sandy-grey slime. It was at such times that I learned the value of the canvas gaiters I was wearing – also of the spare laces I carried, as the part of the gaiter lace that went under the boot was frequently cut by the sharp rocks on the eskers.

As I approached the first of these larger water hurdles, I had almost decided that I had gone far enough for my first trip. I was feeling a little tired and quite hungry (my boiled sweets and chewing gum had long gone) when, with quite a shock, I saw through my binoculars what appeared to be a human figure standing on the skyline of the western edge of the northern esker. I quickened my pace, pausing only to peer through my binoculars to see if the figure had moved. It remained rock steady as I came within about three miles its position. I raised my binoculars again and took a long steady look. I could see a tall thin shape with a round top. I then realised what I was looking at. It was almost certainly a beacon or radar reflector. I had seen a similar object as I had flown over the eastern edge of the Admiralty Islands with Buck Rogers.

With the excitement over, I turned to retrace my steps along the edge of the bay. At one stage, in an effort to find a less soggy way around an area of muskeg, I came across a number of footprints made by large animals. My guess that they had been made by musk ox but even if they had been caribou, it would have made a small dent in the argument often put forward by theorists that the Franklin Expedition could not have survived on that part of the coast as no large animals would have been found to provide food. However, whatever had made the footprints was no longer around. In fact, there was not even the smallest sighting of animal or bird life in the Wall Bay area.

It was late afternoon before I completed my way around the southern curve of the bay. I had decided to take a look at the coastline on my way south rather than try to find my way back along an inland route. Heading south from the bottom edge of Wall Bay, I came to the top of a low ridge and saw laid out before me yet another bay. This bay was not as wide as Wall Bay nor as deeply indented and did not show on the map I had obtained at Yellowknife. In itself, this was not too surprising as the only surveys of the area had been done from the air, and I had seen already how the encroachment of ice could alter the shape of the coastline (a number of small islands off the coast that were marked on the map as 'Position Doubtful' did not actually exist).

Of even greater interest than the bay itself was the fact that, through my binoculars, I could clearly see an esker running towards the ice on the bay's southern edge. At the western tip of the esker, a spit of broken land jutted out into the Victoria Strait – just as shown in my upside-down version of the Peter Bayne map. Any tiredness I may have felt vanished as I strode out along the rocky beach towards the southern esker. After about half a mile, I could see that the beach was parted by a wide stream fed from pools in an area of tundra that ran parallel to the curve of the bay. I had no option but to find a path through this muddy area. I pressed on along a line about two hundred yards from the beach, keeping my eyes down to avoid the mud and muskeg whilst aiming at the tip of the southern esker. It was whilst looking down that I came across a most surprising find. Six large rocks had been placed to form the shape of an arrowhead on a dry patch of ground – and the arrowhead was pointing towards exactly the spot to which I was heading.

With my floundering through the swamp-like tundra having gained a new urgency thanks to the guiding arrowhead, I reached the base of the southern esker in about twenty minutes. I had only just begun to make my way up its gentle slope when, ahead of me, I saw two stone circles; clear evidence that someone at some tim, had erected tents on the spot. To my unskilled eye there was no clue as to their age or their builders (would the Inuit have put their tents on the northern slope of an esker? Perhaps, if it helped to keep the mosquitoes at bay).

There was a pleasing regularity to the size of the rocks used to make the circle that suggested that they had been thoughtfully chosen. I mounted the ridge and walked a few yards along its crest to where the Bayne map indicated a cairn. This turned out to be a hopeless quest as the area before me – looking out to the ice – was a veritable quarry and any number of cairns could have been built and destroyed without leaving any trace. There was one unexpected item that could be seen: a large bole of a tree, silver-grey after years of exposure to the elements, sat upright on a rock shelf slightly above the shoreline. As I was hundreds of miles from the nearest tree I was intrigued as to how such an object had got to where it now rested.

As hopes of finding a cairn faded, I now turned to head eastward along the esker. I passed the tent rings and walked for about 150 yards when my heart leapt into my mouth. There, just on the northern slope of the esker crest were six grave-shaped mounds. Four of the mounds were about six feet in length whilst the remaining two were closer to eight feet. Had I found the source of the Peter Bayne map? There was no sign of the cement that was supposed to have been involved in the burial. And on several occasions, I had seen corrugations in eskers that looked similar to the mounds that I had before me. The tent rings could have been made by Inuits or by any of the other travellers through the area – any of whom could equally have made the arrowhead of rocks nestling in the tundra. I had no intention of despoiling a site that could – just – be of significance. If Canadian archaeologists thought the site worth a visit it would be better left to them.

Leaving the tent rings and mounds behind me I continued heading southwards. I had by now been walking almost non-stop for about eleven hours and was beginning to feel a little tired.

My stomach was also reminding me that I had not eaten since my breakfast of soup and boiled sweets. I was also concerned at the fact that the weather had begun to turn, with the sky darkening and a strong wind getting up from the south. It was not long before I had to lean into a biting wind as I made my way around large areas of tundra that erupted along the coast.

Eventually, in an effort to avoid the patches of muskeg that alternated with the tundra, I drifted inland and was pleased to get the firmer ground of the rocky eskers beneath my feet. Unfortunately, I had so concentrated on placing my feet and bowing before the wind that when I did look up I found myself in a huge saucer-shaped depression. From where I stood, to the rim of the horizon in all directions, there was nothing but a rock desert, the rocks now resembling the porous cinders from an unimaginable furnace. But there was no need for concern, all I had to do was to make my way back to a position from where I could see the ice. I trudged my way toward the seemingly ever-receding skyline which, I felt, was blocking off the view of the Victoria Straits. It seemed an eternity before I climbed the ridge – only to to see yet more rocky ridges ahead of me. I used my binoculars to scan the horizons, but all I could see was yet more rows of ridges. It looked as if the Devil himself had run a garden rake through a vast desert of rock. There was just one directional clue left to me. Although the sky had clouded over and looked dull and threatening, a slightly lighter shade indicated where the sun was. Despite the fact that the sun did not set, and merely circled the sky, the arrival of late evening had to mean that the sun was somewhere in the western half of the heavens. If I was to walk towards the light I would have to reach the coast, then it was simply a question of turning left in the hope that I had not walked past my camp without seeing it.

The next two hours sapped my morale. Each ridge seemed to be always further than it seemed upon first sighting and when the ridge was reached and climbed, yet another empty plain with yet another ridge marking the skyline had the effect of draining the spirit. I was continually afflicted by the strangest desire to lie down and go to sleep in the belief that when I woke up I would be back at home in my warm bed (when I later brought my journal up to date I wrote 'This was a very real sensation' and underlined it). The answer, I found, was to avoid constantly looking at the horizon, instead keeping the eyes down whilst keeping the mind occupied, usually by trying to remember the words to boyhood hymns ('Onward Christian Soldiers' was the most frequently resorted to). Eventually, as I reached the crest of an esker I saw, far off on the horizon, the white gleam of the ice. With proof at last that I was heading in the right direction The wind, which fortunately came from the south and so did not have quite the same edge to it as it would have coming from the west or the north, blew un-impeded across the low ground with a ferocity that took me by surprise. It howled, screamed, and moaned and almost took me off my feet when it gusted unexpectedly. Nevertheless, I could stand any amount of violent wind now that I had the coast in my sights.

After a further hour and a half of crunching towards the west I decided that I could turn south whilst rigidly keeping the coast in view. Even this was not as easy as it sounds. Time and time again the ground would dip, and the ice would disappear behind a ridge causing me to scrabble over the rocks in alarm for fear that I was lost yet again.

Before long I began to recognise the shapes out on the ice, in particular the massive block that could be seen almost west of my camp. My spirits soared immediately and were not even depressed by my encountering a large area of muskeg-riddled tundra. I pressed on over the squelching ground until I heard a sound carried on the wind that brought joy to my heart. It was running water and the clatter of loose rocks. It could only be the stream that flowed past my tent as there were no similar water flows between Wall Bay and Cape Maria Louisa. I was on its northern bank within minutes. Turning left (eastward, uphill and upstream) I trudged on, the tundra making progress very heavy.

Ahead of me I could see an esker and the beginning of rocky ground. I scanned the skyline frequently in the hope of seeing my tent, but the only variation in the barren vista before me was the occasional black rock. Whilst peering hard through my binoculars I caught sight of yet another dark, pointed rock. As it passed through my field of vision, the rock suddenly ducked out of sight behind the esker only to re-appear a second later. My first instinct was to believe that my tiredness was causing my eyes to play tricks. But, again, the rock disappeared only to pop up once more. I now began my approach to the esker with much greater caution constantly watching this bizarre performance through my binoculars. At last, straining through the lenses to make sense of what I was seeing, I noticed that the rock was not quite as black as it had seemed further off. In fact, it had a slight purple tinge. Then, in a flash of understanding and relief, I realised what I was looking at. My bobbing rock was the top of my tent, battered by the wind. My steps gathered pace and I felt my face, quite unbidden, break out into a broad grin as the palpable tension of the moment drained out of me.

Within a quarter of an hour I was up with the curtsying tent, its agitation mirrored by the flags which streamed out as if at the halyards of a full-powered frigate. Although hungry and thirsty, I was far too tired to flash up the stove and cook a meal. Luckily there was some warm water in my flask from the previous morning's breakfast (it had now gone one-o-clock on the morning of the following day), so I drank that whilst devouring a packet of biscuits.

Removing my boots and socks I found that the constant walking over the rocky ground had severely bruised the soles of my feet. Oddly enough, I had not been aware of any discomfort whilst travelling, but they now began to burn with a vengeance. I put on a clean pair of socks and climbed into my sleeping bag in the sure and certain knowledge that I would be asleep in seconds; but it was not to be. Many years ago, when I wore short trousers and long hairy socks, packets of crisps were supplied with a small amount of salt in a twist of blue paper. The salt was sprinkled amongst the crisps and the bag shaken vigorously. I now know what it was like to be inside that bag. The tent bucked and reared as the wind tried to lift it up by getting beneath the ground sheet. When this failed, the wind retaliated by almost flattening the entire structure so that was only inches above my sleeping bag. The whole battering was accompanied a fearful din as the howling of the wind was joined by the machine-gun flapping of the mini-awning that roofed the tent. I buried myself in the depths of the sleeping bag until exhaustion, at last, took charge and I fell into a deep sleep.

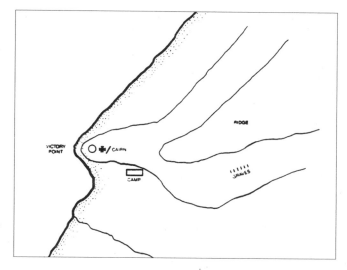

The 'Peter Bayne' map as it appeared in D. C. Woodman's *Unravelling the Franklin Mystery*.

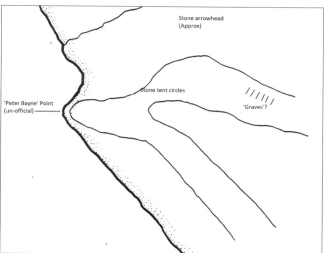

The upside-down 'Peter Bayne' map which may indicate the small point south of Wall Bay.

5

Cape Felix

The three days following my return from the Bayne map foray introduced to me the type of weather that was possible in an Arctic summer. For most of the time the area was enveloped in a dark fog that, at times, restricted the view from the tent to twenty-five yards or less. It was unnerving to walk too far away from the tent and it would only take the slightest deviation from a straight line to have me wandering about in a state of serious alarm until the low, domed shape re-emerged from the mist. I was fortunate, however, in that a wide frost crack ran from the front of my tent directly to the river as a guide when collecting water.

One of the mornings greeted me with a snow shower which, luckily, did not last for long. It was rapidly replaced by heavy rain falling vertically from thick, low cloud. It meant that any attempt to head north, as I intended to do, would have been foolhardy. As it was, I welcomed the chance to rest the bruised soles of my feet.

Whilst waiting for the weather to clear, I used the wire aerial from the useless radio to make a 'stay' or extra 'guy' for the tent. I fashioned it so that I could attach it to any two adjacent supporting rods before anchoring its other end to a large rock. The theory was that, if the wind began to get up I could attach the wire to the weather side, place the rock far enough from the tent to tauten the wire which, in turn, would prevent that side of the tent from buckling. For the most part it proved to be successful, although a shift in the wind rapidly rendered the device more ornamental than useful.

I spent many frustrating hours trying to get something useful out of my short-wave radio. My aim was to receive the BBC World Service, but it was frequently drowned out by static. I could, however, pick up snatches of an Australian radio station, Radio Finland, or Radio Beijing, which would treat me to concerts led by 'People's Worker Singers'. The greatest success I had was in receiving an American programme, 'The Christian

Science Monitor', which kept me up to date with the news despite it being apparently aimed at Africa.

I began to read the Canadian author Pierre Berton's *The Arctic Grail: The Quest for The Northwest Passage And The North Pole 1818-1909*. I found its political correctness grating ('The Indians saved the situation'). It derided the Royal Navy's attempts to find Franklin ('bumbling about in the ice streams'). It was riddled with inaccuracies that included several references to naval officers with 'polished buttons' (never in the history of the Royal Navy) and had lieutenants serving as 1st and 2nd 'Mates' on Royal Navy vessels.

One distinguished captain was mocked for not having been promoted to Flag Rank on his return from the Arctic, a misunderstanding of the system of 'dead men's shoes' of those days. To continue reading this under such circumstances required great endurance – yet, before I left the island, I was forced to read it through four times.

The morning of 17 July put an end to my fiddling with the radio and wading through the book. The sun had burned off the fog and a cloudless sky promised an end to the rain. I rose early and breakfasted on soup, bread and jam, and coffee. Into my rucksack I packed enough food for two days, a thermos flask, a portable stove and gas cylinder, a sleeping bag, a 'bivvy' bag (a bright orange survival bag in lieu of a tent). I also packed my waterproof jacket and its inner liner as I was quite comfortable in my woollen jersey. When packed, the rucksack must have weighed the best part of 40 pounds (half the weight a Royal Marine is expected to carry with ease). When my rifle was included it seemed a great deal more weight than I had intended to carry up to Cape Felix.

Having splashed myself all over with insect repellent, I set off northwards escorted by a dancing cloud of Arctic mosquitoes buzzing in their enraged failure to penetrate my defensive screen. For the first two hours everything went well. I was determined to keep the coast in view and not to be distracted by eskers that ran across my path. I had to keep clearly in mind my aim to get to the very tip of the island, Cape Felix.

My first intimation that all was not to go as smoothly as I had hoped was when I noticed that my accompanying squadrons of mosquitoes had vanished. They had fled before a stiffish breeze which had sprung up from the west bringing with it a layer of sullen cloud that blocked out the sun. By the time that I could see the tree bole through my binoculars standing guard by my tent rings of four days previously, I had donned my inner jacket. An hour later, as I stood on the edge of Wall Bay, I had put on my waterproof jacket, balaclava helmet and gloves. Before me, with the wind now beginning to gain in strength, I could see mist forming over the ice of the bay and on the tundra to its east. Already the esker forming the bay's northern edge was becoming indistinct. For a time I considered the possibility of returning to my base to await for a more promising day: but would such a day ever come? I decided to press on, accepting the fact that I would have to make my way around the ice instead of taking a short-cut across the bay's tundra. I had not long been wading through the cold mud that lined much of the shore of the bay when the fog came down with a vengeance. I seemed to be in a grey, echoing, cave with water running across its floor and a western wall of constantly changing white rubble as I slowly made my way along the edge of the ice. After an hour, I took off my rucksack and sat with my back against a rock as I drank hot coffee from my flask. The rest gave me time to contemplate on the complete and utter desolation around me. The chill dampness that seemed to seep into my body despite my protective clothing, the severely limited horizon that could be hiding unknown horrors and prevented the eyes from grasping at the relief of a navigational feature, the biting wind, and the sickening stench that soured the air of the bay. With the warming drink inside me, I lost no time in getting back onto my feet and heading northwards once again. I had not been walking long after my rest stop when I noticed that the wind had shifted to the north, and it was with great delight that I began to see that the fog was being pushed southwards. Soon, through the clearing mists, I could make out the northern esker less than half a mile in front of me and, before long, was scrabbling up its southern slope hugely pleased to be out of the grip of Wall Bay.

Turning round on the top of the ridge, I could see the bank of grey fog being pushed across the face of the bay with individual plumes of mist being streamed by the wind as if smoke grenades had been scattered in it. By now, I had a clear view to the west and brought up my binoculars to try and find the beacon I had spotted before my retreat from Wall Bay days before. Extraordinarily, there was no sign of the object and I was left doubting the evidence of my own eyes.

Although the wind continued to increase in ferocity, the ground became the best I was to find on the island. The frost shattering of the rocks had produced a level, gravel-like strand that lasted for half a mile north of Wall Bay before dipping down into a wide swathe of tundra.

Cheered by this easy passage, I looked forward keenly to this part of the coast as it was in this area that the American Stephen J. Trafton and his team had located Franklin's summer camp

In a 1991 article published by *The Mariner's Mirror* (v. 77) Trafton had identified the camp as being 'about 1000 feet inland from the shoreline and just north of a series of small braided streams which drain a number of inland lakes'. However, only 'the scattered remains of the torn-down cairn' indicated the camp itself. A further two days of searching by the Trafton team failed to find any additional Franklin remains.

I splashed my way up and down every stream that I crossed in an effort to find the wrecked cairn, but the problem was that the area was studded with random groups of rocks, any one of which could have been the right one. After more than an hour of ranging over the area I decided to return to the southern edge of the tundra and, in one last effort, walk a line parallel to and about a thousand feet from the shore, as far as the eskers forming the northern edge of the tundra.

I found nothing. It seemed clear that the best thing that I could do would be to abandon the search and press on northwards. Just as I came to that decision a thought leapt into my mind. Before he had flown south, Buck Rogers had urged me, 'Use your instinct,' by which he had assumed that over 32 years in the Royal Navy might just trigger a fellow feeling for those others of the Navy who had passed this way 144 years earlier. The thought that came to me was blindingly obvious. If I had been ordered ashore to set up a summer camp I would never had set it up in such a soggy area. In less chilly weather, it would have been infested with mosquitoes, it was difficult to cross, and it was even more difficult to walk more than a few yards without sinking into muskeg.

I made my way to the shingle beach with the intention of starting towards Cape Felix when my eye was caught by a large (3 feet high), black, pyramid-shaped rock that stood in splendid isolation about ten feet from the edge of the ice. It immediately occurred to me what an excellent aiming point such a rock would make when approaching the beach from the Victoria Straits, whether over water or ice. On a coast of almost continuous sand-coloured rocks, the black rock stood out like a beacon. Leaning my rucksack against the monolith, I then tried to place myself in the role of an officer sent by Franklin to set up a camp ashore on this spot (I opted for Lieutenant Gore as his photograph shows him to have been a cheerful sort). There was, I felt, only one order that Gore would have given 'Go for the high ground, boys, go for the high ground.'

In front of me, as I faced inland from the rock, a series of horseshoe-shaped ridges rose gently, one above the other. I set out in a straight line from the beach making my way up the ridges. There were five in all with the top one curving around a small lake. As I walked along the southern crest, I came across a large tent ring. It was to prove an oddly satisfying moment. Could this be Franklin's summer camp? (its first locator, Lieutenant Hobson in 1859, reported that he had found three tents). If it was, why had not Trafton's team found it? Indeed, could it have been built by Trafton and his team only three years earlier? I decided that this latter possibility was unlikely as not only did it look far older, but he was certain to have used modern tents that would not require such a precise stone circle. Slightly to the west of the ring were the remains of what I took to be a small cairn. Even if this cairn was of modest height it would have lined up with the 'marker' rock on the beach giving any sledge or boat an approach line that could be seen from a considerable distance from the shore. On the other hand, the same purpose would have been achieved by the tent itself and my pile of rocks might just have been the remains of a fireplace (Hobson's tents were each accompanied by the crumbled remains of a fireplace).

My delight at finding this spot was rather dampened by the fact that the wind had now brought driving sleet that sliced into me from the north across a darkening sea of rocks (reminding me of the quote from Bunyan's *Pilgrim's Progress* used by Schwatka's second-in-command, Colonel W. H. Gilder, describing this spot as 'the abomination of desolation'). I decided to leave my rucksack where it was as it tended to act like a sail in the wind, pulling me backwards with each savage gust. I removed my thin thermal gloves which had become soaked and ice-cold, pulled up my hood and bent its wire rim into little more than a slit from which I could see the ground a few feet in front of me. Hunched over, and with my rifle across my back, I started off north yet again at about four-o-clock in the afternoon.

The ground I was crossing was fairly easy and provided I did not look up too often (an act which tended to fill my hood with rain or sleet) I was not too disheartened by the rain pouring down the front of my jacket. After little more than a short tramp, I looked up to check my bearings and decided to head for the only clear features in an otherwise flat landscape. These seemed to be the ends of two low eskers, the slate-grey colour of the sky making them appear even darker against the rest of the background. After about a mile of keeping my head down, and with only the occasional directional glance upwards, I became aware that the wind seemed to be easing and the sleet dying away, the sun was even trying to peep underneath the cloud cover. Shaking myself to get rid of as much water as I could, I pulled back my hood and looked upon the strangest sight I had ever seen in the Arctic.

I was within a hundred yards of the eskers I had used to guide my way north. Only they were not eskers.

It was the most beautiful sight I had seen north of the Arctic Circle. Two low-lying mounds, longer than they were broad, lined up north-south, and with a symmetry I had seen nowhere else. It was probably this symmetry which I found the most startling. Whereas everything else that the eye could fall on was totally random in outline and form, here were two features that looked entirely out of character to the rest of the landscape.

Edged by water, and speckled with clumps of purple saxifrage, the place had an almost formal air. My first thought was, 'This is where I would have buried Franklin.' My second was even more to the point; 'This is where Crozier would have buried Franklin.'

I walked over to the nearest and largest of the mounds. It appeared to be about a hundred feet long and about twenty feet high in the centre. The first thing to strike me was that it was made, not of rocks, but of soil. A scattering of rocks and stones lay on the surface of the soil, but the composition was utterly different to the surrounding rocky plain. After splashing through a surrounding area of shallow water, I walked up its side to the top of the mound. There, at the very top, was an area different yet again. A slightly sunken patch, about six feet square (although not regular in outline) was covered in dull red-brown lichen. The possible import of this did not register until I had walked around the shoulders of the mound and reached the northwest slopes. Once again there was a patch of ground covered with the same lichen. But there was a marked difference. Not only was the shape longer than it its breadth, but there remained evidence that it had once been outlined with slabs of rock set vertically into the soil. A closer examination suggested a reason for the lichen being so localised. The area within the edging looked as if it had been surfaced with slabs laid horizontally. With only a thin layer of soil to cover them they had provided a surface for the lichen to flourish.

I then made my way over to the smaller mound. It was about sixty feet long and about fifteen feet high at the centre and surrounded almost entirely by water. An isthmus about twenty feet wide connected the southern end of the mound to a tundra-dotted stretch of esker. This mound had no obvious markings on its surface (apart from the possible remains of tent rings on its northern tip) but, as I looked back at the larger mound, a surprising thought entered my mind. The mounds suddenly looked to me as if they were man-made. Until that point I had assumed that they were natural features, but now the symmetry began to impose itself once more. The largest of the mounds had areas of water on both sides; the smaller had a wide swathe of shallow water on all sides except for its southern tip. The site looked as if it could have been an ordinary tract of tundra until someone had come along and scraped the surrounding surface soil off and piled it up into the mounds. There was no doubt in my mind that almost a hundred seamen, unemployed during the summer months, could have done it with ease.

With my mind still spinning from the possibilities arising from the mounds, I pressed on northwards aiming for a small cairn that I could clearly see about three-quarters of a mile to my left front. The ground could be covered much more easily now that the rain had died away, and the wind had slackened off. To my left, the ice was piled up on the coast even higher than it had been further south, as this part of the island was in direct line with the ice-flow down the McClintock Straits and took the full force of its immense pressure.

The small cairn proved to be little more than a pile of single rocks balanced upon one another. Looking directly eastward, I could see others of a similar type in a line that stretched to the far horizon. Through my binoculars I could see that one of them had been built on the top of a large rock and had the gap-legged look of an Inuit 'inookshuck' suggesting that the whole line was a native construction, possibly built as a navigational aid over the bald terrain.

One thing I could not explain, however, was why the cairn I was standing alongside had a large cross-shaped hole – about three feet deep and the same from side to side – dug just south of it. Peering into the waterlogged cavity, I wondered whether anyone had been up there recently and had tried digging for a buried message. Perhaps it had a natural cause, but I could not think how it could have arisen. The rock markers turned out to be almost certainly direction indicators erected by the Trafton expedition in 1989 – on the other hand, they could have been a line of Inuit inookshucks intended to deflect a herd of caribou towards an Inuit ambush.

One natural feature I spotted almost immediately on leaving the cairn was a most welcome sight. There, grazing peacefully on a patch of tundra, were ten adult musk ox and seven calves, their dark brown shaggy coats covering them entirely except for their eyes, lips, nose, horns and hooves. This sighting was of particular interest as book after book on the subject of the Franklin Expedition trotted out the same tale of a landscape where no animals were ever to be found. Berton, in his *Arctic Grail*, wrote of this area as 'a gloomy, infertile land, barren of game'. Obviously, the herd of musk ox ahead of me had not read his book. I gave the animals a wide berth as, although I would have liked to have seen them retreat into their natural circle of defence with the calves in the centre, I had been warned at Cambridge Bay that they also tend to charge if disturbed. The thought of those heavily horned creatures thundering down on me prompted a healthy respect for their privacy.

I was now probably less than a mile from Cape Felix. As I looked in that direction through my binoculars I got something of a shock. There, quite clear in my lenses, was the beacon I had last seen in the far distance to the northwest of Wall Bay. Now it was standing erect at the tip of the island, about four miles north of the bay. The longer I spent on the northwest corner of King William Island the more I realised that the evidence of the eyes was not always to be believed. As I had stood in the centre of Wall Bay looking up towards the beacon I was actually seeing a type of mirage, a refraction of the light that cast the images of objects and natural features away from their actual sites (the most famous example in the Arctic was Commander John Ross's 'Croker's Mountains' which 'blocked' his western exit from Lancaster Sound). The real thing stood on four red and white legs supporting a large silver disc at its top. As I approached to within a quarter of a mile of the beacon, my attention was distracted elsewhere.

On the north-west slope of an esker were five stone tent rings strung out along the top of a ridge just to the south of a pair of meltwater pools. Three of the rings were large enough to have been the bases of substantial (4-5 man) tents. The remaining circles were too small to have been used as accommodation. To the north of the rings were the remains of a line of three cairns that time, or human intervention, had reduced to piles of rubble. Strangely, as I approached the site, and without any conscious thought process, I felt that I knew what I was looking at.

Once Franklin had become 'beset' in the ice it would have become vitally important to know his rate and direction of drift. With this calculated, the next step would be to find the extent of the ice in the direction of the drift. Then, with the two components together, and with the amount of supplies available to the ship's companies added in, it would be a simple calculation to work out whether or not the expedition could survive a southward drift until clear water could be reached. The amount of remaining supplies would have been easily established. It was also known that Lieutenant Graham Gore, with one other officer and six men, had been sent south in May 1847 – almost certainly to find out the extent of the ice. All that remained was to find out the rate of drift, and to confirm the direction.

The most obvious way to get this information would have been to note the angles of two or three fixed objects ashore. Then, as the angles changed with the drifting of the vessel, the rate and direction could be calculated. But such an option was not open to the Franklin Expedition. All that they could have seen of Cape Felix during the summer would have been a thin, light-brown line on the horizon without a single distinguishing feature. Even the limits of the land could not have been trusted as the piling-up of the ice would have altered the coastal edge.

The answer, of course, was to stand the problem on its head. Instead of measuring against three fixed objects from the ship, why not measure against the ship from three fixed objects? (Two objects would have satisfactory, but three increases the number of angles and provides an in-built check).

The three cairns were in an elevated position, just to the north of the tent rings, and in a line running approximately from east-north-east to west-south-west. The gap between each cairn was about twenty yards. If an imaginary line was taken at right angles from the line of cairns at the centre cairn and extended out over the ice it would head approximately north-north-west (true). The vessels, whose rate of drift was being calculated, would have sat somewhere along this line. A sighting against the same object would have been taken from the remaining cairns and the angles recorded using a horizontal sextant. The movement of the ships through the visible arc could have been observed over time, and the rate of movement calculated.

When, in 1879, Lieutenant Schwatka visited the site, he noted 'We saw some tall and very conspicuous cairns near Cape Felix, which had no records in them, and were apparently erected as points of observation from the ships.'

Modern observations suggest that the rate of ice drift through the Victoria Strait is in the region of a mile to a mile-and-a-half a month. At that rate it would have taken the *Erebus* and *Terror* about six years before they reached open water. One reason why Crozier chose to abandon the ships immediately becomes clear (he would not have known that Victoria Strait has a periodic clearing of the ice – usually once every eleven or twelve years).

Leaving the 'drift calculator' site, I headed for the very tip of the island, under the shadow of the tall beacon. There I was to attempt one of the earliest objectives that I had come to carry out. According to Woodman's *Unravelling the Franklin Mystery*, in 1949 Inspector Henry Larsen of the RCMP had arrived at this spot with the intention of reaching 'two small islets' about a mile north of the Cape. Inuit legend had suggested that this was where Sir John Franklin might be buried. By now, of course, I had very serious doubts about such an idea but, as Woodman repeatedly demonstrates in his book, native myths about Franklin should not be disregarded as being worthless (the official advice about the R. T. Gould map of King William Island of 1927 showing the Franklin-related sites is much more robust on this question – it states that 'The information shown in blue is based upon the various Eskimo reports obtained by these explorers, and probably is not altogether trustworthy').

My more immediate problem was not the existence or otherwise of the islets, but the huge and towering pile of ice blocks that formed a barrier along the shoreline. This barrier did not merely extend, as it did elsewhere, for a matter of a few dozen yards from the coast but could be seen as jagged peaks in the far distance.

I clambered up the base of the obstacle and made my way from the shore between large blocks of ice. I had my rifle loaded and cocked as I had got it into my mind that if a polar bear was to mount an ambush, this was just the place to do it. As I gained height by climbing over the piled-up wreckage, I tried peering ahead through my binoculars to try and find the islets. All I could see was more ice. After a quarter of an hour I had managed to get about two hundred yards from the shore when, with a sudden chill of panic, I realised that not only could I not see any islets ahead of me, I could not see any land behind me. I scrabbled up the nearest pile of ice, my eyes searching beyond the jagged tops of the closely surrounding skyline with every step. Eventually, I found I could see the top of the beacon backed by a thin strand of sand-coloured esker, to the left of the direction in which I had been heading. Even in the short distance I had covered, since leaving the tip of the island I had managed to swing ninety degrees westward. It did not take me long to conclude that this part of the journey would have to be abandoned. I could take refuge in the probability that the islets did not even exist. In his book, Woodman mentions that where these islets are shown on charts (they were not shown on my maps) they were marked as 'existence doubtful'. There were several small 'islands' off the north and west coast of King William Island with the same label that had proved to be non-existent. They had turned out to be 'ice-islands' that had been picked up during the aerial survey of the region. An ice-island is a large amount of soil and rock swept down by glacier ice from the land and carried, possibly hundreds of miles, out on the ice pack. From the air they look like a patch of solid land but, as the ice melts, the rocks and soil fall through to the seabed and the 'island' vanishes.

When I had regained the beach it occurred to me that, as it had now gone ten-o-clock in the evening, I had not eaten since eight that morning nor drunk anything since the coffee I had made in Wall Bay around midday. The desire to keep pressing on, combined with the exciting sights I had come across, had pushed thoughts of sustenance far to the back of my mind. Now they came back with a vengeance. However, there was very little to be done about it apart from sucking on a boiled sweet that I found in my pocket. There was plenty of ice and old snow around which I could use to help my thirst although, even in the Arctic summer, such a temperature-lowering remedy was not recommended.

To take my mind off such matters until I reached my supplies at the 'marker stone' near Franklin's summer camp, I could see, less than half a mile in front of me, a large sturdy-looking cairn on the crest of an esker. It did not take long to reach and as I came up to it, the cairn had all the appearance of a 'recently' built object (by which I understand as anytime in the last fifty years). There was no clue as to its purpose and I did not propose pulling it down as I felt that it was far too recently constructed for me to have any interest in it.

South of the cairn, the terrain flowed in a seemingly never-ending desert of rock shingle. Then again to the south, I could see another, smaller cairn. Could this be the start of a series of signpost cairns leading to the south of the island from its northernmost tip? I set off to find out. This time it took me the best part of an hour before I realised what I had been looking at.

The cairn was part of the chain of cairns (or inookshucks) I had spotted from its westernmost end and, to the horizons on either side of me I could see a line of receding dots as the cairns were strung out to the east and west. They also told me that I was a long way from the track I had followed northwards. I decided to try and pick up the ice-bound coast again by heading southwest.

By now I was beginning to feel tired, but I knew that there was no point in stopping as I had to get to my rucksack and food as quickly as I could. As I clattered across a ridge of dinner-plate sized slabs I was cheered by the sight on the far western horizon of the two mounds on which so much of my hopes rested. Even at this distance, they were clearly etched darkly against a lead-coloured sky, their symmetry out of place in the randomness that surrounded them.

As the faint light of the sun dipped towards its lowest point in the western sky, I could hardly believe my luck when as I reached the crest of an esker and swept the view with my binoculars, I could not only see the ice but also the black rock on the beach against which I had rested my rucksack. Within half an hour I had reached the spot and was sat on the rock, bent forward with my hands on my knees, catching my breath as I looked forward to a warm meal – especially as the temperature seemed to be dropping alarmingly.

My gloves were still wet from the soaking they had received close to this spot earlier in the day, and it was with shivering fingers that I assembled my stove by screwing the long black propane cylinder into the base of the burner. I propped the whole thing up in the centre of a pile of rock and soon had flames at work on the underside of my coffeepot and my first course, in its metallic silver pouch, at the ready. I waited and waited, occasionally putting my hands up to the flames to warm them. Still I waited, but the water refused to boil. Eventually, I had to accept that either the stove was not up to the job, or that the temperature was too low for the amount of heat available. I poured the warmed water into my flask, re-packed my rucksack, put it on my back and headed off south towards Wall Bay. I had decided that the only sensible thing I could do would be to make it back to my base at Cape Maria Louisa as I doubted the wisdom of spending a night out without the benefit of hot food.

Having stuck to the shingle ridge that formed the beach to the west of the tundra below the 'summer camp' ridges, it was not long before I was striding along the closely packed chippings that led to the northern edge of Wall Bay. The view, as I looked down into the bay, was anything but inviting. Spirals of water vapour streamed up from the innumerable meltwater ponds scattered amongst the tundra inland from the curve of the bay ice, and the sickly-sweet stench of corruption soon assailed my nostrils. To the east I could see a horizon formed by a dark esker marking the far edge of the tundra, ahead of me the thin line of the ridges that awaited my arrival on my way south. To my right I could see the shingle beach that ran around the edge of the bay with its huge isolated blocks of ice and the grey-white fingers where the ice reached beyond the beach itself. Over the whole scene nothing moved apart from the drifting clouds of water vapour. A dense, heavy silence lay as cheering as a coffin lid. I decided that rather than go the long way around the beach, I would strike out across the tundra about a mile to the east of the furthest curve of the bay. It was close to one-thirty in the morning as I slid down the esker onto the gloom of the tundra – not a time noted for the spirit to be at its most buoyant.

For a while, the going was not too difficult as I was able thread my way through small streams and ponds using residual shingle that had probably worked its way down from the northern esker. But soon the gravel-like base ran out, and I was reduced to stepping from one clump of tundra grass to another, carefully balancing all the time to avoid toppling into the mud and muskeg that waited in all directions. In was not long, however, before a treacherous patch of greenery revealed itself to be a camouflage for the malevolent muskeg and my leg disappeared up to my knee. I found myself having to heave in the opposite direction in order to avoid sprawling in the foul stuff or finding myself staggering into one of the meltwater ponds, the smaller ones by now freezing over. My particular nightmare was of falling over and putting my arm out to stop myself, only to see it sink into the ice-cold grey slime. I had to concentrate on every single step I took, keyed up with the expectation of having to leap to the next clump of tundra grass in the hope that it was solid. The situation was not improved when it became clear that there was nowhere I could stop and rest. To linger for more than a few moments would prompt a sinking feeling as the ground began to collapse beneath me. I had no choice but to keep moving on (my adopted cheerful doggerel, 'Keep on keeping on', had never seemed so appropriate). My legs, weighed down with mud, began to lose all feeling as in the clearer skies of the east a huge silver moon rose above the eskers.

Even on the largest maps, the distance across Wall Bay does not seem very far. It is probably less than two miles, and yet it was to be two and a half hours after I entered it before I was able to drag myself up the slope of the southern esker where I found a flat rock, no bigger than a pub ashtray, on which I could rest a single knee. I slipped off the rucksack and gulped in the sharply cold air as I fought for breath. I had intended to make a coffee with the warm water in my flask, but my chattering teeth insisted that I drink the water straight from the flask. It was a glorious moment to feel the warm liquid going down inside me but within seconds I was shaking violently. It was clear that there was no better answer to the cold than keeping moving, but I found that even the simple act of putting on my rucksack was almost impossible. Eventually, by getting on all fours and ignoring the pain from the rocks pressing into my knees, I could lever myself upright. I soon realised that I was unlikely to get more than a hundred yards at a time and so made targets of largish rocks at about that distance as aiming points, standing upright at each position until the shivering forced me to get moving once again.

I continued this way for about half an hour when, looking up, I realised that the white sheen of the Victoria Strait had disappeared, and that I had not got a clue where I was, or in which direction I was heading. This, I thought, was just about as bad as it could get. But I was wrong. Turning round to see if I could pick up a bearing from the direction I had just come my heart plummeted to my boots as I saw a monstrous pewter-coloured fog bank just about to fall upon me. Pathetically, I pulled out a sheet of paper and a pencil from my top pocket and I wrote in wavering script 'Franklin might be in mounds 1 mile north of summer camp' and tucked it back into my jacket.

Within minutes of seeing the fog, even the smallest hope of making any progress seemed to be dashed. Then a most peculiar affect began to make itself felt, something I had never known, or even imagined, before. My mind started to talk to me. The words were clear and their meaning unmistakable. I was told that if I did not survive – so what? In a hundred years, who would care less? A lot of people had died on this coast; it would be very easy to join them. All I had to do was to lie down and go to sleep – I might even wake up in the morning and all would be well.

It was extraordinary how real this phenomenon seemed, and how easy it would have been to follow the trend of the messages. I tried shaking my head and putting thoughts into my mind to drive out these dangerous notions. I thought of my home and my family, of my friends. Nothing could hold back the poisonous thoughts for long enough to triumph over them. Then, with my knees ready to buckle, an answer pushed its way through the seductive voices.

When I had left England, the chief controversy ranged around the question of the appointment of women priests by the Church of England. My own view of the matter may be described as a gentle lean towards being in favour of the idea. After all, this was the end of the 20th century; women had shown themselves to be as good (and as bad) as men in a vastly expanded field of endeavour, why not in the clergy? Consequently, I decided to mentally advance the argument to show that the time had come for women priests, both as a means of detaching the mind from the painful feet and legs, whilst at the same time crowding out the all too real whispers gently lulling me to surrender.

Concentrating as hard as I could to the exclusion of all outside (and internal) distractions, I began to reason with myself in the cause of women priests. It proved to be a very interesting experience. Time and again I would find long forgotten sections of the Bible imposing themselves, snatches of Sunday school teaching would suddenly appear, and links in chains of thought would suddenly forge themselves into whole new ideas. As I reached the limit of my knowledge or ability to construct an argument, my mind would spin back to the beginning and I would examine the whole debate in the most minute detail once again. I would repeat the same sequence of thought time and time again as, bent over by the weight of my pack, and with my legs operating as if by remote control beneath me, I crunched and squelched my way down the northwest coast of King William Island.

I could not even guess how many times I ran the ever expanding debate through my mind as the night hours slid by, but it was approaching eight o-clock in the morning (almost twenty-four hours since I had set out) when, whilst plodding heavily through a patch of tundra, the sound of water running over rocks broke into my mechanically churning mind. It could only be the Cape Maria Louisa river. Stooping under the weight of my rucksack, I turned eastward and began to climb upwards to where I knew my tent would be. As I spotted the purple and green domed oasis in its sea of rocks, only one thought now kept repeating itself again and again through my mind: 'Bloody marvellous'.

With infinite slowness, my mind, now seeming to be dead of most emotion and effect, directed my fingers to undo the straps of my rucksack, allowing it to fall to the ground. My rifle clattered down alongside it as, with every inch of my body aching, I unzipped the front of the tent and crawled inside. It took an immense effort not to just curl up and go to sleep, but I knew I had to get some heat inside me. I pulled off my mud-caked boots and gaiters, turning the boots upside down in case of rain and then, with aching slowness, I pumped up the pressure in the stove and, before long, had a pot of water on the boil. The resulting shepherd's pie, followed by cherry sponge pudding washed down by coffee, might not have been the most elegant banquet in the world, but few were devoured with greater relish.

And what of my self-imposed, possibly life-saving, debate on women priests? Having scoured the most remote corners of my (admittedly, limited capacity) brain, I became implacably opposed to the idea. (Not long after my return from the Arctic, I told a bishop of my experience in hauling myself clear of Wall Bay, the beguiling voices, and the mental struggle. He listened politely before telling me that he felt the time had come when God had decided that women should play a more 'inclusive' role in the Church. This gave me the chance to give the most gloriously pompous reply: 'And to whom do you think I was talking?')

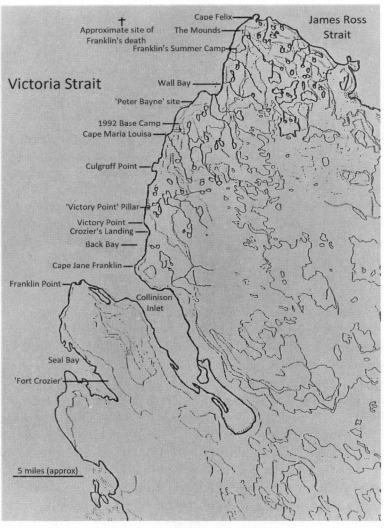

Map of the north-western corner of King William Island.

6

Crozier's Landing

The next two full days after my return from Cape Felix were taken up with resting my tender feet (the soles had, once again, taken a hammering from the rocky terrain) and my aching frame. Taking one of the large cardboard ration boxes, I filled it with rocks and made it into a passable seat so that I could rest elsewhere than inside the tent. The hard, bright sunlight brought with it clouds of mosquitoes that had to be kept at bay by regularly drenching myself in insect repellent. My flags draped themselves limply on their aerial-flagstaff and the temperature reached eight degrees at midday as the only clouds in the sky hung far to the west. Yet again I tried to tune my short-wave radio to the BBC but ended up with American 'hell-fire' preachers for company. When all else failed, I was forced to resort to churning through Berton's uninspiring chapters about nineteenth-century American explorers who, with their 'win at all cost' attitude tramped towards the North Pole leaving a trail of discarded fingers and toes littering the snowscape.

By the third day, I was beginning to feel much more cheerful when the weather decided to take a hand and turned violently for the worse. Back came the wind and the fog (I never ceased to be amazed that both these elements could exist at the same time) whilst heavy downpours of rain provided variation. During one of these nights, I had dropped off to sleep accompanied by a screeching wind, and with my external tent anchor (wire aerial and heavy rock) doing its best to prevent the tent buckling. I awoke with some alarm to find the top of my head (the only part protruding from the sleeping bag) being splashed with icy rain. In the grey light I could see that the wind had changed direction and that the tent was buckling hard over. This had the effect of bringing the small awning covering the netting at the top of the tent to an almost vertical position causing its edge to be lifted by the wind, allowing the rain in. At the very least, I would have to get out of the tent to re-position the anchor.

As I did so, I noticed a large green mosquito net that had been provided by the Canadian military store. Having got dressed in my waterproofs, I went out clutching the net and, with the anchoring wire repositioned and tautened, I tied rocks into the net's four corners and threw it over the top of the tent. I sacrificed some of the inner space of the tent as the rocks weighed the sides in, but it put an end to any further risk of rain spraying the inside of the tent. The next day I had to empty the tent, trusting that a stiff breeze would dry a number of dampened articles of clothing whilst I used a pair of thick woollen sea-boot stockings to wipe dry the tent's groundsheet.

As a counter to the sheer volume of noise brought by the wind, there were times when a thick fog would bring a totally deadening silence enlivened only by the occasional loon whirring overhead. I would stand outside the tent keeping perfectly still just to listen for the next sound, any movement I made over the ground producing an unnaturally loud din.

One sound had become unexpectedly diminished. During one of the days I was weathered-in after my first visit to Wall Bay, I built myself stepping stones across the adjacent river in the belief that when I headed south I would need them to get across the fast flowing waters. However, as the days passed, the volume of water coming down grew less and less until a whole host of stepping stones and easy fords were revealed. I assumed that the draining of the overflow from the lakes and ponds beyond the eastern ridge had begun to slow down. There were now places where I could jump across to the other bank where only days before, I could not have pole-vaulted.

The fifth morning after my return from Cape Felix proved to have the promise of a sparkling clear day. The air was sharp, I felt fit, and the level hook of Franklin Point beckoned beyond the glinting ice of Back Bay. I packed my rucksack with survival bag, inner and outer jackets, and a flask of hot water. For food, I took packets of biscuits, boiled sweets, and bars of chocolate. After a breakfast of chilli con carne and bread and honey, I turned my luminous tape direction indicator arrow round, loaded up, crossed the river, and began the walk southward.

The view directly to the south was one of a vast plain of sand-coloured rocks. It had looked as if it was level, but on closer examination, proved to be a series of gently rising and falling folds. Despite this I was still surprised to find that, after an hour of walking, I could still clearly see my tent through my binoculars. When, at last, it looked as if I was to lose sight of my camp, I searched out a large rock and made a pile of smaller rocks on its top. When this was to about to disappear behind a ridge, I built another marker, and continued marking my trail as the ground over which I was travelling became more ridged. This time I was determined not to get lost.

Walking southward was proving to be considerably easier than the northern journeys. The weather continued fine; my chain of rock piles would provide an easy return, and the ice, although disappearing behind the occasional ridge, remained on my right. After a little over three hours of walking, and with the seventeenth marker just about to hide behind an intervening esker, I built my final rock pile on the crest of a ridge overlooking the outflow of watery mudflat that was Culgruff Point.

This patch of land made its mark in the history of Arctic exploration as the final base of Commander James Ross in his sledge journey down the coast in 1830. From this position, he decided that he and Thomas Abernathy, the second mate of the *Victory*, would leave their three companions for one final push to the south. His food supply was dangerously low, and the *Victory* was many miles away on the east coast of the Boothia Peninsula, but determination overtook his caution. I looked around, more in hope than expectation, for any sign of their stay, but could find not a trace.

From Culgruff Point I had determined to keep closer to the beach as I believed that this would have been the route taken by Ross. In fact, an esker ran almost parallel to the ice and, apart from the occasional patch of tundra, was to prove an easy path to follow. Just as was the case with most of the eskers on the coast, its composition varied from chippings of gravel-like proportions to sections piled with boulders. The soles of my feet were beginning to make their presence felt, and I welcomed the chance to splash through the soggy tundra to cool my boots down.

After a further three hours walking, I found my path barred by a river. The river itself came as no surprise as it was clearly shown on all the maps. What was surprising was its size. Fed by a number of large lakes to the east, the river had cut itself a respectable gorge through the eskers and now flowed wide and deep into the notch it had made for itself in the ice. During his journey down the coast, Dr Beattie had decided that the river could not be crossed (he and his party were carrying all their camping equipment including an inflatable boat) so he took to the ice to walk around the obstacle. It took him two hours before he made his way back to the shore beyond the southern bank. I, on the other hand, was much more lightly laden and felt that I could manage to get across using the many rocks that protruded from the racing waters.

I stepped off, watched by a gaggle of loons sunning themselves on the edge of the ice. My route was carefully planned and involved wading upstream before picking up a series of rocks which went in the opposite direction. Using the old naval principle of 'one hand for the Queen and one for yourself' (or, in this case, 'foot') I made my way steadily across, at times having to lift a large rock out of the stream and put it ahead of me to bridge a gap too wide to jump. The crossing only took about ten minutes but provided a few diverting moments. Once I had gained the southern bank I made a pile of rocks to indicate the start of the return route.

The low beach esker continued on southward and with it the easy travelling. My target was Ross's 'Victory Point', which my map showed to be the northern 'corner' of Back Bay (in fact of 'Irving Bay', which was not named on the map) and was about an hour's walk from the river. Clutching a copy of the sketch drawn by Ross, I could easily see the great sweep of the bay as it appeared (falsely) to continue round to the tip of Point Franklin, now less than eight miles from me across the ice. Although I was still more than a mile from Victory Point (or 'Point Victory' as some maps have it), it seemed to me that I was already close to the place where Ross had made the drawing. As I looked around, the sight of a brilliantly white snowy owl sitting on a three-foot tall column of rocks on the crest of an esker to my left caught my eye. Following a hunch, I walked up the ridge towards the rocks, the owl taking off and flying to a new perch inland. Now that I was up at the level of the rocks, the scene – with the exception of the rock column which appeared to be about six feet tall – began to imitate the sketch more and more closely. I moved slightly to the left so that the feature assumed the same position with regard to Franklin Point as it did on the drawing. I could not escape the conclusion that this was the spot on which Ross had stood at midnight on 30 May 1830. The view was identical except for the fact that no Union Flag braved a southern breeze, its flagpole wedged among the rock pile. I then moved closer to the rocks themselves and was staggered to see that the snowy owl's perch was not just a random pile of rocks but a number of slabs placed carefully on top of each other. Was this indeed Ross's Victory Point?

The actual position of Victory Point had caused considerable debate amongst academics over several decades. The vast majority of the studies in the matter were done by poring over books, documents, and maps, the only method available to the scholars and researchers. I could only go on the evidence before my eyes. The time it had taken me to reach my pile of rocks south of the wide river, on the journey from Culgruff Point, was almost identical to that taken by Ross and Abernethy to reach their Victory Point; we both claimed three hours (although Ross claimed the distance to be over ten miles, I measured three and half). The view that presented itself to me as I stood behind the Snowy Owl's perch was unquestionably the same as in Ross's sketch. I was at least thirty feet above the beach level and about fifty feet from the western edge of the esker. Intriguingly, just hidden from view behind the rocks, a small section of the beach jutted out into the ice. The triangular peninsula, no more than fifty feet wide at its base and jutting out a similar distance, could have made a distinctive shore-line feature quite capable of being named 'Victory Point' although it could have been hidden from Ross by ice. Furthermore, although often described as a 'cairn', the formation in front of me was clearly a pillar made of flat slabs placed on top of each other. The document found by Hobson and McClintock in the cairn at Crozier's Landing noted:

> This paper was found by Lt Irving under the cairn supposed to have been built by Sir James Ross in 1831 four miles to the northward, where it had been deposited by the late Commander Gore in June 1847. Sir James Ross' pillar has not however been found and the paper has been transferred to this position which is that in which Sir James Ross' pillar was erected.

Did Gore deposit the paper in the column of rocks in front of me, only to have Irving later describe it as a 'cairn'? Captain FitzJames (who wrote this part of the document) preferred to describe it as a 'pillar' just as Ross had described it, but, in its absence, assumed – for no clear reason – that it had once stood where the ships' companies came ashore.

On his return, Ross submitted his charts to the Admiralty who, in 1836, issued the chart 'Arctic America', Sheet II, No 261. Shown on this chart is the section of the northwest coast of King William Island visited by Ross. On it a position is marked as 'Cdr. Jas. Ross Furthest 1830'. It looks to be exactly the spot marked by my pillar of flat stones.

Back on the beach esker, I made my way south and reached the 'official' Victory Point after about three-quarters of an hour. From this point the main bay curves dramatically eastward before returning westwards at Cape Jane Franklin and the entrance to Collinson Inlet (the presence of which Ross was unaware of when he named both the Cape and the far Point). Also, from the same position on the northern tip of Back Bay another, smaller bay, 'Irving Bay', serves as a gentle indentation to the larger bay. Running parallel to this bay a twenty-foot-high esker follows its entire curve. As I walked along the beach past this esker, with much of its western slopes still covered in snow, I came across a number of tent rings, some too small to have been used for accommodation so may have been for store tents. As was the case elsewhere, it was extremely difficult to assess the age of these remains, or whether or not the Inuit made them. I continued on for about half a mile to where the esker ended abruptly in its southward march before it turned sharply inland. Just at the point where the crest turned I could see the remains of a cairn. I scrambled up the ridge towards the cairn, almost stopping dead in my tracks as I caught sight of something that appeared to have fallen from its inner cavity. There, looking most unnatural in those barren surroundings, lay a box measuring about twelve by eighteen inches and four inches deep. I approached this item almost without breathing and treading as gently as I could in case it might disappear. But it was not to be a great new Franklin Expedition find. Instead, the cairn had been built by a number of Canadian Franklin enthusiasts led by a Commissioner for the Northwest Territories. They were known as the 'Franklin Project' and this title was along the top of the box followed by a list of the names of those who had been involved. In an eminently practical approach to the area, the box itself contained emergency supplies for anyone who found themselves in difficulty whilst passing along the coast. I placed it back at the base of the crumbling cairn in the fervent hope that it would never have to be used.

Looking around to take a closer inspection of my position, I realised, with a deep sense of awe, exactly where I was. This was Crozier's Landing, the very spot where Captain Francis Rawdon Moira Crozier RN had come ashore with the 105 survivors from the deserted ships. Once they had mustered and organised themselves, they had set off south towards 'Back's Great Fish River' on the mainland of North America. None were to survive the journey.

A few yards to the north of the 'Franklin Project' cairn, I came across the base of the great cairn the men from the *Terror* and *Erebus* had erected. Its identification was only possible by the use of the description and photograph from Trafton's article in *The Mariner's Mirror*. Lieutenant Hobson had come across this rock marker on 6 July 1859, 133 years earlier. He had broken the cairn down to find a metal canister containing a message giving the only information known about the fate of the Franklin Expedition. He had then re-built the cairn before setting off to return to the *Fox*, arriving almost on the point of death from scurvy. His departure from Crozier's Landing was closely followed by the arrival of the commander of the *Fox*, Captain Leopold McClintock RN – the last Royal Navy officer to visit the site before I stood on the spot.

There was yet another mournful reminder of the awful tragedy surrounding the events of 1848. About three hundred yards south of the esker tip, on a strand of almost sand-like gravel, I could see a gently decaying cairn looking forlorn and sad despite the bright sun that cast a lengthening shadow from its once carefully arranged stone blocks. This (according to Trafton) was the site of the grave of Lieutenant John Irving RN after whom the shallow bay was named. The grave seems to have been missed by both Hobson and McClintock (and by Beattie, who having camped at Crozier's Landing for two days, searched for the Irving grave without success) and was not found until the American, Lieutenant Schwatka, came across it in July 1879. Schwatka was able to identify the scattered bones by means of a medal awarded as a mathematics prize at Greenwich Naval College. The medal had, fortunately, been overlooked by the natives who (so it has always been assumed) had broken into the grave in the search for much valued metal, cloth and wood.

The American collected the bones and eventually arranged for the remains to be returned to their homeland for interment. With the grave now empty, Schwatka built a cairn on the spot. In 1930 the Canadian, Major L. T. Burwash, dismantled the cairn and re-built it. Inspector H. A. Larsen of the RCMP did the same during his visit in 1949.

Later, during his 1994 expedition to the site, the leading defender of Inuit folk memory, David Woodman, cast serious doubts on the Schwatka/Burwash/Larsen/Trafton cairn. In the near vicinity of Crozier's Landing, he came across an open trench that copied, almost exactly, the sketch produced by Schwatka's companion Heinrich Klutschak. The feature certainly had more of the look of a grave, but there were some objections. Where was the cairn supposedly built by Schwatka? Furthermore, according to Woodman's measurements, the grave was less than a foot deep, and only about four feet long. On the other hand, the cairn I could see from Crozier's Landing, was built on the shingle – not even within sight of the sort of large rocks sketched by Klutschak.

Larsen had also come across another cairn on high ground behind Cape Jane Franklin, so I set off to walk the two miles to the cape. Again, the going was good, and I soon discovered that the ground eastward of the cape was yet another sea of rocks. After reaching as far as the northern coast of Collinson Inlet, still frozen solid throughout its length, I turned and ranged over the numerous ridges for over an hour but found no sign of a cairn. It was early evening as I returned to Crozier's Landing.

Standing once again on the ridge, with Irving's grave to my left, I looked out across the ice conscious of the deep silence that lay upon the site of so much hope, yet the source of so much tragedy. All around me the landscape held names of Royal Naval officers. Irving Bay after Lieutenant John Irving; Back Bay after Midshipman Back who had accompanied Franklin on his earlier overland expedition and was later to lead an expedition in search of his old commander; Collinson Inlet, whose name honoured Captain Richard Collinson of HMS *Investigator* who so nearly came across the clues to the disaster as he sledged up the Victoria Strait in April 1853. And finally, the great promontory of Franklin Point named by Ross after his friend in 1830. Ross also took the opportunity to name the northern corner of Collinson Inlet after Franklin's wife, Jane, who in later years was to earn by her determination not to accept the loss of her husband until it was rendered beyond doubt a place high on the list of Arctic luminaries. I cocked my rifle and fired three rounds in salute to them all, the officers, the seamen and the Marines, the wives and the searchers. The explosions echoed around the rocky eskers and rolled out across the ice. Eerily, as the sound of gunfire died away to the west, a shiver ran down my spine as the sound of three sharp cracks echoed from the far-off horizon – my salute had been returned.

With the weather remaining bright, and with only a very gentle breeze from the northwest, I decided that on my return journey, I would take to the ice and make my way around the mouth of the river north of Victory Point. I stepped easily onto the ice directly from the gravel beach and clambered over the low hummocks that lined up a few yards offshore. The ice on this part of the coast had not been subjected to the same great pressures that applied further north and as a result there were fewer large ridges. Walking over level ice about one hundred yards from the beach proved to be an uplifting sensation, marred only by the slight churning feeling in the stomach when open cracks were encountered. When this happened the decision had to be made whether or not to try and jump across or whether to search for a narrower gap. Much depended upon the landing area, if it sloped back towards the crack it was better to leave it and try elsewhere. The full thickness of the blue-white ice could be seen for several feet below the water. Even further down, the brown rocky seabed could be seen in startling clarity. Most of the surface was covered in frozen snow that crunched with a pleasing directness beneath my boots. I could see no other tracks other than those in my wake, and the only signs of life were a pair of loons who paddled up and down a wide crack between the river's mouth and my position. It took less than half an hour to achieve the shore to the north of the river. Three hours later I was back at Culgruff Point.

So far, the journey back to my base had been little more than a pleasant walk enlivened by the detour on the ice. I expected to be in my tent by half-past one or two o'clock and was beginning to look forward to a hot meal. The sun, although getting low on the horizon, remained bright. There was a little chill in the air, but not enough to cause me to take my woollen inner jacket from my rucksack. It was a perfect time for me to pick up my chain of markers that would guide me across the expanse of otherwise featureless landscape. From the shore I could see the first of these small rock piles and made my way up the gentle slope towards it.

I had only reached the third of my markers when, despite a detailed scanning of the horizon through my binoculars, I found that I could not see the fourth. Taking this as being little more than a nuisance, I reasonably assumed that if I kept going in more or less the same direction I would be bound to come across the next marker or failing that, the following one. Two hours later I was still walking and not a single marker had come into view. Even worse, not only had I lost the ice, but I found that I had been walking along a shallow depression edged to the east and west by shingle eskers. When I walked up to the crest of the western ridge I was appalled to see nothing but lake-strewn tundra. The sun was behind me as I crossed the depression to the eastern ridge. From its crest I could not only see yet more lakes and tundra but also, most surprisingly, a ridge of high ground. The only such feature that I knew of was the isolated ridge that could be seen on the eastern horizon from my tent. I realised that this had to be the same ridge and I was looking at it from a point about three miles to the south.

It looked as if I had walked down a rocky limb that jutted into the tundra plain. I could try retracing my steps but the risk of wandering even further away from my true route was much too great to be seriously considered. At least with the tall ridge I had some kind of fixed reference point. Somewhere to its west lay my camp. It was clearly impossible to go directly in that direction as there was one huge lake and many smaller ones, all threaded together by streams blocking the way. Between the high ground and me I could make out several stunted eskers which could be used to cross the otherwise soggy ground.

After using the last of the warm water from my flask to make a mug of coffee I set off. I stepped off onto the coarse tundra grass. Where the ground was firm there was no difficulty. In fact, some stretches were no worse than scrubby moorland and reminded me of the terrain I had known further to the south of the island. Other stretches were not unlike English water meadows that could be splashed through with care. All, however, were subject to the sudden eruption of muskeg waiting to claim an unwary boot. Muskeg also tended to line the many streams that had to be crossed. I found that the best method was to find a spot where a number of grassy clumps broke the surface of the fast-flowing waters. I would plan a route, then, with a run, jump from clump to clump in the hope that they would support my fleeting passage and that my landing on the far side would not hide a patch of muskeg. Some of the time I was successful. The distance to the high ground (which I had nicknamed 'Spion Kop' as I was now into my third reading of a book on the Boer War I had brought with me) was extended considerably by the need to roam up and down streams trying to find a suitable place to cross and it was close to half-past two in the morning before I began to ascend its lower slopes.

Much of the western slopes of the hill were covered in snow and using the sides of my boots, I found it easy to make my way up. I had no intention of trying to reach the top, which I estimated was about three hundred feet above the flat tundra. Having reached about a third of the way up, I searched the horizon through my binoculars and was delighted to see to the north the continuous line of a low esker running to the west where it joined another that veered to the south. It was as if the tundra had been given its own ramparts. I pressed on, leaving the heights to my rear as I struck out northwards. It took me three hours of wading around lakes, ponds, and meltwater pools before I could walk up the blissfully dry slope of the northern esker. The passage had none of the horrors of Wall Bay and required little more than a head-down slog, but as I began the westward trudge along the crest of the esker I started to experience a new emotion. A wave of anger at my situation flooded over me. I had taken what I thought to be sensible precautions to avoid getting lost. I had marked my path and ensured that the ice was always no more than one ridge away. Yet it had gone wrong. If I was to break a leg or was otherwise rendered immobile any searchers would assume that the indicator arrow by my tent would show the direction I had taken – southward. But there I was, miles to the east (or even northeast). It was strange how my inner fury increased with every step I took until I decided that getting worked up over events I could do nothing about was something I should resolutely avoid. Silently raging at everything around me proved to be a new and unwelcome sensation, so I decided to occupy my mind by, once again, mentally argue the case for and against the ordination of women as priests in the Church of England. In the circumstances south of Wall Bay, forcefully organising the mind to think logically had been a defence against exhaustion and cold. Now it was an attempt to calm a seething fury. As before, it worked brilliantly, with the debate passing from one side to the other in a continuing cycle of repeated argument. Within an hour I had reached the south-leading esker with my anti-female priesthood stance confirmed.

There was still some way to go before, at last, I could see the shimmering ice of the Victoria Strait. The old problems of no reference points on which the eyes could rest frequently meant that horizons were much further away than originally thought. Once the far crest had been reached, a morale-sapping vista of yet more barren rock unfolded. The only answer was to mentally withdraw from the surroundings and limit the ambition to placing one foot in front of the other. I eventually struck the coast just south of the tent rings I had found when looking into the Peter Bayne map (my western esker had been forced to swing to the northwest to avoid a deep salient of lakes and tundra). Then, sticking firmly to a ridge that ran parallel with the shoreline, I ran into a strip of tundra that lined the northern bank of the Cape Maria Louisa river.

I was less than half a mile from my tent – after more than twenty-six hours of almost non-stop walking.

Some years later, I was reading William Henry Gilder's account of being with Schwatka on his expedition (*Schwatka's Search: Sledging in the Arctic in Quest of the Franklin Records*). Although not mentioned by name in the script, the hill I had climbed part-way up appeared on a map showing the expedition's route after Schwatka climbed to the top. On the map, it had been given the name Helmholtz Hill – probably after the nineteenth-century German philosopher, physicist, and physician Hermann von Helmholtz. Gilder's account:

The following day Lieutenant Schwatka and I took Toolooah (a dog sledge driver) with us inland, and sent Frank and Henry (the other two white men on the expedition) down the coast towards Victory Point. From the top of a high hill, about six miles from camp, we had an uninterrupted view for many miles in every direction, and swept the entire field with a spy-glass – but saw nothing like a cache or cairns. It was all a barren waste, with many ponds and lakes, some still covered with ice, and others, being more shallow, were entirely clear, as in the case with most of those near the coast. A few patches of snow could be seen here and there on the hillsides. We had to cross one deep snowbank before reaching the crest of the hill, and upon our descent came upon a depression in the snow, which Toolooah recognised as a bear's igloo. A few patches of white wool near the entrance confirmed his opinion. I crawled in as far as I could, to see in what sort of a house the polar bear hibernated, and found it very much in size and shape like those of the Inuits. The only difference, as far as I could see, was that this was dug out of a snowbank, instead of being built upon the surface and afterwards buried by the drift.

Bearing in mind that Hermann von Holtz was also honoured by having an asteroid and craters on the Moon and on Mars named after him, my wearied choice of 'Spion Kop' (a British Army defeat) was clearly overshadowed.

7

Up and Away

The morning of my fourteenth day on the island brought with it a return to a thick fog which meant that no aircraft would be coming in that day to collect me. At any time from now on, I could expect to hear the welcome drone of an engine, although I felt a sharp concern about the lack of an obvious landing ground in my immediate area. I had two complete meals left from my rations which, it seemed, worked out just about right for my being airlifted out on the sixteenth day. Again, my fuel – little more than an inch in the bottom of its container – looked as if it had been nicely judged to see me off the island.

The following day proved to be crisp and clear with the merest wisp of cloud. I used my time waiting for the sound of an aircraft to lay out a large anchor design using dark rocks against a patch of light-coloured gravel chippings close by my tent. I put the letters 'R' and 'N' just above the anchor flukes and the date '1992' at the bottom. At the top, just above the anchor ring, I built a cairn to the height of about four feet. Perhaps, I thought, in years to come, when Arctic tourism has become fashionable (as it will) my marks on the landscape might provide an interesting diversion as they follow a future 'Franklin Heritage Trail' (the stone anchor and cairn were found twenty-six years later – still in pristine condition – by an American expedition in 2018).

By the early evening, a fog bank had rolled in and put an end to any hopes of leaving.

I was lying propped up on my sleeping bag at about eleven-o-clock the following morning (having been driven inside by squadrons of mosquitoes) when the peace was shattered by the sound of an aircraft engine flying low from the west. I tumbled out of the tent to see a small red and white aeroplane banking to the north almost overhead. I jumped up and down and waved my arms vigorously, but the aircraft continued in the direction of Cape Felix. I dived back into the tent and brought out my binoculars. Through them I could see that the aeroplane was undeviating in its direction – away from me. Soon it was reduced to a tiny dot which then disappeared. Surely they had seen me? My flags fluttered in a strong wind and my cairn and anchor could surely not be missed from such a low-flying craft. I waited, constantly scanning the northern horizon, and trying not to be concerned. Suddenly, there it was, coming back, at first a dark dot, then a shape with wings. Within minutes it was flying low overhead, the roar of its engine filling the landscape with its alien sound. Round and round the aircraft flew in ever increasing circles, then levelled out as if to land south of the river. It then banked away and headed off towards 'Spion Kop' returning via the great esker that had brought me back from those heights. For three-quarters of an hour the aircraft circled and flew along the eskers then, with a final low sweep over my camp, it headed south. Within minutes it had gone. A heavy silence returned to Cape Maria Louisa.

The following day, the last biscuit from the ration packs was eaten. All I now had left was fifteen 'one-cup' packets of hot chocolate, a packet and half of 'Fishermen's Friends' (strong throat lozenges) and a quarter of a bottle of rum. I rationed myself to one cup of hot chocolate in the morning and one in the evening. I boiled one coffeepot of water for the first drink, saving the rest of the hot water in my flask for the evening. I had a throat lozenge whenever I felt particularly hungry and had a small tot of rum before retiring to my sleeping bag.

Whilst the days slowly passed, I could do little but inspect the estate by walking around the surrounding eskers. I dreaded the thought of being away from the camp if the aircraft returned and stayed close to the tent. Apart from ploughing through my ghastly reading material once again, there seemed to little I could do to fill the long hours. What I eventually found I could do was to study my immediate environment more closely.

Even from a short distance away the eskers and their connecting plains look entirely barren but – especially at the edges of the tundra and by meltwater ponds – rare, brightly coloured jewels could be found. These were small clumps of flowers that glowed against the green-brown tundra grass or the honey-coloured rocks. The most common flower on the northwest corner of King William Island was the purple saxifrage, low-lying and clinging to rocky fissures, the purple of its petals often dulled in lustre by being close to the bright orange-yellow of jewel lichen. Small cushions of moss-campion and Alpine forget-me-nots provided specks of pink and blue along the banks of streams and lakes whilst Arctic dandelions grew alongside flat, scrubby, Arctic willows on the tundra. The tundra also played host to the pristine white and gold mountain aven. The aven's near cousin, the Arctic poppy, was most often found struggling desperately to exist on the most barren stretches of the eskers, usually a single bloom with a root that burrowed for twelve inches, or even deeper, through the dry rocks in the search for moisture.

No insect was so numerous, or so detested, as the Arctic mosquito. The message that there was some fresh blood available must have spread far and wide for they turned up in their legions to torment their target. The options for dealing with them were limited to using copious amounts of insect repellent (which smelled like something retrieved from a blocked sewer) or retreating under cover of the tent. To help in staying outside, Bob Martin had given me a hooded green mesh 'bug jacket' which needed to be regularly soaked in an insect repellent elegantly named 'Bug Off'. The garment lacked a certain elegance but remained effective until the liquid repellent required replenishment. A far more welcome but extremely rare insect visitor was a small butterfly similar in size and colouring to the English small tortoiseshell. I only saw the butterfly four times, but I also came across a hairy caterpillar, dark brown with two orange spots along its back, and a silken cocoon attached to the upper side of a flat rock. Whether they were part of the same butterfly family I had no way of knowing.

I knew, both from my studies and from my experience on the south of the island that butterflies could be expected. But I was surprised to come across what I took to be a moth. My limited experience in such matters had taught me that as a rule butterflies settled with their wings upright (except when drying them out) but moths kept their wing outstretched. When the sun was at its highest, and there was little wind, I had often seen a tiny black object swooping and diving about my camp. Then, one day as I was standing outside my tent, I saw one of these objects land on it. It had dense black delta-shaped wings about three-quarters of an inch from tip to tip. After several sprints across the rocks in pursuit, I managed to see one land in its natural habitat and was intrigued to at how it flattened itself against the rock, wrapping its wings around any corners just as if it was using them to hang on. In fact, it was quite clear what it was doing. In clinging to the rock in such a manner it looked exactly like a shadow, making it practically invisible. I managed to capture two specimens of my 'shadow moth' which I later gave to the Northern Heritage Centre for their opinion. The insects ended up at the University of Victoria's Department of Biology and proved to be very poor travellers. Upon inspection of the mangled remains the experts declared themselves to be unsure but thought them 'most likely adults of Sympistris zetterstedti – Lepidoptera: Noctuidae'. (Exactly as I had thought all along).

For a long time, the only mammals I saw were the lemmings which darted for cover as I approached their area of tundra and, far off, through my binoculars, I could see seals basking in the sun. Almost invariably these were the common ringed seals but on one occasion, I saw the much bulkier form of a bearded seal lying on the ice, its head up on the lookout for predators.

One clear day, when standing idly outside my tent, my heart leapt into my mouth. From the corner of my eye I had spotted something moving over the esker to the west of my camp. I brought up my binoculars and was overjoyed to see an Arctic fox trotting along the ridge. The animal, grey furred on top with white legs and underparts, stopped once to have a look at me (probably distracted by my flags) allowing me to see its grey face with its white stripe down the centre. After a few minutes it disappeared over a northern ridge, its brief presence having raised my spirits. I was to see it just once more as, to the south, it came under attack from a pair of birds, the fox rearing up to fend off its tormentors.

The birds were jaegers (from the German for 'hunters'), white and grey with swallow-like tails. They were one of the few birds I came across in any numbers during my travels. Occasionally, a jaeger would hover low above my head, presumably to check whether or not I was any threat to its nest (or, perhaps, to decide if I was good to eat), but I was never attacked. Arctic terns also came to have a close look (I always thought they came to see if I was alive). On two occasions, a flight of whistling swans flapped overhead in a stately manner as they headed to the south, and I once saw the chin-strapped erect heads of Canada geese as they walked across the tundra. Loons could normally be heard rather than seen, their haunting cry penetrating the sunlit night air, and eiders, so common to the south of the island, only made their appearance once.

Through my binoculars, I watched a bird of unknown name bringing food to a nest of chicks. The bird, about the same size as an English thrush with brown body and a red-brown head and neck, would be greeted on its return by a loud chorus of cheeping of the nest's occupants. I decided to have a closer look and, as my shadow fell across the nest, half a dozen yellow bills opened wide with the same shrill pleading. Then, as if by magic, and with one accord, the gaping beaks closed, and the extended heads shrank back in an effort to disappear from view. They were a little late in reacting, but I could understand the survival principle at work.

On more than one occasion I found myself scanning the bare ridges through my binoculars when I felt my heart almost stop as I saw a tall figure standing still on the top of a far-off ridge having not been there only a moment before. It looked for all the world like a lone Indian brave watching the wagon train rumble through his territory. With heart racing, I broke into a run towards the stranger, stopping to get my breath back and took another look through the glasses – he was still there. As I got closer the binoculars were no longer needed, I could quite clearly see the head and body of someone I assumed was either looking in amazement at another human wheezing across the eskers to reach them or was, possibly, facing in the opposite direction and unaware of my presence. Then, with the figure apparently still a long way off, it sprouted wings and lazily glided away. My stranger had turned out to be a snowy owl, and the lack of perspective and the effect of the light had made it seem much further away and much larger than it really was.

My favourite bird during my stay proved to be the russet-throated knot. I found them settled on lakes and ponds, particularly towards Cape Felix, refusing to be disturbed as I walked past. As I watched, they would spin around on the surface of the waters, their action probably designed to disturb the bottom and thus churn up food. I fondly imagined that they were doing their bit to keep the mosquito population down. Nevertheless, after six days without food, even the friendly knot was beginning to look like a tasty morsel.

After my breakfast of hot chocolate on the twenty-sixth day of my stay on the island I took stock of my supplies. Ten days after the last visit by an aircraft, I was down to three packets of hot chocolate and half a packet of 'Fishermen's Friends' (the last of the rum had disappeared the previous night). Just to the west of my tent, I had arranged the fluorescent tape to spell out to any passing aircraft that I was out of food and fuel (I had about a quarter of an inch of fuel left).

The day itself was bright and sunlit with just the hint of a breeze from the north and, as I stood outside the door of my tent at about eleven-o-clock, I fancied I heard a new sound on the wind. I did not act immediately for there had been many occasions when I had been fooled by the wind into thinking that I had heard the sound of an aircraft. I stood still, nevertheless, and strained to listen. There it was again – a definite engine noise. I grabbed my binoculars and circled the horizon. Nothing could be seen. I listened again. The south, that was where it seemed to be coming from. Looking hard in the direction of Franklin Point, I caught sight of a tiny black dot against the blue of the sky. It held to its course and grew larger by the minute.

As the aircraft came closer I could see a pair of floats suspended beneath its fuselage. This sight caused me some concern as I knew that there was no chance at all of the aeroplane coming down on the sea, and the lakes that lay to my east seemed far too shallow. The aircraft kept to its course heading directly for me and soon, in a flash of bright yellow, roared low over my head. The Beaver (as I later found out the aircraft to be) banked, turned, and flew over me once again as I waved at the pilot. This time he turned eastward and cruised up and down the lakes and meltwater ponds that flooded the tundra as far as the horizon. I watched from the ridge of the eastern esker as he flew back and forth rarely more than a few feet from the surface of the shallow waters. After a number of passes he gained height and began to head southwards. Was this going to be yet another abandoned attempt to lift me out? About two miles to the south, the aircraft turned and, steering in a north-easterly direction, began to lose both height and speed. Down he came, lower and lower, until he disappeared behind a ridge a mile to my south-east. I held my breath as, through my binoculars, I looked for the aircraft to re-appear. But nothing emerged, and the sound of the engine had stopped.

As fast as I could, I began to cover the ground between me and the aircraft-obscuring esker. I splashed through the river, jumped across patches of muskeg, and squelched over tundra until, breathless and dizzy, I scrambled up the esker's western slope. There, before me, in the middle of a long rock-studded lake, sat the aircraft, its side emblazoned with the words 'ADLAIR AVIATION'. The pilot was standing on the port float prodding the bottom of the lake with a wooden paddle.

'Good morning!' I yelled, using my voice for the first time in twenty-six days.

'Hi!' came back the reply, 'how are you?'

'All the better for seeing you.'

'OK' shouted back the pilot 'I've got a bit of a problem. There's only inches under my floats. I can only take you and any really necessary kit.'

'Right! Don't go away.'

I was not about to debate the point. If it meant abandoning the tent and other items I had on loan then so be it. I could always throw myself on the mercy of the Chief of Staff and plead of circumstances beyond my control. I made my way back to my camp and packed a bag with my cameras, notebooks, and items of personal loan such as my flask, 'bivvy' bag, and rifle. Regrettably, the tent itself, which was an equally personal loan, had to remain. I bundled into the tent the fluorescent tapes and orange survival bag which I had used as message carriers, threw in the stove and fuel can, and zipped up the tent hoping no-one would ever come that way and find my large bag of dirty laundry.

The last items I put in my single bag of important items were my white ensign and 'sledge flag'. Then, despite suddenly feeling sick, dizzy, and exhausted, I headed off at my best speed towards the waiting aircraft, full of secret fear that the pilot would grow tired of waiting and take off without me.

But he was still there, in the centre of the lake, as I breasted the ridge. I made my way down to the edge of the water, lowered the bag to the ground, and bent over breathing heavily. I could hardly answer as the pilot shouted across 'I'm sorry, but I can't get in any closer. Looks like you'll have to wade out.' With a weak wave I signalled back that I understood.

After many days of trying to keep my feet dry, I was now required to do the opposite. There was no steeling of the mind, no stiffening of the spine or upper lip, just a stepping-off into the icy water. In fact, as the water poured into my boots and crept up beyond my knees it concentrated my mind and helped to shake off any cloudiness that had clogged my brain. In probably less than a minute I was standing alongside the aircraft float reaching up to shake the hand of the pilot. 'Hi!' he said once again 'Mike, Mike Carr-Harris. Hand me your bag, then I'll get you up.'

I rolled myself onto the float, shook off the excess water from my sodden limbs and, at the invitation of Mike, climbed into the aircraft and sat in the passenger seat on the right of the pilot's position. Mike was concerned about the number of rocks just lying below the surface and had spent his time whilst waiting for me in manoeuvring the aircraft to a position which would give the best hope for a hazard-free run for take-off. As he was settling into his seat, he turned to me and said, 'How have you been getting on for food – we've all been a bit concerned?'

'I haven't eaten for ten days,' I replied.

'Really?' he said. Reaching down he pulled out a wrapped package. 'I've some sandwiches, do you want one?'

'Yes please. What are they?'

'Garlic sausage.'

'Er, no thanks. I'll wait 'till we get to Cambridge Bay.' (A chap, after all, can only lower his standards so far.)

I then enquired if he had plenty of fuel. When he replied in the affirmative, I asked him if he could fly north beyond Wall Bay as I wanted to take a photograph of the 'burial mounds' from the air. Having no problem with this request, Mike readily agreed and switched on the engine, operating another lever at the same time. The propeller spun round once and stopped. Again Mike switched on. Again the propeller jerked into life and promptly died as I pretended to be totally unconcerned. On the third attempt the propeller spun round twice, paused for a fraction of a second, then the engine burst into glorious life. Moments later the aircraft eased forward and was soon skimming over the surface of the water and we were airborne, free of the island. Almost immediately, Mike banked the aircraft to port so that I could have a final photograph of the campsite. Then, turning north, we headed towards the mounds. In the few brief minutes it took to reach Wall Bay I was staggered to see how different the terrain looked from the air. Everything looked flat, even smooth, and could not possibly cause a problem to anyone walking across it. Dr Beattie has written that when walking on this part of the island the walker covers two kilometres for every one kilometre of coastline. I not only agreed, but also added a considerable number of extra miles from my meanderings. I had done close on a hundred miles, and the deeply ridged soles of my boots were, in parts, worn smooth. Yet below me the ground looked as if a round of golf could be played on it.

The island had not finished with me. As we approached Wall Bay, a huge bank of grey fog rolled down its northern edge. There was nothing Mike could do about it, we had to turn back. As he banked over the ice, I could clearly see the silver-grey tree stump marking the place where the Peter Bayne map had led me. The flight down the coast revealed just how much the ice had been piled up where it met the land. In parts it looked like a mini-mountain ranges following the coastline. We flew past Victory Point and Crozier's Landing, past Collinson's Inlet, still frozen solid over its entire length, and Franklin Point.

At Cape Crozier, the westernmost tip of King William Island, we turned westwards and were soon over the Royal Geographical Society Islands with McClintock Point to our starboard.

By now the ice in the Victoria Strait was beginning to break up and great stretches of open water could be seen as we reached the northern edge of Jenny Lind Island. Then we reached the great landmass of Victoria Island and soon had the hunched bulk of Mount Pelly fifty miles away on the horizon.

Below us the tundra, dotted with lakes and ponds, rolled past with the occasional musk ox running for non-existent cover from the sound of the aircraft's engine. There seemed to be no eskers or rocky plains, just the mud-coloured Arctic shorn of its winter coat. Before long we were banking over an arm of Cambridge Bay, reducing speed and losing height. No ice remained to impede our 'landing' as the Beaver touched down with the *Maude* quietly decaying to our starboard side. Once safely down, Mike manoeuvred the aircraft towards a wooden jetty where a well-muffled figure paced up and down waiting to moor us alongside. When the lines had been secured, I stepped down from the aircraft to find that beneath the thick clothing of our welcoming party was a young woman with fair hair, an engaging smile, and a buoyant enthusiasm for anything to do with Sir John Franklin and the North-West Passage. Her name was Joanne and she was the daughter of Willy Laserich, the owner of Adlair Aviation. Whilst bombarding me with questions, she drove Mike and I to the RCMP Detachment where I would pick up my bag of clean clothes and personal effects. (The Constable on duty proved to be Canada's only RCMP man to suffer from a dramatic sense of humour failure. When I told him I had been forced to abandon the RCMP radio he paled, stared at me in disbelief and I am certain would have locked me up if I had not told him I would discuss the matter with my 'big buddy', his Chief Superintendent. I exaggerated for effect).

Joanna dropped me off at the Arctic Island Lodge where I booked in for the two nights before the next flight south to Yellowknife. Feeling highly conscious of my crumpled and grimy appearance in such pristine surroundings, I made my way to my room and was soon in the shower scrubbing off the effect of almost a month without a decent wash. Feeling pounds lighter, I telephoned my wife who was almost delighted to have been woken up in the early hours of the morning to be informed that she could put the insurance certificates away. I then dressed in wonderfully clean clothes and made my way to the restaurant for the first food I had tasted in ten days. I was determined to have something with cheese. Like Ben Gunn, I had missed cheese more than anything else. As I walked into the dining room, I was delighted to see Buck Rogers (still eating revolting peanut butter and jelly sandwiches). It turned out that he had been in the first aircraft with Frank O'Connor and he explained that severe crosswinds had defeated their attempts at landing. However, they had not been too concerned about me as they had seen me 'still standing'. When another opportunity to get to me had failed, Bob O'Connor had got in touch with Adlair Aviation to see if they could help. They could, and Mike Carr-Harris had undertaken the four-hundred-mile round trip.

With my desire to taste cheese satisfied, I returned to my room and was soon snuggled down, feeling extremely weary. But a most uncalled-for reaction settled over me. I found my mind determined to keep me awake as if in fear of it all being a dream, and when I woke up the following morning, I would find that I was back in my tent on the island. It seemed hours before I could shake myself clear of this feeling and finally fall asleep.

After breakfast the next day, I had to prepare myself for what could be the most awkward moment of my stay in Canada. There was no way I could avoid making my way over to the Adlair Aviation offices to discuss the cost of my flight from the island to Cambridge Bay. I had just enough money left to get back to Calgary, so I would have to try and persuade the aircraft's owners to accept my word that when I arrived back in England I would take out a bank loan and settle their account. Not knowing where their office was, I called at Willy Laserich's home where I met his wife, a bright, cheerful English lady who had lost none of her Londoner's wit and talent for making a stranger feel at home.

Willy, it turned out, was out of town so his son, Rene, came out onto the porch to talk as his mother provided us with cups of tea. I told him of my financial situation but, before dealing with the problem, he asked me to tell him what I had done on the island. When I finished my tale, Rene put down his cup and spoke the words I least expected to hear. 'Well', he said, 'I reckon you can have this one on us.' He then handed me a baseball cap with a design on the front stating that 'I flew with Willy's bandits.'

The Canadian 'frontier spirit' had struck again. There was no demand for like payment, no quid pro quo, just an honest desire to help where possible. Any protestations of gratitude were brushed aside in a continuing discussion about the fate of the Franklin Party. My tread was considerably lighter as I left than it had been on arrival.

The next day, having said farewell to the good people of Cambridge Bay, I took a flight back to Yellowknife, luxuriating in a gin and tonic as I flew south over the Arctic Circle. Colin Perriman had arranged for me to stay at his accommodation whilst I was in town.

There were three people I had to see before I moved on. The first, the Chief of Staff, proved to be out of town so I asked to see the Duty Officer who turned out to be an Army major. I told him what I had been doing and how much I regretted having to report that I had abandoned all the kit I had taken on loan from the Canadian military. I grandly told him that the responsibility was entirely mine and that I was prepared to respond to any action that he felt was necessary. He responded to all this pompous nonsense by telling me not to worry about it, it was 'no big deal' and that the military would get 'in touch' if there was any 'settling-up to do'. (They never did.)

I then called at the Prince of Wales Northern Heritage Centre and reported my findings to the Director, Doctor Charles ('Chuck') Arnold. He listened carefully and got me to draw a few diagrams. He then began to expand on the potential of what I had come across and seemed to become more enthusiastic the more he explored the different possibilities that might arise. He was particularly excited by the possibility that if my mounds proved to be the site of Franklin and his men, the expedition's camera might also have been buried with them. Eventually, with mounting eagerness he was discussing the likelihood of a team of professionals taking a look at my findings the following year. In return, I thanked him and his team for the help they had given me and presented him with two dead moths.

Finally, I made my way over to Aero Arctic's head office and main base, where I was able to tell Bob O'Connor of the discoveries that he had been chiefly instrumental in achieving. Having come to know him, albeit briefly, I was not in the least bit surprised when he insisted in glossing over his contribution to concentrate on the outcome of my visit. He, his team, and of course, his pilot, Buck Rogers, had quite literally got me off the ground. It was very difficult to ensure that he understood the extent of my gratitude.

Before leaving Yellowknife, I was interviewed by a reporter from the *Baltimore Sun* who just happened to be in town. When I read his account, I was considerably encouraged by what he had written: 'Lieutenant Coleman is a kind of latter-day Victorian, with a spade-shaped beard, a voice that is a friendly drum, and a manifest sense of English rectitude and exceptionalism.' These comments show clearly that the truth is safe in the hands of the press, and newspapers can always be trusted.

After calling in at Calgary to pay a brief call on my friends in the 'Calgary Light Horse' I arrived at the airport to await my RAF flight home. I made my way to the departure lounge which I found to be packed with lots of immaculately dressed soldiers of a Dragoon regiment waiting for the same aircraft. For the first time since my departure from King William Island, I felt somewhat less than tidily dressed. My clothes were hanging off me, and with my hair and beard of such a length that I looked not unlike a grumpy musk ox. Shortly after my arrival, a broadcast announced that 'Officers are to board the aircraft.' To surprised looks on the part of the soldiers, I headed down the long embarkation tunnel. As I entered the aircraft and made my way between the seats, a female squadron leader Loadmaster ran down the aisle after me. 'Excuse me,' she said looking with marked disdain at the shuffling, carrier-bag carrying, tramp-like figure in front of her, 'This is a military flight.'

'Indeed, Ma'am,' I replied with years of experience in tackling RAF officers, 'And your flight is graced by the presence of the Senior Service. My name is Coleman, Lieutenant Coleman.'

Her mouth opened 'Ahh.'

I nodded towards the seats, 'Anywhere?'

'Oh, yes. Of course.'

I picked a window seat with masses of legroom. Not a single soldier came to take a seat anywhere near me, and the Loadmaster looked after my every need for the entire flight. I have no idea what they thought I had been up to, but they clearly believed I should be approached with caution and fed lots of sticky buns.

I arrived home to settle down to write my report of the events of the past few weeks. I sent a copy to Doctor Peter Wadhams, the Reader in Polar Studies at Cambridge University. His interest and encouragement (supported by his wife, Pia Casarini whom I had met at the National Maritime Museum) was to lead me, and others of a like mind, on a remarkable further unfolding of the Franklin Story.

8

Mustering the Hands

My return from my 1992 visit to the Arctic had resulted in the same round of lectures, slideshows, and after-dinner speaking as my earlier visit. Although the Captain Scott disaster had replaced the Franklin saga in the forefront of the national mind more than eighty years ago, the mid-Victorian tragedy easily aroused wide interest. And chief amongst the questions I was asked was 'When will you be going back to investigate further?'

There were three barriers to be overcome before I could answer that question. In the first case I had to get permission from the Royal Navy. I wrote to the Commander, my immediate superior, who applied on my behalf to the Director of Naval Recruiting. It was not long before a reply arrived from the Commander containing the direction 'Go for it'. Next I had to obtain the permission of the Canadian Authorities, not merely to re-visit King William Island but to do so with an accredited archaeologist who could supervise any digging, in particular at the site of the mounds I had come across the previous year.

Finally, and as always the most difficult, there was the vexing question of funding the project. Despite contacting a wide range of organisations, the raising of cash proved to be extremely difficult. 'Recession' was nearly always given as the reason why the money was not available. In the end, my funds came from interested individuals (including £5 from a ninety-year-old lady who claimed Franklin as a distant relative) and from two handsome donations by the Boston and the Spalding Royal Naval Associations. In the end, I had mustered enough for about four nights' accommodation in the Arctic where weather can force an unwelcome delay. As for the huge costs that can be involved in travelling both to and through Canada, and the topping up of my Arctic kit, it looked as if another scrounged lift from the Royal Air Force, and a visit to the bank manager was on the cards.

However, by the beginning of 1993, a disturbing effect began to make itself felt. There was a marked cooling of enthusiasm from the Canadian authorities – i.e. the Prince of Wales Northern Heritage Centre at Yellowknife. When I spoke to their resident Arctic archaeologist, Margaret Bertulli, it was as if she was trying to hint to me that there were problems on the horizon without being able to commit herself to a definite statement. The Director of the Centre, when I spoke to him, appeared to have lost all interest and was trying to distance himself from any involvement.

There is, however, one single factor that no enterprise can exist without and that is luck.

For me, the luck can probably be traced back to the moment at the National Maritime Museum when I met for the first time Maria Pia Casarini, Arctic historian, expert on the life of Jane, Lady Franklin, and wife of Dr Peter Wadhams, then Director of the Scott Polar Research Institute (SPRI). I had visited SPRI a number of times on my return from my 1992 visit to King William Island, and had always found support for my attempt to return once again on the Franklin trail. When I informed Peter of my plight, he suggested that it might be possible for him to take up the planning of an expedition under the general umbrella of SPRI. Whilst he did not have access to immediate funding, he felt that he might just have enough contacts in the academic and polar world to get the thing off the ground. Of course, I leapt at the chance. I did not care who was in charge, as long as the project continued to roll. Peter was one of the leading figures in polar research in the country (which meant, in practical terms, in the world). He had visited both polar regions nearly thirty times and had been awarded the Polar Medal by the Queen. Unfortunately, little of this seemed to impress the Canadian authorities who dragged their feet; even at one stage sending a fax message to Pia that any visitors would be welcome to make a day trip to King William Island – probably a world record distance for a day trip. Peter, however, persisted with the result that his message finally got through, that we were a serious outfit who would carry out a professional search of the area and report our findings directly to the authorities at Yellowknife. They were happy with this on the additional understanding that should there be any digging it should only be done under the supervision of a trained Arctic archaeologist. I would have liked to have had Margaret Bertulli with us as in the past, she was not merely highly competent in her field, but would also have provided some splendid diversionary (and hot) debates on current Canadian theories regarding the Franklin expedition and the role of the Royal Navy in Arctic exploration in general. Unfortunately, she was to be involved in excavating the finds made by a Canadian journalist at Erebus Bay on King William Island the year before. That expedition had relocated an undisclosed number of skulls and other bones and part of what might be a sledge or boat on the path taken by Franklin's men when they abandoned their ships in 1848.

With archaeologist or not, Peter found that the task of raising funds for the expedition to be more difficult than he had at first imagined. Then, one evening, I had an excited telephone call from Pia. An American supporter of SPRI had made a visit to the Institute and had expressed a keen interest in becoming involved in our proposed search for Franklin – *and he had his own aircraft.* Shortly afterwards Peter telephoned to say that a Frenchman had also expressed interest and that he (Peter) would like to arrange a meeting with all involved at SPRI. The American was conveniently passing through England on his way to South Africa, the Frenchman was available at the same time, and Peter, Pia and I were free to get together at SPRI, so a meeting was arranged.

Prior to the meeting I had hoped to contribute to the funding by writing to the producers of the cough lozenge, 'Fisherman's Friend'. I explained what had happened, and how their product had helped me through the ten days of extreme shortage on a remote Arctic island. They graciously replied with a single packet of 'Fisherman's Friends'.

Our American benefactor was Dan Weinstein. A tall man of fifty-nine years with a shock of grey hair that he insisted was actually deep brown and only turned grey when it fell to the barber's scissors. He had a liking for alligator-skin cowboy boots and came fully equipped with an attitude chiefly expressed in his regular phrase, 'Well, let's do it!' Dan's business life had been spent very successfully in the field of metal production. He had retired from the daily grind but had retained just enough economic interest to allow him to follow his chief interest, polar research. This study had taken him to both poles, and he had taken a particular interest in the regulated opening of Antarctica to visitors. With his generosity and enthusiasm, Dan embodied all that was best in his countrymen and was to prove a mainstay of the expedition.

Shortly after my arrival at SPRI, the Frenchman walked through the door of Peter's office. Thirty-three year-old Pierre Sauvadet had recently sailed a small yacht through the northeast passage and had set up his own foundation for polar exploration. His chief ambition was to build a ship in which he could emulate Nansen by drifting through the polar sea embedded in the ice. Pierre appeared to be a friendly individualist who laughed easily. He had a reasonably good command of the English language so, in an effort to foster good relationships, I offered to let him teach me French whilst we were away and in return, I would teach him cricket. My offer was rejected with a shudder. He did ask, however, if I would help him with his English. This I gladly offered to do and, when he told me that he had to get his equipment 'preparated', I was pleased to inform him that the correct word was 'propagated'.

After all the introductions had been made, we moved into the Institute's library to discuss potential programmes etc. But, no matter how we tried to approach the subject it soon became clear that there was a serious risk that we would be embarking on the enterprise with very severe underfunding. Our best hope turned out to be Pierre, who assured us that he would be able to raise the money by making an arrangement with a photographic agency.

Dan also helped to ease the load by suggesting that his aircraft would be available to get us as far as Cambridge Bay. He would have liked to have gone further and taken us to our target area on King William Island, but the small retractable wheels of his Super King Air could not be fitted with the tundra tyres needed to land on that broken and boggy terrain.

There still remained the problem of finding an Arctic archaeologist with a permit to dig during the 1993 season. Fortunately, Dan had contacted Professor James Savelle of McGill University. Not only was he a very senior archaeologist who had already worked on the Franklin story as part of the Beechey Island exhumations but was known to have a permit. As we stood in his office, Peter made a telephone call to 'Jim'. After the usual pleasantries, the telephone was handed to me and I found myself describing the mounds and giving my thoughts on what they might contain. The Professor did not seem overly excited about the possibilities but after a few questions regarding the geological composition of the site, agreed to accompany us on King William Island.

A few weeks after the meeting, Peter telephoned me to say that Pierre seemed to have pulled off a stunning deal. A television production company had agreed to fund the entire project providing they could have the sole filming rights. This was splendid news and was readily agreed upon. At long last I and the rest of the team could relax and prepare ourselves for the forthcoming expedition. But nobody said life had to be that easy. With less than a month to go, Peter had to inform me that the television company had either changed its mind, or that Pierre had got it wrong in the first place. Now it seemed that the television company would only agree to make the film (if they felt that its substance was strong enough) and offer it for distribution throughout the television world. Then, having recovered their expenses and their profits, they would return to the expedition any remaining money that the film might make.

This arrangement promptly deprived us of our initial funding required to get us up to the Arctic. Peter flew over to Paris to try and put our case but, despite his best efforts, could secure no more than a promise for a modest assistance with the cost of hiring aircraft.

Through Peter's own contacts, he was able to get a small but important, number of free air-hours from the Continental Polar Shelf Project, and after a telephone call from me, United Distillers offered us a sum large enough to cover Peter's, Pia's, and my airfare to the US. Their interest was triggered by the useful fact that Felix Booth, a nineteenth-century gin magnate, had helped Captain John Ross into the Arctic in 1829. In fact, we were about to base ourselves conveniently on Cape Felix just to the west of the Boothia Peninsula.

Nevertheless, in the end, Peter, Pia, Pierre and Dan were forced to underwrite the bulk of the expedition's costs in the hope that any profits from the photographic agency and the television film would be enough to reimburse them. Their kindness and courage made me cringe in the thought that my own contribution was very small indeed, but they dismissed my concerns and assured me that but for my earlier work, the project would never have been considered.

There was just one other task to be performed before leaving. Both the photographic agency and the film makers wanted to spend some time at SPRI filming Peter, Pia, and myself prior to our departure; in my case they wanted me to be in uniform. I found that appearing in front of a camera caused me little difficulty. I was quite happy to talk to the lens (actually, a fraction to the right), but I was not so keen on being filmed in conversation, an exercise I found to feel entirely stilted and bogus.

The occasion did, however, give us the opportunity to meet the cameraman who would be accompanying us on our trip. His name was Maurice. He was about six feet tall, stockily built, and looked quite tough. He had filmed in most parts of the world and had been involved in expeditions climbing Mount Everest and sailing around Cape Horn. Part of his youth had been spent in the French Navy and he had a passable command of the English language. As I left the Institute to return to Lincolnshire, Maurice shook my hand with the words 'See you in Cambridge Bay.' The expedition, it seemed, was up and running.

The author, Peter Wadhams, Pia Casarini-Wadhams, and Dan Weinstein at the Scott Polar Research Institute with a celebratory pre-expedition cake.

9

Northward Ho!

After a routine flight out to Newark Airport, New York, Pia, Peter and I were met by Dan. He was accompanied by forty-year-old New Zealander Max Wenden who was to become the sixth member of the expedition. Max's bearded features always made him look as if he had just clambered off the polar ice at the end of an epic journey. This may have had something to do with the fact that he spent most of his recent winters in Antarctica. He had been amongst the first to greet Ranulph Fiennes and Dr Stroud at the end of their cross-Antarctic walk. He was also a skilled polar ('Bush') pilot and would act as Dan's co-pilot on our flight north. However, it would be as the expedition's logistics expert that he received unstinted praise. Among his great delights was studying wildlife at close quarters, and in trying to imitate my flat East Midland accent (which was a bit rich coming from someone who liked camping in a 'tint' and eating 'fush').

Having collected our baggage, we climbed into an airport transport and made our way to a corner of the airfield where Dan had parked his aircraft (named 'Jonolyn' after Mrs Weinstein). The Super King Air was resplendent in the blue and yellow livery of the University of Michigan, Dan's favourite football team. It was twin-engined and could carry eight passengers in the kind of luxury we could only have dreamed of in our recently vacated jumbo jet. Whilst Max stowed our luggage, Dan showed us around the inside of the aircraft which proved to be a model of space-saving design that would have been the envy of even the most elaborate mobile home.

The flight to Dan's home town of Jamestown took about an hour and fifteen minutes. On arrival we unloaded the aircraft and boarded Dan's capacious van. The vehicle was painted in the same colours as the aircraft in case there was any doubt as to where Dan's footballing loyalty lay. Dan lived in a Tudoresque pile surrounded by trees and lawns. Vivid red cardinals and brilliant blue jays could be seen swinging from bird feeders, whilst squirrels rampaged among the branches of tall trees and chipmunks held sway at ground level. We met Dan's wife, the original, and previously sole, owner of the name 'Jonolyn' until Dan bestowed it upon his aircraft. Like Dan, she offered us the freedom of her home and its comforts before our journey north, which was somehow like taking a condemned man to see lunch being prepared on the morning of his execution.

As there was plenty of daylight remaining, Max was keen to get on with packing the expedition's stores, a task he had been working at before our arrival. I decided that I could be best employed as his assistant and spent much of the remainder of the day packing large plastic containers under Max's direction. He was a fanatic for getting everything in the right order, at the right weight, and under the right label. Tents had to be unpacked, spread, and checked for wear. In particular, he was concerned about my information that tent pegs would be no use at all on the King William Island rocks. He had hoped that the tents would have snow-skirts at the base of the fly-sheets so that they could be weighted down with rocks but as the skirts were non-existent, we spent a long time extending the guy-ropes and adding extra guys for securing to boulders.

I spent the following morning trotting behind Max and Jonolyn as they made the final purchases for our supplies. Max, in particular, was quite brilliant in working out both what was needed and how much. He also was insistent in getting as many 'freshies' as possible as he was a great believer in fresh fruit and vegetables. Some would have to be purchased further north, but they would be almost certainly a great deal more expensive and the more that could be obtained in Jamestown the better. On our return, Max gave me the job of carefully weighing each package and recording the weight as he was very concerned about the total weight we would have to carry in the aircraft. As I weighed the boxes and noted down their weights and contents Max completed another job suggested by Dan. As an American, Dan firmly believed that there was no point in suffering pointlessly, whether on expedition or not. To this end he had a wooden board prepared with a hole in the centre. Over this Max fitted a luxuriously padded toilet seat. Needless to say, the toilet seat was blue and yellow with the legend 'Go Blue' emblazoned on the underside of the lid.

Whilst Max and I prepared the stores, Dan, Peter and Pia busied themselves rigging aerials and testing the radio. After a modest success in contacting Dan's aircraft (which had been flown to a distant airport for a minor modification) they turned their attention to giving a trial to the contents of a large chrome box that we had brought out with us. This was the Proton Magnetometer, an instrument which would tell us whether or not any cavities existed below the surface of any potential gravesites.

It could also be used to detect metal below the surface. After a long series of trials, Peter declared himself to be happy with the instrument's operation.

With the approach of late afternoon, Dan, Max and I flew to Buffalo to meet Pierre. His arrival not only completed the team, but also gave Max considerable trouble as the New Zealand accent is incapable of distinguishing between 'Pia' and 'Pierre'. Eventually, Pia resolved the problem by allowing him to refer to her by her 'Sunday-best' name, 'Maria-Pia'.

The final task for the day was to load the aircraft. Yet again this was a job for Max to oversee as he alone knew the full picture of what we were carrying and the capacity of the aircraft. It proved to be a very difficult task. The design of the plastic containers allowed them to be stacked when empty, but they lost their flexibility when full and piled against each other. Max forced anything he possibly could in between the gaps whilst locking everything down with securing straps. Our personal baggage took up most of the forward part of the aircraft's cabin whilst the magnetometer had to be wedged into the centre aisle. If Dan and Max were to get to the aircraft's cockpit they would have to climb over a pile half the height of the cabin.

There had been a series of intriguing coincidences as the expedition began to pull itself together. The newly appointed First Sea Lord's family name was 'Bathurst', a name reflected in the Canadian Arctic's Bathurst Island and Bathurst Inlet. The new Commandant General of the Royal Marines bore the name 'Ross', a pre-eminent name in Arctic exploration and geography. A current naval officer's promotion list carried both McClintock and Hobson (discoverers of the fate of the Franklin expedition) and as we left Cambridge for Heathrow airport, we had followed a car bearing the registration plate 'RAE', the name of the Hudson's Bay Company factor who made the first contact with the Inuit that was to prove the direction in which the Franklin expedition had gone in 1846. As we breakfasted on Jonolyn's blueberry muffins on the morning of our departure from Jamestown, the local radio put out Vaughan Williams' tribute to Captain Scott, 'Symphonia Antarctica'.

So well had Dan and Max organised our departure that, within minutes of our arrival at the airport we were airborne and heading north over the Great Lakes. As we flew over Lake Huron we passed over Penetanguishene, the site of the Royal Navy's most northern base during the first half of the nineteenth century, and the place where in 1825 Franklin had noted '7pm. The distressing intelligence of my dear wife's death has just reached me.'

Our first port of call was to be Thunder Bay on Lake Superior, a harbour close by Fort William where Franklin and Surgeon Richardson had left Lieutenant George Back to organise the rear party as they headed towards Cumberland House, 1500 miles away on the Saskatchewan River. The name 'Back' was to figure prominently during the next few weeks of our expedition.

Whilst taxiing towards the Canadian customs post, the aircraft was joined in formation by two large fuel tankers. As I opened the aircraft door (I had been appointed official 'door monitor') and descended the steps I was greeted by the two tanker drivers, smart, polite, young men in baseball caps. Would I care to fuel whilst I was there? I handed the request on to Dan who found himself in the position of having to decide which of the two he would use to fill up his aircraft. In the end he decided by tossing a coin. An amiable agreement which had the loser raising his cap and wishing us a 'nice day'. Once through the simple formalities of customs, we went to the airport restaurant for lunch. Pierre was not impressed. With his hamburger in his hands he looked at me and said, 'I sink we 'ave already left civilisation.' As we prepared to leave Thunder Bay, the local Air Traffic Controller asked Dan where we were going. He turned out to have a great interest in Franklin and wished us well on our way.

We then began the long haul to Churchill, an ex-Hudson Bay Trading Post situated on the western coast of the bay itself. Churchill had earned a name as a good place for polar bear spotters as so many bears gathered there during the breeding season that a polar bear 'gaol' had been established to protect the townspeople from their unwanted attention when they wandered into Churchill itself. Once it was safe to do so, the bears were taken out into the wild and released. Churchill airport proved to be something of a disappointment. On the final approaches to the landing strip, we saw a series of pre-Cambrian-shield rocks decorated with the unwelcoming wreckage of an aircraft. On the ground, in temperatures above 30°C, the airport seemed to have been abandoned and an air of decay hung over the whole area. The main building was totally deserted with the restaurant empty apart from a tired-looking Christmas tree and a few drooping decorations. Outside, against a backdrop of rusting gas storage tanks, a few Canada geese nibbled unenthusiastically at the warm tundra grass whilst Hudson Bay glinted in the background like the Mediterranean off Morocco. We all felt less than cheered when Dan told us that our next, and final, stop, Cambridge Bay, had reported that they were fogged-in and there was a possibility that we might have to remain at Churchill. After some consideration, and with a plane-full of fuel, Dan decided to try for Cambridge Bay. Should we fail to get in there due to the poor weather, we could swing westwards and head for Yellowknife. If Yellowknife proved to be out of operation we would have to return to Churchill. With our fingers crossed we took off and were soon flying over the vast barren lands with innumerable lakes that glowed red, purple, blue, green and burnished gold through the polarised windows of the aircraft. As the long hours passed at 27,000 feet we read, chatted, or slept. Pierre spent some time sewing the badge of the European Space Agency onto his Arctic jacket. He gave us all one, but I refused to sew mine on as the designer had contrived to portray the Union Flag upside down.

It was with some relief that Max came on the aircraft's broadcast to tell us that the weather was improving at Cambridge Bay and Dan had decided to head north. We next heard Max announce that we had just crossed the Arctic Circle at 101 degrees, 20 minutes West, nine hours after leaving Jamestown. Shreds of fog still hung around the land below us, but there were ample gaps for Dan to take the aircraft through to a landing on the gravel runway that welcomed visitors to Cambridge Bay.

The only change that I could see from my visit the previous year was the sleek lines of a jet aircraft wearing the colours of Willy Laserich's 'Adlair'. As we eased ourselves alongside this splendid new addition to the Arctic's aircraft, I got out of my seat to lower the door and steps. There, waiting for us, was a crumpled figure in a battered baseball cap, who came forward, shook my hand, and, through his own personal cloud of diesel fumes, said 'Hi! I'm Wilf.' This was Wilf MacDonald who Dan had arranged through Willy Laserich to look after our requirements whilst at Cambridge Bay. Taking on unusual tasks seemed to be Wilf's role in life. We found out later that not only was he the hamlet's 'Mr Fixit', but he was also the Fire Chief and the Coroner. He had booked us into one of the two local hotels (unfortunately, the wrong one) and Max soon had our personal baggage out of the aircraft and into the wreck on wheels that Wilf proudly told us that he had arranged for us to use as we wished during our stay. The 'Wilf-mobile', despite its disreputable appearance and horrendous rattling and banging, proved to be a good friend to us over the next two days, especially in the preparation of our stores and the supply rearrangements between the village and the airport, which were some distance apart. That evening, with the exception of Dan who had had a long day at the aircraft's controls, we made our way over to the Arctic Islands Lodge where the French film crew was staying. In the hotel's restaurant we came across Maurice, the cameraman, who introduced us to his sound technician, Francois Bouet, who had little in his favour apart from being darkly handsome, witty, and an expert in mountaineering, skydiving, and scuba-diving. Fortunately, his limitations were further constrained by a modest skill in the English language. As a group we chatted away for most of the evening in a most amiable manner, but I saw Peter and Maurice getting into a huddle to one side that resulted in a disturbing number of shaking heads.

When we left the film crew and returned to our own hotel, we learned what the problem was. The French television production company had promised Peter during his visit to Paris a number of aircraft hours. Written confirmation of this had been expected before we left England. It had not arrived and now it turned out, almost on the very eve of our departure for King William Island, that Maurice had brought no confirmation with him. Although Maurice insisted that a fax had been sent with the required information, Peter felt that to press on without some form of written evidence of good faith would be irresponsible.

It seemed that, at the eleventh hour, everything might be called off. I offered to take a risk and try to fax my bank manager to see if I could raise a loan but, as was pointed out, in reality the amount I would be likely to raise would have little effect on the ouitlay required. Pia was in favour of continuing but with an implacable Latin logic she was to display on more than one occasion, demanded that if the television company were not prepared to come up with the money, they should be left behind.

The next morning Maurice insisted that he had neither the means nor the authorisation to pay for aircraft time yet was adamant that his backers had approved the expenditure for aircraft hours. Peter, in an attempt to avoid a Pia-style ultimatum, agreed to carry on with preparations for the remainder of the day if Maurice would get back to his people and demand that they faxed the confirmation through.

There were a number of tasks we had to get through that day in order to leave on the morrow. I helped Max to finish our shopping (mainly 'freshies') from the hamlet's Co-operative store and from the 'Northern' (ex-Hudson's Bay Company). We then repaired as a group to the Royal Canadian Mounted Police Detachment to complete a 'Wilderness Report'. The last to arrive was Pia who entered the police station covered in snow. Outside, the air was full of heavy snowflakes whirling about in a stiff wind. It was a stern reminder that the Arctic summer was drawing to a close. It was equally a reminder that we could well do without. Luckily for us the snow did not settle and proved to be the last we were to see for some days.

Wilf had arranged our weapons. One gun was a straightforward pump-action shotgun that fired a solid shot. The other was a well-worn, but quite useable, .303 Lee Enfield of the design that I had used on first entering the Royal Navy, and similar to the one I had used the previous year on King William Island. We were also provided with flares and 'bear-bangers', light projectiles that could be fired in the direction of a threatening polar bear, the loud aerial explosion being intended to scare the animal away. With regard to the use of the weapons, there was a general feeling among the group that we should try the guns out before we left for King William Island. Amongst the six of us, five had had very little weapon training. Dan had had some during his time with the United States Air Force many years earlier, Max had no need of such items in the Antarctic, and Peter, Pia, and Pierre could not muster more than half a dozen firings of a weapon in their entire life. For everybody's safety it was decided that I should give some basic instruction in the use of the rifle.

We all climbed aboard the Wilfmobile and drove the short distance to the town rubbish tip. A bitterly cold wind had got up and scythed through us as I stood at the front and gave the standard gunnery lecture about 'never, never, never point a gun at anyone.' I then explained the principle of the bolt-action rifle and demonstrated with one round into a convenient bank of soil about fifty yards away. One by one the shivering listeners shuffled forward to try their hand at loading and firing. All went well with the men. Then it was the turn of Pia. Her light frame, muffled in a bright red Arctic coat, stepped forward and I handed her the 10lbs of loaded rifle. For a moment it looked as if her left arm would be too short to reach far enough down the rifle barrel to hold it securely, but she leaned forward, pulled the butt into her shoulder as instructed, and fired the weapon using the gloved middle finger of her right hand. The bullet whined over the bank and disappeared towards Victoria Strait. She had not blinked an eye at the recoil or the detonation and gave us all an early demonstration of the fact that we had onboard an Italian woman who would be recognised by the rest of the team for her grit and determination. Loading another round into the breech she fired the gun once again, this time hitting the bank, still using her middle finger to squeeze the trigger.

With the weapon training session completed, the party took the Wilfmobile around the eastern arm of Cambridge Bay to get a close look at Amundsen's *Maude* as she lay half-over on her starboard side about twenty yards from the shore. Although it was only a year since I had last seen this historic vessel she seemed to be in an even more parlous state with the water higher up her exposed ribs. She remained a decaying monument to a noble past.

For me, the bay also held a ghost ship. In 1852 Captain Richard Collinson had brought HMS *Enterprise* to winter in the Cambridge Bay harbour. Sledging parties had been sent northwards up the Victoria Strait but although they passed within fifty miles of the coast of King William Island, they were unaware of its existence and missed, by that distance, the chance of being the first to discover the fate of the Franklin Expedition.

Joanna, Willy Laserich's daughter who a year earlier had been the first to meet me on my return to Cambridge Bay, paid a visit. She was kind enough to comment on my improved appearance since she had last seen me (the weight I had lost had soon been replaced). Apart from such a welcome sight, my meeting with Joanna was timely as Dan felt we ought to contact Willy. I found out from her that her father was in town and ready to meet us. As I returned to the restaurant after telephoning to make an appointment, I noticed a changed atmosphere in our group. It was the news that we had all been waiting for, the television company had sent a fax confirming their financial contribution. Although limited in its scope and effect, the aircraft hours that could be purchased would help to ease the strain on those who had risked their own finances in funding the expedition.

At Willy's house, with introductions completed, Willy and Dan soon fell into a conversation verging on the nostalgic about the number and types of aircraft they had flown. Willy was particularly interested in Dan's Super King Air, whilst Dan was full of admiration for Willy's latest jet aircraft. With some effort, Peter managed to steer the conversation back to our forthcoming requirements and our intention to land the party near to Cape Felix. At this Willy turned to me and asked me if there was a good site for landing a Twin-Engined Otter near the Cape. This simple question caused me some consternation as I knew next to nothing about flying and was reluctant to commit an aircraft to a landing on that rocky terrain solely on my experience of walking over the ground. The best I could come up with was that, in my highly unqualified opinion, there were several wide eskers and fields of small rock which might be suitable for a landing strip. Several options were put forward for getting the group onto King William Island. The matter had become more complex than at first thought as the weight of eight people (the team plus the film crew), all their stores, and a four-wheeled ATV with trailer was too heavy for a Twin-engined Otter. Eventually, Willy came up with a solution that would, at least, carry our plans forward. Dan would fly Max, Pierre, and me (the 'Advance Party') to Gjoa Haven on the southeast coast of King William Island. Willy would also come along to have a look at the Super King Air in operation and with Dan flying low and slow over the region around Cape Felix, take a close look at potential landing sites. Once at Gjoa Haven, the Advance Party would pick up a Beaver aircraft on floats (the same one that had plucked me from the island twelve months earlier) and try to make a landing on a lake near Cape Felix or, if any of the sea was clear of ice in the same region, 'land' just offshore. This arrangement was possible because a lone canoeist had just paddled down the Back River and at Chantry Inlet at the mouth of the river had triggered off his personal rescue alert. The Beaver had been sent out to pick the canoeist up and would be at Gjoa Haven.

As we were about to leave Willy's house, I was delighted to see Mike Carr-Harris's friendly smile enter the room. It had been Mike behind the controls of the Beaver when it landed and picked me up the year before. Again, he commented on my improved appearance with the remark that the last time he had seen me I had looked 'about seventy-five years old' (younger than I felt).

Extra fuel was obtained, batteries for the film crew's satellite link, and a four-wheeled ATV with an accompanying two-seat trailer emblazoned with the legend 'Girl Mobile' (the trailer was changed as soon as Willie swapped it for a more practical model). The group also felt that as we were using three tents, we should really have an extra gun. Wilf came up with a rifle of obscure origin that looked as if it had been dredged from the bottom of the bay. Willy rejected the weapon out of hand on our behalf and produced a splendid bolt-action Mauser and the ammunition to go with it. (Local opinion was divided as to whether we would have to use our guns. Some said that the Cape Felix area was 'infested' with polar bears, others said that none had ever been reported. I had only ever seen footprints.)

The Advance Party were to take a tent, sleeping bags, a radio, a stove, and food for eventual transport in the Beaver. The Otter was to carry half the remaining stores with the other half being stowed away in Willy's airport building.

Dan Weinstein. American pilot (with his own aircraft), polar enthusiast, and key sponsor of the expedition.

10

Mounds and Rings

Dan eased the aircraft nose down to bring us just below the cloud cover. We had left Cambridge Bay 30 minutes earlier and now beneath the clouds, we had the solid ice of Victoria Strait stretching from horizon to horizon. Within minutes we could make out the thin sandy outline of the northwest coast of King William Island and, before long, the great gash of Collinson's Inlet could be clearly seen. Just to the north, Back Bay bowed inwards towards Crozier's Landing and, as the aircraft swung to port, the coastline northwards past Victory Point was revealed. We flew on steadily with Willy crouched behind Dan and Max as he prepared to look for a landing place at Cape Felix. Dan gently brought the aircraft lower and lower and reduced our speed as we passed Cape Maria Louisa. I looked for my anchor and cairn from the year before in vain, but, with a leaden feeling in my stomach, I soon recognised the malevolent bulge of Wall Bay. The bay appeared free of fog as we flew over, and even the ice had retreated well beyond the shoreline leaving a wide swathe of glittering blue water. But nothing could disguise the broad streaks of grey-brown muskeg and sky-reflecting streams that scoured the land to the east of the bay and had produced the worst hours of my previous stay on the island.

The aircraft was soon passing the angular ridges of Franklin's summer camp and I began to search the coast keenly for a sight of the mounds that I hoped would contain evidence of the Franklin expedition – even, possibly, Franklin himself. To my consternation, I felt the aircraft begin to bank to the right and I realised that we were passing around Cape Felix. I had not seen any sight of the mounds which I had always thought would have been easy to spot from the air. I kept this concern to myself as Willy looked hard at the land below us. Suddenly he was satisfied. He had spotted a bay to the east of Cape Felix that was clear of ice, which meant that the Beaver could come in on its floats, and the vast expanse of yellow-brown esker to the south suggested a number of possible landing sites in the area for the Otter. We gained height and headed south towards Gjoa Haven, the thin layer of clouds beneath us stretching like a shroud over the silent, haunted, land.

On arriving at the settlement, we found we had arrived shortly after the Beaver had returned with the canoeist who had triggered off the alert at Chantry Inlet. I met the rescued adventurer inside the small airport cabin. He was bearded, about my age, and dressed in what appeared to be green 'combination' underwear. He had left his home in Calgary with the intention of canoeing down the Back River taking no more than three weeks. It was, however, to be more than five before he found himself in trouble at the mouth of the river and after one day without food (and with no chance of obtaining any), he had triggered off his alarm signal. Of course, I scoffed at his paltry single day without food and boasted of my ten days of enforced fasting the year before – choosing to ignore the splendid achievement of his solo journey down the Back River (Captain Back had noted more than eighty rapids during his 1834 expedition down the 'Great Fish' River). Dan took the opportunity to join in the standard frontier spirit that can still be found throughout the north and offered the canoeist a lift back to Cambridge Bay along with his equipment. Willy tended to take a philosophical view about the use of his aircraft for such rescue missions. If he could get paid, all well and good, but it was far more important that lives should be saved than losing sleep over the cost of a 'couple of barrels of fuel'.

Having loaded our kit onto the back of a truck, the Advance Party was driven through the small hamlet to the shore of the tiny inlet where Amundsen had spent two winters during his voyage through the North-West Passage 1903-1905. The spot where I had launched a sledge journey over the ice of Simpson Strait to Todd Island three years earlier was now blue, and its open water sparkling in the harsh sunlight.

Hard up against the shore, the tip of its floats resting on the sand, floated the Beaver which I had last seen alongside the jetty at Cambridge Bay after it had lifted me off Cape Maria Louisa. Crouched at the tip of the port wing was the baseball hatted D.J. 'Smitty' Smith busy re-fuelling the aircraft. Clambering down from the wing and along the port float he jumped ashore and, grinning widely, shook me by the hand. As I introduced him to Max and Pierre he began to tell them grossly exaggerated stories about my last visit to the region. At last, however, we managed to get our kit loaded into the aircraft. Smitty asked me to cast off for him as the others climbed onboard, and it was with a last push of my boot the Beaver drifted out from the shore. As a pilot, Max was interested in the flying of a floatplane, so he sat in front with Smitty whilst Pierre and I sat in the rear. After some minutes of cruising up and down the inlet, Smitty pushed the throttle forward and eased the aircraft into the air, our line of take-off taking us directly out over the Simpson Strait. The aircraft then swung through 180 degrees and headed to the north.

After a few minutes of flying we reached a height of twelve hundred feet and Pierre fell asleep as the lake-riddled tundra unfolded beneath us (we were soon to learn that hibernation was a natural state of affairs for our gallant Frenchman). Far from feeling tired, I kept my eyes glued to the drab scene below in the hope that I might spot the anchor I had marked out with rocks in 1990. After about half an hour of this, I raised my eyes to the middle distance where they met the extremely surprising sight of a large grey four-engined aircraft apparently speeding directly at us. Before I had time to say anything, the other aircraft roared beneath us and, as it appeared in the port window, I saw it sway from one side to the other as if in shock. Over our head-sets came 'What the...?' followed by a rapid apology resting on the fact that the last thing the other pilot had expected to see was a Beaver on floats in the middle of King William Island. Being a charitable sort, Smitty avoided asking the pilot of a submarine-chasing Orion aircraft how many submarines he expected to find at five hundred feet over the tundra. When the other pilot learned that the Royal Navy was onboard and what we were about to undertake he wished us the best of luck in our search.

In response to a request from me, Smitty edged the aircraft over to the west so that we would be able to take a low look at the coastline from Collinson Inlet northwards. We ran down the length of the frozen surface of the wide inlet and banked around the sand-coloured plain that was Cape Jane Franklin. Ahead lay Back Bay, followed by the scar of the river I had forded just prior to climbing up to my 'Victory Point'. I then asked Smitty if he could move slightly inland whilst I searched the eskers below for sign of the narrow river that reached the shore at Cape Maria Louisa. Then, suddenly, there it was and, tucked up against the easternmost esker on the northern bank were my cairn and anchor. Smitty dipped his starboard wing and we turned in a tight circle around the spot as I looked down at what had been my backyard for almost a month a year earlier.

Having indulged my nostalgia, we continued northwards over Wall Bay (still fog-free) and the 'summer-camp' site. With a mounting urgency, I scanned the shoreline for sign of the mounds but had seen nothing as for the second time that day the Advance Party flew around the red and white radar reflector at Cape Felix. It did not take Smitty long to spot the small bay just to the east of the Cape. Ice formed a barrier about two hundred yards out from the beach and ice floes lined the bay's edges to the east and west, but there was a good southerly breeze blowing and Smitty soon had us lined up for a landing into the wind. As we swooped down I could see that the choppy waters suggested that the wind was of some strength. It was not, however, strong enough to cause any difficulty to the 'landing', which was carried out perfectly, the aircraft's engine pulling us right up to the rocky beach. Pierre and I stepped out onto the floats with mooring lines which we secured to large rocks in order to prevent the Beaver from being blown back out to sea. We then unloaded the aircraft, piling our stores up the beach well away from any risk of being blown away by the aircraft's propellers. As we were doing this, Pierre threw my sleeping bag ashore only to see the wind blow it back to settle on the icy waters between the floats. As I fished it out with an extended radio aerial, Pierre's English vocabulary took on a new dimension (fortunately, the bag's cover prevented too much damage being done).

With all our kit ashore, Smitty spotted a new, and unwelcome, problem. The Beaver had coasted ashore between two sandbars that just failed to break the surface of the water. Instead of just drifting out and using the engine and propeller to turn the aircraft, he would have to turn it around where it had settled and power his way through the narrow gap. This proved to be easier said than done. As the aircraft was pushed off the beach it tried to float away under the pressure of the wind and, invariably, drifted towards the underwater obstacles. I was given the job of hanging onto a mooring line to try and prevent this but, as the rest turned the machine, its broadside-on fuselage caused even more wind resistance which resulted in me being dragged down the beach in a shower of rocks. For a while it looked as if Smitty was about to become a non-volunteer member of the expedition. This, probably, had also occurred to Smitty who with an air of desperation climbed back into the aircraft and after short time rummaging around, re-appeared with a pair of thigh-high waders. Now able to get into the water, Smitty could hold the aircraft clear of the sandbars as I held it to the shore and the others pushed on the floats to turn it round. It still proved to be a struggle fighting the stiffening wind for control of the Beaver, but to Smitty's immense relief we managed to get the aircraft turned around to face out to sea. He then told us what we had to do next. Max and Pierre were to hold on to each of the floats to keep the Beaver lined up with the narrow exit between the obstacles. The most interesting task, however, was to go to me. I had to hang on to the aircraft's tail plane to stop the wind blowing the machine off the shore before Smitty was ready. I would know he was ready when he pushed the throttle forward. Should I continue to hang on beyond this point, I would become an unwanted appendage with a tendency to join the ice floes drifting out in the bay. With the engine racing, Max and Pierre battled to keep the Beaver lined up with the gap between the sandbars whilst I heaved back on the tail plane. Suddenly the engine gave a much louder roar as Smitty stepped on the gas. I had just uncurled my fingers as the aircraft leapt forward and skimmed beautifully between the twin hazards. We stood and watched with relief as Smitty taxied the aircraft almost to the edge of the ice. He then turned into the wind and came straight at us framed by a plume of spray. The Beaver lifted off well clear of the beach and flew directly over our heads giving us a shower of crystal droplets to accompany Smitty's 'thumbs up' out of the cockpit window.

We stood and rested from our exertions as the sound of the Beaver disappeared to the southwest. But Max did not intend that we should stand around for too long. He began to unpack the tent and the radio while I won the job of rigging the radio aerial. As we did this, an Arctic fox trotted over a low southern ridge and made his way down to the beach. It seemed as if he was blithely unaware of our presence until he was about twenty-five yards from us. Then, in an almost comic reaction, he froze for a second before ducking down to try and hide behind the small rocks on the beach. Realising the sheer inadequacy of this he gave a yelp and sprinted back over the ridge. I fondly imagined that it could be the same fox who had so cheered me during my lonely hours the year before but, as I was soon to discover, there seemed to be a great deal more Arctic foxes around Cape Felix than there had ever been at Cape Maria Louise. More meat supplies for Franklin's trapped ships?

We made ourselves a warm drink from flasks of hot water as Max tried unsuccessfully to raise the incoming Main Party. Max tried to adjust the aerial in an attempt to make contact. I made my way around a small arm of the bay towards an object I had seen through my binoculars. After a short walk the object turned out to be a tripod on which had been attached a plaque in commemoration of a visit to the area by soldiers of Princess Patricia's Canadian Light Infantry Regiment. They had searched the ground in vain for evidence of Franklin during the summer of 1967.

I had just arrived back at the tent when Max signalled me to keep quiet. There, faintly above the wind, could be heard the sound of an aircraft engine, first to the northwest and then swinging north. Raising my binoculars I scanned the sky above the ice and picked up the shape of the Otter heading directly for us. Max picked up our hand-held VHF radio and was soon in contact with the pilot. Where, he wanted to know, did we want him to land? Before a decision could be made the aircraft was roaring in low over our heads and banking to circle around to the east of us. All I could suggest to Max was that the aircraft should try to land somewhere to the southwest of us. He passed the information on with the sensible addition that the pilot should land at a site he alone felt to be best. But the 'Adlair' pilots are nothing if not obliging, and the aircraft was soon making passes over a spot to our southwest.

I climbed to the top of the nearby ridge and watched as the Otter made its final run and disappeared behind a far-off shingle esker. The southerly wind brought the sound that Max recognised as final braking as I caught sight of the tip of a tail plane peeping over the top of the ridge. The aircraft had come down about three miles from our position.

Returning to the tent we packed up everything except for the tent itself. After about half-an-hour we heard the unmistakeable sound of an ATV being driven towards us and a sweep with the binoculars revealed Dan, with Peter sat behind him, towing the trailer and heading in our direction. By the time he breasted the ridge, we had the tent stowed in its bag and alongside the rest of our stores ready to go into the ATV's trailer. Dan told us that the aircraft pilot had opted to remain on the ground until it was confirmed that the Advance Party was safe and well, so once the stores had been packed into the trailer and Pierre had climbed onboard as pillion passenger, he set off on the return journey.

Max, Peter and I, free of any load to carry, spent the next half-hour walking over the eskers towards the flag-like tail plane of the Otter poking up over the rocky ridges. How different it was to walk those barren reaches in company with other people – especially Max, who despite his wealth of Antarctic experience retained an almost innocent delight in everything around him. Every bird's flight was worth following, every differently coloured rock was worth investigating, and every silent reflection in a meltwater pool was worth a moment's appreciation.

Walking up the gentle slope of a rubble-strewn esker we heard the engines of the Otter rumble into life and by the time we had reached the top we could see the aircraft trundling along the ground before turning into the wind. With a lurch and a furious revving of its engines the Otter rolled forward and within an incredibly short distance had left the ground. Max and I joined the others just as the last sounds of the aircraft faded from earshot.

The camp had been established to the west of a low ridge about a mile from the ice of Victoria Strait and about the same distance from Cape Felix. The television crew had already set up their tent on the best, most level stretch of rocky ground and were busy erecting their aerial through which we all would be able to reach the outside world through voice and fax. The rest of us decided to join them on their level strip, and Pierre and I helped to put up the tents as Dan and Max formed the large plastic containers into a kitchen area. Pia and Peter took the ATV down to a large pond between us and the shore to top up our water tanks.

They had not been long returned before Max had a brew on the go. Dan took on the responsibility of positioning the imposing toilet he had brought. For some reason known only to himself, Dan placed the imposing structure exactly on the skyline of the ridge behind us, whereas a few more yards would have taken it entirely from view. The rest of us thought this was somewhat eccentric and the television crew was reduced to hysterics but, as without Dan's initiative in bringing the padded contraption we would have been reduced to a lonely tramp into the wilderness, we accepted his decision. Later the toilet was voted the best seat on the island, both for comfort and for its view across the eskers to an imposing cairn on the next skyline.

With the tents secured to heavy rocks and the thermal mattresses inflated (vital on this terrain) we gratefully accepted the first of many excellent meals cooked by Max. As we dined I was asked on several occasions whether or not we were close to the mounds. All I could come up with was that they had to be somewhere to the southwest. My plan was that we should make for the beach and simply head south. If we did that, I felt, we had to come across them before we reached the site of Franklin's summer camp. Although I occasionally experienced the leaden fear that I might not be able to find them, I reasoned that if I had found them once without actually looking for them, I should have no difficulty in locating them this time. Nevertheless, I kept glancing in the direction of the low-lying sun in the hope of recognising something of the terrain, but without success.

As it had been a long day we decided to turn in after our meal ready for an early start. Before everyone actually took to their tents, I went along and checked that each tent had a weapon just in case those who believed that the area was 'infested' with polar bears turned out to be correct. The best idea seemed to be to give one rifle to Dan and Max whilst Pierre and I had the other. This left the shotgun for Pia and Peter with its relatively simple pump-action cocking and loading. However, Peter felt that as he had not actually fired the gun, he ought to put a round through it just for the experience. After warning the remainder of the team about what we were about to do, Peter and I walked a short distance from the camp and I showed him how the weapon worked. Although he had never fired a shotgun in his life, he understood that such guns had a fairly hefty recoil. What neither he nor I realised was that the use of solid shot increases the 'kick' significantly, and his pull of the trigger had him rocking back as the ground just a few yards to our front exploded with the impact of the lead ball. He looked quizzically at the gun and said slowly 'OK ... um'. It was clear that Peter was unlikely to leave the Halls of Academe to become an international mercenary.

As we walked back to the tent, a horrific thought flashed through my mind. Peter and Pia's tent was the middle one in a line of three. If either of them opened fire to fend off a menacing bear, the chances were that each of the other two tents could become the unwilling recipient of a solid lead shot. With this in mind, I heavily suggested to Peter and Pia that if they were molested during the night, they should only fire through the roof of their tent and depend upon the rest of us tumbling out to come to their aid. Luckily, this procedure was never put to the test.

The following morning proved to be a fine, clear day. Over breakfast, the talk was of little else other than our impending task of finding the mounds and getting to work on the supposed gravesite. I had no plan to re-locate the mounds apart from simply striking out for the coast and then following the beach southwards. With this in mind, I stood apart from the rest and looked out towards Victoria Strait, the ice a brilliant white tinged with amethyst in the morning light. My eyes had only just begun to follow the coastline a few degrees to the south when I stopped, turned, and ran the few paces to my tent. Snatching up my binoculars, I focused on the spot I had just seen. There, two far off dark humps against the frozen sea, were the mounds! Within seconds everyone was peering in the direction of my pointing finger. Our camp had been pitched about two miles from our most pressing target. With the sun in the west, they had been lost in the glare of the ice, but the morning sun had lighted the sandy shades of the eskers against which the darker grey of the mounds was clearly visible.

Little time was lost in preparing to make our way over to our immediate goal. It was decided that the film crew should take the ATV as they were burdened with camera, tripod and batteries whilst the remainder of us set off together on foot. It took less than an hour to come up to the mounds. Most of the journey was over frost-shattered rocks with a few areas of soggy tundra to be circumnavigated or splashed through, depending on the footwear.

I approached the mounds in the grip of a suppressed mix of fear and excitement centred on the pit of my stomach. Would the others believe my theory? Would they think I had wasted their time and money? Would I get off the island alive? In an effort to cut short the waiting, I directed Peter and Pia around the northern shoulder of the large mound and led them to the gravesite. There it was, unchanged since the year before. The flat slabs lining its length with the remnants of the edging upright slabs still standing exactly as I remembered them. Peter looked at the depression in the ground and said 'Yes. Very interesting. Yes. Looks good.' Pia responded with a surprising Anglo-Saxonism and gave me a hug. Dan, Max and Pierre all agreed that it looked as if we had something important enough to be examined closely. Even Francois came up to me with his arm outstretched and said, 'Let me shake zur 'and of a famoose man'. Only Maurice, the cameraman, put a damper on the occasion by insisting on a pantomime in which Peter and I repeatedly walked up to the grave at the edge of which Peter would congratulate me and shake me by the hand. Eventually Peter was freed from his starring role to take the ATV back to the camp and pick up the magnetometer whilst the rest of us rambled over the area looking for any clues that might tie the site down to the Franklin era. But apart from some interesting tent rings close to the small mound, nothing was turned up.

The rest of that cold but sunny afternoon really belonged to Peter and Pia. Using nylon cord they made a half-metre grid over the depression and erected a mast-like structure a short distance away. Then Peter, who had to ensure that he was not wearing any metal (I had to lend him a pullover as his warm clothing tended to have zips), placed the tip of a rod-like probe on the surface of the soil. After a couple of seconds he would intone the reading from a meter strapped to his chest. Pia would note down the figures and mark the position in the grid. As Peter approached the depression, the readings, which consisted of two numbers, varied in value but remained constant in value order i.e.. a high number followed by a lower number. After three or four readings outside the gravesite, the tip of the probe entered the depression. The wait for his reading seemed incredibly long to the small band of watchers.

Then it came. The number values were reversed. 'We have an anomaly,' announced Peter in an attempted matter-of-fact voice that failed to hide his excitement. There was another anomaly, then another, followed by yet another. The procession of inverted numbers continued until Peter lifted the probe out of the sunken area. The readings then reverted to their previous value order. Another sweep was made across the gridded area. Again, when the probe was placed over the depression, a succession of anomalies appeared. Eventually, even Peter threw off his scientific detachment and exclaimed 'Wow! There's something really weird here.'

As the readings continued, I felt a great feeling of relief come over me. The grave (if that was what it was) might not contain the body of Franklin, or even a body at all, but the complex electronics of the proton magnetometer assured us all that there was something below those moss-covered slabs. I seemed to have been carrying the secret fear of being the cause of wasted time and money for a lifetime, but now it looked as if my hunch had paid off.

I found the slow progress of the magnetometer down the length of the depression, despite its apparent confirmation of my hopes, to be a long and wearisome process. I kept getting the urge to push Peter to one side and start ripping up the slabs. By late afternoon, I suggested that I might go north along the shore to try and find the site of another possible grave I had come across the previous year. Max, who had spent much of the day wandering off to explore the region around the mounds, offered to come with me. For someone as active as Max, the plodding demands of scientific accuracy held the very minimum of attraction. We set off towards a crude cairn that topped a ridge about a mile towards Cape Felix.

It was always a pleasure to march alongside Max. He never failed in his appreciation of the scene around him and seemed to be forever drawing my attention to seals that watched us from afar, pointing out the colour contrasts in the tundra patches, or declaiming on the values of the sparkling, crisp air. Even his patience was slightly tested, however, as we approached the rock pile when we came under the close attention of a colony of Arctic terns. These beautifully streamlined, gull-like birds did not like intruders and showed their displeasure by diving down at our heads whilst emitting ear-piercing shrieks. For some strange reason they took a particular dislike to Max's royal blue headgear, a bizarre cone-shaped affair with earflaps that made him look like an extremely scruffy Tibetan monk.

Having pressed on through the aerial bombardment, we were just about to come level with the cairn when Max suddenly grabbed my arm and hissed 'Look!' At first I could see nothing until Max whispered, 'Behind the rock'. There, about twenty feet ahead of us, just poking above a small low-lying rock, was the head of an Arctic fox. Its dark eyes glittered as, with ears rigidly erect, it watched from behind its protective barrier. For a while nothing could be heard but the screeching of the terns to our rear as we stood perfectly still, neither the fox nor us making a move. Then Max, slowly raising his camera, inched his way forward whilst I remained where I was. The fox watched intently as Max drew level with the cairn and reduced the distance between them to no more than ten feet. At this stage, Max felt he dared not go closer for fear of sending the animal scuttling off. He began to bring his camera up to his eye when an extraordinary thing happened. The fox got to his feet and, with a caution to match Max's, began to creep towards him. We both held our breath as the small grey and white creature, its curiosity defeating its natural fear, placed itself between Max and the cairn. It looked up at Max's face from less than three feet away as the New Zealander took a photograph. After lingering for a few seconds more, and without the slightest degree of urgency, the fox turned and loped off to return to its rock from where it continued to watch us. Our brief contact with another creature had provided a rare and magic moment that would stay with me for as long as I live.

Giving our newfound friend a wide berth, Max and I continued on along the beach. I found the oval of rocks marked at one end by a pillar of slabs that I had taken to be a possible grave. It was almost directly to the west of our camp and would be easy to find once again should Peter decide that he would like to run his magnetometer over it. We now turned out back to the low sun and made our way back to the camp. As usual Max was keen to get a brew on the go and to get a late supper ready for the others when they returned from the mounds.

As it happened, they were not long behind us. The magnetometer survey had continued to produce 'anomalies' over the gravesite. There was no doubting the fact that there had to be something under the slabs, but there was little now that could be done until the arrival of Professor Savelle and a colleague late the following afternoon. Peter had decided that he would like to carry out the whole exercise once again to check his findings. I decided that I should stay with him and Pia to show a little more loyalty to the effort they were putting in, but Maurice was to have other ideas. He was going to make me a star.

The morning was spent in the company of the film crew who had me standing and sitting on the ice of Victoria Strait whilst waving my arms pompously as I described the arrival of Franklin's ships and discussed their probable fate. I had to walk up and down the beach, straight at the camera, and away from it. But the nearest I came to displaying an appropriate artistic temperament was when Maurice demanded that I sing 'Lady Franklin's Lament' at the top of my voice as I strode along the rocky strand towards his unblinking lens. I tried to explain that an Englishman would not behave in such a manner (I was supposed to be alone in an empty landscape), but Maurice insisted. There is unlikely to have been a worse sound on the island since it first rose clear of the icy waters.

It was past midday before I was released from my labours and allowed to make my way back to the camp. Maurice continued to roam the edge of the ice hoping to catch some seals or loons with his camera whilst Francois experimented with his sound equipment. He had let me listen through headphones to the crystal-clear and bell-like chiming of the melting ice as his microphones picked up the haunting sound from beneath a dripping ice floe.

Peter had by now finished his second survey of the grave using the magnetometer. His finding of the previous day had been confirmed, and he was even able to add more detail to the 'picture' he had worked on overnight. The Proton Magnetometer was convinced that beneath my lichen-covered slabs lay something worth investigating.

With the mounds looking as if they might reveal their exciting secret, Dan was keen to visit the site of my 'Drift Calculator' of the year before. There was at least a couple of hours before the Canadian experts were due to arrive so Dan, Max and I set off the mile or so northward from the camp. Peter and Pia stayed behind to concentrate on their calculations whilst Pierre – yet again – took to his sleeping bag. The walk was an easy one with the area almost devoid of tundra and with just one large meltwater pond to skirt around. The site was not difficult to find despite the column of stones which had led me to it twelve months earlier having toppled. The series of tent rings and rock piles were given a surface search for any signs that could confirm a link with the Franklin era, but nothing was seen. What was found, however, was that the tent rings I had recorded on my previous visit were not the only ones. Several others continued along the ridge to the east. Among them, and most exciting of all, was the rubble outline of a rectangular tent of the same dimensions as the Royal Naval tents of the mid-nineteenth century. Was this to be the proof that I had been seeking? We would soon find out as we were about to be joined by two men who had made a close study of such matters. Before our return to the camp, I re-erected the stone pillar, crowning it with the fist-sized, grey pre-Cambrian rock that had topped it when I had last seen it.

We were about a quarter of a mile from the tents when we heard the drone of an aircraft engine. In no time at all the orange and silver Twin Otter was circling for a look at the landing site and, by the time we had reached the camp the pilot was bringing the machine down within yards of the tents. As the propellers gave their final spin, the door was opened by the flight engineer and a ladder let down. Two bearded figures alighted and introduced themselves as Art Dyke and Jim Savelle. There was no time for small talk as the pilot, a late middle-aged man of grim visage who clearly wanted to spend as little time on King William Island as possible, demanded brusquely that we get the aircraft unloaded immediately.

'Happy Harry' was to lighten our lives once more before we left the Arctic. According to his urging, we bent ourselves to the task of emptying the Otter as fast as we could. The passengers had each brought a four-wheel ATV, which were removed from the aircraft by simply hauling them out and allowing them to drop the six feet to the ground with a crash that sounded fatal but appeared to do them no damage at all. With tents and kit discharged from the aircraft's belly, we dragged the baggage clear as the pilot rolled the Otter forward, turned it into the wind, and took off with barely a glance in our direction.

Luckily, the passengers he had dropped off were of a more pleasant disposition. Art Dyke had spent a number of years in the Arctic undertaking a variety of tasks for the Canadian Geological Survey. He looked like a bespectacled Viking with his long gingery beard, but proved to have a pleasing personality and was always tolerant of total ignorance in others in matters of geology. Jim Savelle, on the other hand, treated every query about archaeological matters like an assault upon his personal honour. In archaeological matters, the archaeologist was king, and any challenge to orthodoxy was considered a heresy worthy of nothing more than being crushed into oblivion.

He seemed to have more letters after his name than he had in it and was currently the Professor of Archaeology at McGill University's Department of Anthropology. However, despite his imposing title and position, and his habit of throwing up barricades to block any light trying to enter his black arts, Jim proved to be a splendid companion. He had been with Owen Beattie on Beechey Island when the bodies of Franklin's men were disinterred from their icy tombs and, on another field trip had seen a friend gripped by the head in the jaws of a polar bear and dragged off. On that occasion, Jim found he had to aim his rifle at a spot on the animal where it would bring the beast down but not injure – or even kill – his colleague. Fortunately, he was successful and as a result of the incident he evolved a system of tent arrangement that allowed each tent to be overlooked by others as a safety measure. Also, should a polar bear get into the camp, there would be plenty of ways in which it could leave without feeling trapped and thus becoming more dangerous.

Whilst Max and I helped Art to put up their tents, Peter brought Jim up-to-date with the events that had happened so far and discussed with him the shape that the magnetometer had revealed beneath the slabs. Both Jim and Art seemed intrigued by the findings and as soon as their tents were up they expressed an eagerness to get over to the mounds to have a look for themselves.

We set off in a gently fading light towards the southwestern shoreline, the expedition members generally walking together whilst Jim and Art took off on their ATVs in a wide sweep to the east, which would allow us to get to the mounds ahead of them. We had been there for a few minutes only when the two Canadians roared along the beach from the south and drove their machines up the side of the large mound (Max nicknamed them the 'Arctic Hell's Angels' from their habit of tearing about the eskers at high speed, but I preferred to think of Jim, with his beard, green jacket and red hat, to be more like Hell's Garden Gnome).

Dismounting from their ATVs, Jim and Art walked over to where we stood in a small knot around the depression in the ground. They stood at the lower edge and looked along the length of the broken slabs. Art was the first to speak. The mounds he could deal with easily. Although they seemed regular in shape, even accepting their symmetry, he had not the slightest doubt that they were simply made of 'marine till' and had been formed when the island dragged itself from beneath the water. But, as for the gravesite, he had never seen anything like it before, and he felt that it was unlikely to be natural.

Jim spoke in measured terms, choosing every word carefully. He also had never seen anything like it. Yes, it was in the right position and, yes, the shape and apparent internal collapse were all good indicators. He was not, however, prepared to voice an opinion at this stage. Nevertheless, he was keen to get to work on the site the following morning.

As we returned to camp, Peter came up to me and said that, in his opinion, we had just heard a classic case of an academic being 'wildly excited'. They just had different ways of showing it.

The following morning saw a cluster of people standing around a long depression on the lower slope of a large mound, their bright clothing providing a rare flash of colour in the bleak landscape. Over their heads, the deep blue sky arched from honey-coloured eskers to sparkling blue-white ice. Untroubled by even the slightest breeze, they looked down at the ground in front of them. The whole of the gravesite had been marked-out by a grid of white nylon cord. At the lower end, Jim Savelle was sketching the positions of the slabs and rocks that lined the surface. He was helped by Art, who measured the distances between various points as indicated by the archaeologist. It was a slow process that had the rest of us stamping our feet in chilled impatience, but we knew that if this was to be an important site, the job of exploring beneath its surface had to be done with care and skill. The skill level took an immediate dip when, having completed his preliminary sketch, Jim called me forward to start removing the layer of moss and lichen from the lower slabs. Conscious of the television camera covering every one of my moves, I tried to act as if being involved in an archaeological dig was, for me, a perfectly normal event. I examined each trowel-load of moss as if I expected a naval button to pop out at any time. Of course, none did, and by the time I had cleared the surface of the first grid square, I was glad to hand the trowel back to Jim, who cleared the next square with an elegance born of experience. Soon, everyone else was having a go whilst Jim brushed any remaining grains of soil off the level slabs.

Eventually, the whole paved area within the depression was cleared and Jim embarked upon a second sketch. It looked as if three large slabs had been used to cover the 'grave' with smaller ones placed on top to cover any gaps. The large slabs had cracked in several places but there was no disguising the fact that they had once been whole and of a similar size.

With his second drawing completed, Jim lifted the smaller slabs. A small moment of excitement was provided when a white feather was revealed beneath one. This, we felt, was clear evidence that the slabs had been deliberately laid. How else could a feather get underneath? Jim and Art were not, however, quite so convinced and explained that the feather could have found itself beneath the surface through the actions of lemmings.

Then it was time for Jim to start lifting the larger slabs. Starting at the bottom he prised each angular piece clear of its neighbour. Beneath each one was a patch of cement-coloured granular 'till'. Once two of the grids had been cleared, Art bent down to take a closer look. It did not take him long to come up with some disturbing news. To his trained eye it seemed that the soil beneath the slabs was far too compacted to have been disturbed at any time in the reasonable past. A few more flat pieces of slab were lifted, each revealing more of the same bland foundation. Art picked up one of the spades we had brought with us and began to dig just above the lowest part of the grave. The soil seemed to have an almost buttery consistency as it gave way to the spades edge. Again Art expressed his opinion that the soil had never been disturbed when, with a sound that sent a surge of excitement through the standing observers, the spade struck something hard. The geologist did not join in the expectant silence. Looking up, he said in a resigned voice, 'Permafrost'. At eighteen inches below the surface he had hit permanently frozen sub-soil.

Another hole was dug higher up with the same result, then Dan took up a spade and began to dig near the middle of the site. Again compacted soil covered nothing but the rock-hard permafrost. We all then knew the worst. Whatever the site was, it was apparently not a grave. A heavy silence fell over the scene as dashed hopes had, at last, to be faced squarely. I felt as if I had glimpsed the towers of Camelot glittering in an evening sun, only to find the following morning that I had been looking at a glue factory.

I was not to be allowed any wallowing in dejection. A tap on my shoulder had me turning around to be faced by the unsympathetic lens of Maurice's television camera. "Ow do you feel now, eh' demanded the Frenchman. I spouted some platitudes about not minding and tried to be cheerful, but my stomach felt as if it was lined with lead. A number of excellent and worthy people had spent a lot of money on my conclusions and now they had been let down.

It was just at that moment when I was made aware of the exact quality of the people who had accompanied me northwards. Extraordinarily, they seemed to be concerned more about how I felt than the waste of their personal investment. I was patted sympathetically on the back and arm and even given a consolatory kiss by Pia. One of the academics (possibly Peter) even went as far as to suggest that, even if it had been a grave, the discovery of a body would have been unlikely to have advanced an explanation of the Franklin mystery.

Having transported all our equipment back to camp, we discussed the morning's work as Max organised lunch. After some thought, Art came up with what he thought might be an explanation for the apparent 'grave'. He felt it could be a rare example of a 'megaclast'. In this case, a large rock is shattered by the intense cold along lines of strata giving it the appearance of a sliced loaf of bread (there were innumerable examples of smaller rocks on King William Island which had suffered the same effect). Then, if the slope beneath it was at exactly the right angle, the 'slices' would, over decades – possibly centuries – separate and slide down the slope to form a level pavement. The same movement, or possibly the movement of ice, would cause smaller slabs to be pushed up against the larger slabs and tipped on end, thus giving the impression of a vertical lining around the site. If meltwater was to be channelled beneath the slabs, enough of the fine till could be washed out to cause a depression into which the heavy horizontal slabs would gradually sink. As for the false lead given by the Proton Magnetometer, Peter felt that this could be accounted for by lemming burrows beneath the slabs. I, however, not having seen any such burrows, believed that the machine's error was more likely to have been caused by air gaps between the different levels of slabs.

This miserable day, having started badly, continued to go downhill. After lunch and the 'gravesite' post-mortem, we de-camped and headed off to the site of the 'Drift Calculator'. As we wandered among the tent rings Jim pointed to each one in succession chanting as he went 'Native ... native ... native'.

But what about the line of three cairns that made up the drift calculator itself? 'Native food caches'. How did he know? 'Animal bones'.

I soon realised that my imagination and flexibility of reasoning was about to be battered against academic dogma. I could not understand why any natives would have built three food caches, in a line and approximately 25 feet apart, alongside their tents. Surely caches were intended for remote locations, not on the front doorstep? And, even if they were caches, what would have stopped Franklin's people from adapting them as a means of checking the drift of their ships? Could British sailors have eaten the animals? Jim's answer was always the same: 'Show me the evidence.'

In an attempt to demonstrate the futility of my objections, Jim explained the key design feature associated with a native tent ring.

He pointed out that the floor area was divided into two halves, a 'paved' area for day-to-day activities (cooking etc) and an unpaved area for sleeping. At one tent he bent down and picked up a rectangular piece of whalebone. 'Native carving,' he explained before replacing the item carefully back into the depression its removal had left. I tried to counter this by insisting that Franklin's people could have easily adapted native methods when ashore. There were plenty of examples ranging from Parry and James Ross to McClintock that this was commonly done. And what about the whalebone? To my eyes it looked suspiciously regular with its squared-off corners and straight edges – it could have just as easily have been done by one of the seamen from the *Erebus* or *Terror* (such bone and ivory carving was carried out on all Arctic ships). Again, Jim's only reply was the stultifying 'Show me the evidence.'

I felt that by taking another tack I might breach Jim's dogmatic barriers. Taking him to the rectangle of rocks I had earlier taken to be the outline of an Admiralty Pattern tent I tried to use his own arguments against him. Here was a site that did not have any associated animal bones, it was not circular, and did not have a paved area. By his own reasoning it could not possibly a native tent. But it was all to no avail. The mere fact that it did not look like a native tent was not the same, in his opinion, as proving it to have been a tent used by Franklin's people. He argued that if British seamen had used the site, there would be some evidence of the fact remaining – something like a button perhaps. I tried to counter this by saying that all evidence suggested that any site visited by the natives would have been stripped of any items they thought valuable and that, in reality, meant anything made of metal or wood (nineteenth-century Franklin searchers frequently found buttons in the possession of native Eskimos). Anything thought of as valueless by the natives, such as paper, would have long since vanished through natural decay. But none of this made the slightest impression on Jim who refused to budge an inch. I did not feel that I had been outgunned, but I clearly lacked the firepower to penetrate the armour of Academe.

My delight was wholly unconfined when, some years later, during a visit to the Map Room of the Royal Geographical Society, I chanced upon Admiralty Chart No. 5101, a chart of King William Island produced by Lieutenant Commander Rupert T. Gould in 1927 when he was working at the Admiralty Hydrographic Office. Gould – noted for his practical and mental attention to detail – had assembled all the then known discoveries found on the island onto the chart. Just to the south of Cape Felix, he recorded a cairn discovered by the American Franklin searcher, US Cavalry officer Lieutenant Frederick Schwatka. Schwatka had identified the cairn as an 'Observation Cairn'. Furthermore, in his own account of the Schwatka Expedition, Colonel Gilder noted:

> During the year and a half that the *Erebus* and *Terror* were frozen fast in the Victoria Strait, the officers had probably surveyed the adjacent shores very carefully, and had undoubtedly made observations that were highly important. Especially would this be the case with their magnetical observations, as they were right upon the magnetic pole. We saw some tall and very conspicuous cairns near Cape Felix, which had no records in them, and were apparently erected as points of observation from the ships.

Although I still believed that the cairns I had seen lining the shore were used as a triangulation base *towards* the ships, the single cairn built on high ground just to the south of Cape Felix may have been seen from the ships, and could have been used as a means of confirming the measurements taken from the lower cairns. However, whatever activities had taken place at the site, Gilder recorded that they found, close by Cape Felix

> ...what appeared to be a torn-down cairn, and a quantity of canvas and coarse red woollen stuff, pieces of blue cloth, broken bottles, and other similar stuff, showing that there had been a permanent camping place here from the vessels, while a piece of an ornamented china tea-cup, and cans of preserved potatoes showed that it was in charge of an officer.

Close to Schwatka's 'observation cairn', Lieutenant Hobson had earlier reported finding three tents containing

> ...evidence of a visit from a party belonging to the *Erebus* or *Terror*. A quantity of small articles were found in and about the tents. In the smallest of the three (which I supposed to have been an officer's tent) a packet of needles, fragments of a small red ensign, a good deal of that, two eye pieces of small telescopes and fragments of the stockings, mitts, with [illegible] on the others. Inside the sleeping gear the remains of blankets frocks sox (sic) and common cloth trousers, mitts, stockings, with also a large bank of twine, a good deal of old clothing was strewn about the place.

Would Hobson's and Schwatka's finds have been enough evidence to have satisfied the academics?

That evening there was a general feeling that, despite the day's setbacks, there was no reason to let the morale of the expedition be affected.

During an earlier walk down to the beach with Pierre, I had been with him when he had spotted a large wooden pallet of the type used when loading trucks. How it had reached that remote shore was unclear, but it was possible that it had come all the way from northern Siberia, across the Polar Sea, and down the McClintock Strait. Whatever its history, it was decided by all that it would make a splendid bonfire, so Dan and I took the ATV and its trailer down to the beach to collect it. Back at the camp, Francois and I got rid of a lot of frustration by smashing the pallet to pieces and before long the wood was blazing away majestically. In the meantime, Max had cooked a superb Arctic char that Jim and Art had brought with them. This salmon-like fish, although plentiful throughout the Arctic, was rarely to be found to the south, and could prove to be a profitable future export from the region to the jaded tables of European hostesses. To help with the proceedings I produced a bottle of West Indian rum that had been given to me by the Royal Naval Association in Boston, Lincolnshire. Pia produced her guitar and she and Peter entertained us to some of his self-composed songs. The only jarring note to the evening came when there was a group demand that I contribute to the singing. Fortunately, a veil of amnesia has long dropped over the highly forgettable performance.

On the following morning it was decided that I should lead everyone to the site of Franklin's summer camp. The 'V'-shaped ridges had been easily spotted from the air during both of my flights over the coast and I had no doubt that I should be able to reach them with little difficulty. As they were little more than five miles to the south, we thought that we would veer inland and approach the site from the east. This route would give us a chance to see a little more of the island than we would have done by keeping strictly to the coast. Jim and Art took their ATVs whilst Peter and Pia drove ours, all four intending to have a look for a cairn found by Schwatka in 1878. The cairn had been located in 1989 by Stephen Trafton. Inside, he found a pencilled message from Schwatka recording his fellow countryman's visit.

The foot party set off in the wake of the roaring ATVs. Our route took us over a wide sea of rocks dotted with patches of tundra and the occasional small lake. We were well wrapped up as this was to prove the coldest day on the island, low clouds having blotted out the sun. After about an hour of easy walking we came across a wide but shallow depression surrounded by a ring of low ridges. Much of the southern half of the area was filled by a lake on whose northern shore we could see Jim and Art studying a group of tent rings. When we had joined them we learned that this was the camp set up by Trafton in 1989. It was clearly a good spot with plenty of fresh water and possible shelter from the worst of the winds. It was also somewhere close to the middle of the Franklin search region for that part of the island.

Looking up from the spot at the surrounding ridges we could see a number of stone columns obviously erected as beacons to help his exploring parties return to their base. I then realised that most, if not all, of the similar stone pillars I had seen the year before and had taken to be native constructions were, in fact, markers put up by Trafton's people. This was felt by all of us to be something of an irritant as their existence had frequently led us to investigate these stone piles in the belief that they might hold some significance for our own explorations. No-one objected to the principle of rock beacons being used in such a way but, it was felt, a map should have been produced by the builders showing the marker's locations, which would have not only avoided any errors in assumption, but could also have been put to practical use.

As this matter was being discussed I felt the flush of hypocrisy fall about my ears. Whilst joining in the general condemnation of this lapse by the American researcher, I suddenly realised that the year before I had tramped my way from Cape Maria Louisa to just north of Victory Point erecting rock markers as I went and I had not bothered to mark them down on any map ('He that is without sin among you, let him first cast a stone').

From the Trafton campsite, we began to swing in a south-westerly direction and, after less than an hour's walking, I began to recognise the easterly edge of the group of ridges and furrows that marked the site of Franklin's summer camp. Again, the site had been first found by Schwatka and located by Trafton in 1989. It was also known that Trafton intended to return to the site at a future date to take a closer look at particular areas he felt needed closer examination. We felt it to be important that we should not cut across his intended investigations, despite the fact that we had no actual knowledge of his intentions.

As we climbed up the gentle slopes towards the highest point, I had no difficulty in finding the tent ring I had first seen the previous year. Soon the others had found more rings and I even came across a series of features that were about six feet long and two feet wide surrounded by regularly spaced-out rocks. One site, separated from the others, even appeared to have a flat slab carefully placed at its head.

Having walked over the area in company with Dan and Max, I then made my way down to the beach where I had no difficulty in finding my 'marker' rock from the year before. I now knew (thanks to Art) that it was of pre-Cambrian origin and had probably been dropped at this spot by a vast ice-sheet that had retreated north before the dawn of pre-history. What I could guarantee being not quite so ancient was the arrangement of three small rocks at the foot of the larger black one.

These I had arranged on one miserable day twelve months earlier to support my propane stove in the futile hope that it would produce some heat. These three stones, along with my cairn and anchor at Cape Maria Louisa and my south-trending markers, were the only surviving evidence that I had once walked these shores.

The sound of approaching ATVs brought me back to the present with Jim and Art bouncing over the rock-strewn ridges. They had reached the Schwatka cairn but had lost Peter and Pia on the way back. This was serious news: our colleagues were towing the trailer in which was our lunch.

Whilst we waited for the sound of our own ATV, I took the two Canadians to the features I had found both this and last year. Jim rapidly dismissed my tent ring, and the other tent rings found by Max and Dan, as being of native construction. My experience at Cape Felix had taught me the futility of trying to debate the tent ring's origins with him so I went on to point out the grave-like sites I had spotted. This time it was Art's turn to be dismissive. To his trained eye the features were clearly natural. Only one caught both their attentions: the arrangement of rocks with a slab at its head. This, they both felt, did hold some promise. But again, it was felt that it was likely to have been located by Trafton and may well be on his return agenda. Accordingly, it was left for his future visit.

By now the sky had turned a leaden grey which matched the dark mutterings from the surly troops who were beginning to feel the cold and the absence of lunch. It was just at this point I realised what the problem was. Looking southwards through my binoculars, I found myself looking at the dark malevolence of Wall Bay, my particular bête noire of the previous year. Although the ice had retreated beyond the beach, the area still had an air of foreboding as an iron-grey fog bank filled the eastern half of the bay. I could not gaze upon the scene without recalling my frightful journey across its bogs and ice-streams. It had only been a couple of hours or so out of a lifetime, but they were hours in which every second could be recalled in painful detail. I felt perfectly justified in blaming our current weather deterioration on our proximity to Wall Bay.

Turning away from the grisly panorama, I walked back over the ridges to where a mutinous rabble was discussing what they would do to Peter and Pia if they did not turn up with our lunch in the very near future. I continued walking for a few more yards and then trained my binoculars northwards (I had the only pair of binoculars in the party). I ranged over the grey eskers but could see nothing of our missing team-mates and was just about to turn and look eastwards when I caught a flash of bright red about two miles away. Calling out to the others that I had seen Peter and Pia, I kept my glasses trained on the tiny red dot as it disappeared and re-emerged behind the intervening ridges. Unfortunately, the patch of red seemed to be passing from right to left rather than closing with us. Then the red speck turned and started to head inland once again. Clearly, Peter had no idea where we were. Once Max had grasped the situation, he took immediate and appropriate action. I heard a sharp crack behind me and saw the brilliant red of a signal flare arcing through the sky. Within a minute, I was able to confirm that the ATV was now headed in our direction and before long we could clearly hear the rattle of its engine as Peter inched it towards us (or so it seemed to us stamping our feet and tucking our mittens under our armpits in an attempt to keep warm).

All thoughts of exacting revenge on the academics was pushed to one side (until it was their turn to buy the next round of drinks) as we broached the flasks of hot water and made ourselves warming drinks. Peter was particularly insistent that we should all make the effort to get over to the 'Schwatka' cairn as it was one of the two confirmed Franklin cairns in the Cape Felix region. Accordingly, when we had lunched on the salami sausage and chocolate bars provided by Max, we decided to return to camp to get a warm meal before heading eastwards to the cairn site. In order to get things moving, I drove the ATV back to camp with Pia as pillion passenger. On our arrival, I left her to get the stoves working and the water on the boil whilst I went back to the foot party to see if anyone wanted a lift. As I juddered my way over a particularly uneven stretch of esker, I was joined by an Arctic fox which kept running in front of my machine whilst turning its head and yelping at me. Eventually tiring of the game, the animal simply sat down where it was, causing me to swerve around it. With its ears and eyes alert as it watched me vibrate noisily past the tip of its nose, it would have been unaware of how close it had come to being the first traffic victim on north King William Island.

When I reached the party toiling over the ground I found that they were, in fact, enjoying their stroll and were not seeking a lift back to the camp. So I relieved them of their rucksacks and other burdens and went back to help Pia prepare the food. After a substantial and warming meal and after an hour or so's rest we set off in a south-easterly direction to have a look at the cairn site. Almost as I could have expected, now that we had left the vicinity of Wall Bay, the weather had improved and the dark clouds had given way to a light, high cloud base which did nothing for the temperature but restored a brightness to the scene. Peter set off astride the ATV intending to show us the way, whilst Max went off ahead with the film crew and I brought up the rear with Dan and Pia. Again the passage was quite easy over seas of rocks stretching from horizon to horizon and whilst we enjoyed a leisurely stroll, Max and the film crew marched on until they were about half a mile ahead. At this point they came alongside a stretch of tundra and were about to be joined by Peter when, rather than skirting the tundra, he elected to drive the ATV through it. I happened to be watching through my binoculars when I saw him come to an abrupt halt in the middle of the green-brown patch. I knew immediately what had happened. Unfortunately, Peter had chosen to go through a patch of tundra that had retained a significant amount of water and had found an especially liquid area of muskeg. I watched as Max, wearing high boots, waded out to see what he could do to extract Peter from his difficulty. Not to be outdone, the film crew, with typical French élan, took off their boots and socks, rolled up their trousers, and with a gasp that could not be heard at our distance but could be well imagined, stepped into the freezing bog and waded out to the stricken machine. We, of course, had speeded up in an attempt to lend a hand, but were too far away to be of any immediate use. Ahead of us we could see the party heaving and pushing on the ATV and slowly but surely it was being freed from the grip of the grey sludge. By the time we arrive on the scene, the machine was being hauled up onto the more secure rocky ground and Peter was promising very large gin and tonics for the entire party in compensation for having nearly lost the ATV. It had not been a good day for our chief scientist's bar bill.

Before we were due to arrive at the cairn site, we had been promised another more gruesome object. Pia, who had already seen it, guided us over a ridge from the top of which two things met our eyes. Firstly, there was a superb view of the frozen waters of James Ross Strait, the first time I had seen them from ground level. Just in front of us, the coastline buckled inwards to form a deep inlet to meet the rising ground on which we stood. The second sight sat astride a large rock at the head of this inlet. It was the partly mummified body of a young polar bear. Birds and foxes had clearly gnawed at the side that was turned towards us as the entire skeleton was revealed. The other side, which had obviously been laid against the ground, retained its white fur. The body had been found by Jim and Art who had propped it up in the bizarre position we now found it. They were of the opinion that it had either become separated from its mother, or that its mother had been killed (we were to see a number of large polar bear skins hung out to cure as we passed through the Arctic). Unable to fend for itself, or simply due to a lack of food in the area, it had then probably starved to death. Although it had to be described as 'young' it still looked large enough to constitute a threat if encountered in the open. This thought was strengthened by the large and menacing teeth that grinned from the half-exposed skull. It was these teeth that were to prevent the animal's death becoming totally pointless. There was, apparently, a researcher at the Scott Polar Research Institute studying the structure and ageing of polar bear teeth, so, with the help of his Swiss Army knife, Max removed the fore part of the bear's muzzle in the cause of science.

It was less than half an hour's walk from the sad sight of the dead bear to the ridge dominated by a cairn. Unfortunately, the cairn which dominated the scene was not the one built by Franklin's crews and found by Schwatka, but a modern one probably built by Trafton. Time and the elements had reduced the actual cairn we had come to see to a pile of rocks no more than two feet high. Inside the modern cairn, a plastic cylinder had been found in which a sheet of paper listed the members of Trafton's 1989 expedition and a small red, white and blue 'Explorer's Club' flag. These items were replaced after we had added another sheet of paper listing our names.

Using my binoculars and standing beside the cairn, I could see the position of the Franklin cairn recorded by Gould on high ground just to the south of Cape Felix. It followed, therefore, that the same view could have been obtained through a standard naval telescope of the mid-19th century. This led Peter to conclude that by lining up the cairns a continuation of the resultant line would lead directly to the spot at which – at the time of the cairn building – Franklin's two ships were beset in the ice. I liked the idea, as it could be used to support my theory of the drift calculator. The two theories, if used together, could be used to pinpoint the area where the ships were 'beset' in the ice on 12 September 1846. However, I also felt that the cairn served another, more practical purpose. It seemed to me that it would be in the interest of the morale of Franklin's sailors to let them off the ships as much as possible. During the summer months in particular, allowing the seamen to stretch their legs would have been an important recreation. But there would have to be some basic rules to prevent the men wandering out of reach of any searching parties if any went missing. I could well imagine instructions being issued stating that any search parties would only cover the area bounded by Cape Felix to the north, the summer camp to the west, and the cairn over on the east. By leaving that area, the individual or group would have knowingly placed themselves at risk. If the Franklin Expedition's experience of becoming easily lost in that region matched my own (and supported by the experience of others) it would have been common sense to declare a southern boundary. Of course, neither my idea nor Peter's was mutually exclusive, there was no reason why both could not be valid at the same time.

There was to be no party that night. Not simply because of a lack of rum or material for a bonfire, but it had been a long day. I had escaped a portion of the footslogging thanks to my ride on the ATV from the summer camp. Dan, Max, Pierre and the film crew, however, had walked for more than twenty miles over what was essentially large gravel, and welcomed the opportunity for an early night.

The drone of an aircraft engine hammering against the early morning air told us that we were about to lose the company of Jim and Art. Their tents had been collapsed and stowed, baggage packed, and ATVs drained of fuel. They had both expanded our knowledge of things Arctic and I in particular had been given a stern lesson in how to deal with academically inspired partiality (demand that they prove their own their negatives). Jim had made his own minor discovery on the island when he had picked up a rusting smoke grenade canister and found inside a list of members of Princess Patricia's Canadian Light Infantry Regiment who had taken part in the 1967 Canadian Army search of northwest King William Island.

We all helped with the loading of their kit, the ATVs being run up planks and man-hauled into the body of the aircraft. They were to return to their summer base at Hazard Inlet to continue their research work into Thule Eskimo sites. Before they arrived there, they would be calling in at Resolute Bay, and so our television crew hitched a lift northward.

Two hours later, the sound of a second Otter told us that it was time for us to leave the island. Only one tent remained erected in case our weight proved to be too much for a single journey, and someone would have to stay behind for a few hours (Max especially would have welcomed more time to get closer to the island's wildlife). The Adlair crew, however, ensured us that we could all lift off together now that we had temporarily lost the television crew. With Dan and I having carried out the last job on the island – collecting the red fuel containers that lined the rubble runway – the aircraft engineer closed the door on our forbidding, yet fascinating, estate for the last five nights.

About twenty minutes after our take-off, and to the south of the ice barrier that sealed off Victoria Strait, we saw below us a Canadian Coastguard vessel. With supreme irony it turned out to be the *Sir John Franklin* taking the explorer's name through the very straits he had strived for.

At Cambridge Bay, Dan checked over his aircraft whilst Max superintended the transfer of stores that had been left in the care of Willy Laserich. The RCMP was informed of our arrival, and Peter took care of the administrative details regarding flight costs etc. Pierre and I helped Max and were given the job of topping up the 'freshies' whilst Pia taxied us around in the 'Wilfmobile'.

Our visit to King William Island had not turned out as we had hoped. At best we had simply shown that a pair of academics believed that my mounds were nothing more than a natural phenomenon (of which more later). We had walked over the known Franklin sites and seen the features that remained. A little had been learned about Arctic archaeology and geology (also, of which a lot more later). But although there had been little in the way of physical adversity, we had come through as a team in the face of disappointment and not allowed failure to be seen as a setback. We were in good shape to tackle the next stage of our journey.

11

Haunted Shores

During a visit to the Scott Polar Research Institute to discuss the problems of financing the 1993 expedition with Peter, I was introduced to a most extraordinary individual. Whilst voyaging as guest lecturers on a ship travelling through the North-West Passage in 1992, Peter and Pia had met Wayne Davidson at Resolute Bay where he worked as a 'Weather Technician'. They had been intrigued by his stories of strange grave-like sites and stone circles at a remote spot in the High Arctic. Of even greater interest, was his description of the remains of a boat which he claimed was not of native origin and was connected in some way to the Franklin disappearance. Wayne had come to England to try and raise money to mount an expedition to have a look at the sites in the company of experts who might be able to cast some light on the mystery. I was a little saddened by his naive belief that the Royal Navy would be eager to fund a search for one its most illustrious captains and had to explain that 1993 was not a good time to seek finances from that quarter. He, in turn, was appalled that there was not a giant wave of support for his ideas from every part of England. He had grown up in the belief that Franklin was one of England's leading heroes yet, much to his dismay, he was rapidly learning that most people had never even heard of him.

It was not an easy task to try and deflect Wayne from his purpose as he could be intimidating both in appearance and manner. He was well over six feet tall and obviously very strong. He was dressed in the type of plaid shirt much favoured by traditional lumberjacks and wore a pair of faded jeans held up by braces. Beneath an unkempt mass of black hair, his pale face seemed almost skull-like, an effect heightened when he grinned and revealed his teeth. From deep sockets, his dark eyes burned with the fervour of an Old Testament prophet. His French-Canadian ancestry strengthened the very first impression I had of him. To me, he seemed the exact embodiment of the type of trapper or 'voyageur' who would have been familiar to Sir John Franklin during his travels overland through nineteenth-century Canada.

It was when Wayne began to talk that the most unusual effect of all made itself apparent. Whenever he got onto the subject of the Franklin Expedition, and its possible involvement with the native community, facts would come rushing out at all angles, great soaring leaps of logic would take place, or a new subject would be introduced in the middle of a sentence. It was for all the world as if his brain worked in a different gear to his tongue. But there was no doubting that he was an extremely intelligent and knowledgeable man with a wealth of experience in the Arctic. The trick was to somehow apply a brake to his enthusiasm so that ordinary mortals could follow his reasoning.

Fortunately, Wayne had brought his own braking mechanism over to England with him in the form of his friend, Martha Idlout. Martha was an Inuit from Resolute Bay, and Wayne had lived with her and her two children for seven years. Like many of the native women we were to meet, she seemed to be better educated than most of the men of her race and was capable of holding her own in any conversation or company. A tiny figure when compared to Wayne, she would step in with a few well-chosen words when he tended to go over the top.

I had brought some photographic slides of the mounds on King William Island for Wayne to have a look at. When he had seen them, he told us that in his opinion the type of lichen that covered much of the slabs was possibly too slow-growing to have spread undisturbed since 1848. However, he had no idea what the 'grave' itself was and had supported the view that it should be investigated.

We then had the opportunity to look at the slides Wayne had taken at the site of his marker and tent rings. From a grey, snow-covered landscape, a narrow black slab reared up with roughly spherical rocks on either side. A disappointed Wayne watched me closely as I looked at the slide. He had expected me to come up with an instant naval explanation for the arrangement of rocks. He had seen the liberal use of cannon balls on naval memorials and graves and believed that the two rocks at the foot of the black upright were meant to represent similar objects. I was not convinced. There were more slides of rock piles and stone circles but as the ground was mainly under snow, if was difficult to be clear about exactly what we were seeing. As a result, Wayne, Martha and I retired to a room with a blackboard where I drew out a plan view of my Cape Felix 'drift calculator'. This drew forth a great wave of excitement from Wayne as he was of the belief that some of the markers on his site were also possible drift calculators or even direction indicators. He then asked me to sketch on the board the structure of a mid-nineteenth century

Admiralty Pattern tent. I had to admit that my knowledge on such matters was extremely vague and could only be based upon the contemporary illustrations I had seen, but when I had completed my drawing, there was an excited banging on the table as Wayne told me that there were signs at the site that matched my sketch exactly. I was not terribly impressed, however, as I felt that anyone would have come to the same conclusion that I had. However, it now meant that Wayne and I got on well together and, because I was 'a naval officer' (and, therefore, apparently trustworthy), he would let me into the secret of the site's location. The markers, rings, and boat site were at Back Bay on the northeast coast of Prince of Wales Island, a hundred miles south of Resolute Bay.

Regrettably, there was a slight political edge to the story. The Resolute Bay natives had known the sites at Back Bay for generations as they had used Back Bay as a hunting campsite for many years. For some reason, never made clear to us (but possibly becoming clearer later), there was a strong feeling among the natives that the Canadian authorities in the region should not be involved. It was only the fact that Wayne had lived amongst the people for seven years and had become accepted into their community that had allowed the local Elders to give him permission to pass on the location of the site. Even then, the knowledge was to be limited to those Wayne felt secure in informing.

This gave us something of a problem as we knew that any application to do a proper search of the ground would require the presence of an Arctic archaeologist authorised by the Prince of Wales Heritage Centre at Yellowknife. Yet, if we were to respect the wishes of the local people, Yellowknife could not be informed. During the succeeding days, Peter discussed the problem with Dan and Pierre and a compromise was worked out. If anything of importance was found at Back Bay it would be inevitable that it would get out into the press, either as a result of the television crew in their professional capacity, or through scientific papers or reports published by the team members. Obviously Yellowknife would get to hear of the visit at some stage. It was decided, therefore, that the expedition would visit the sites simply as guests of the Resolute Inuit. No digging or tampering with the sites would take place beyond a surface search and, on completion, the expedition would report its findings to Yellowknife. It would then be up to the authorities at the Heritage Centre to deal with any political problems should they consider the matter worth pursuing.

Our arrival at Resolute Bay (now often shortened to 'Resolute' on Canadian maps) revealed our staging post to be a cheerless, barren place surrounded by lumpy flat-topped hills of a universal brown colour. Despite its forlorn appearance, however, the place was of great interest to me in its role as a mustering point for the early Franklin search expeditions. Just to the south of the harbour lay the flat, cliff-encircled Griffith Island (named in 1819 by Parry after Midshipman W. Nelson Griffith of HMS *Griper*). Between the island and the mainland, the frozen seas had provided the anchorages for the 1850 Horatio Austin Expedition and the 1852 Sir Edward Belcher Expedition. Sledge parties had gone out under the leadership of such men as Captain Erasmus Ommanney and Lieutenants Leopold McClintock and Sherard Osborn. The graves at Beechey Island had been found, and the ship's company of HMS *Investigator* rescued.

Somewhere beneath the icy waters lay the remains of Belcher's abandoned ships *Assistance* and *Pioneer*. To the west the *Intrepid* had succumbed to the ice as the deserted HMS *Resolute* after whom the bay had been named drifted eastward past Griffith Island in ghostly silence on a 1200-mile journey out through Lancaster Sound and down Baffin Bay. Having been picked up by an American whaling vessel, she was refitted by the Americans and returned to the Royal Navy. In exchange, a desk was made from the *Resolute*'s timbers and presented to the President of the United States. It remains in use in the Oval Office to this day.

The area seems to have had the same appeal to the members of the search expeditions as it had to me. One of the *Assistance*'s officers wrote during the early 1850s that the place 'is perhaps one of the most dreary and desolate spots that can well be conceived'. I could well agree as under an iron-grey sky and slashed by an icy wind we unloaded our overnight baggage and threw it in the back of the transport that was to take us the five miles from the 'Base' to the Resolute Bay hamlet. The native settlement had followed the establishment of a military base and weather station. A small number of families had been forcibly moved from the Pond Inlet settlement on Baffin Bay in 1953 (by tradition, to establish Canadian sovereignty in the area) and the population was now in the region of 200.

Our accommodation was to be at a commercial establishment that gloried in the fact that it was frequently used by expeditions visiting both the geographic and magnetic North Poles, both of which must have seemed welcome after a stay at the 'hotel'. The place was run on lines not dissimilar to Colditz or a 1940s seaside guesthouse. A succession of orders barked at us from the walls, 'All footwear and outer clothing is to be removed and left in the lobby area,' 'When not in use bathroom doors are to be left open,' 'All lights are to be switched off when not in use.' I came to the attention of the female head guard as I held the outside door open to let Max pass through and was rewarded by a screamed 'Ernest! Close that door!' The staff's eerie habit of closely watching their guests at all times led Dan to nickname the place the Bate's Motel. Fortunately, there were two good reasons to leave the place and go out into the miserable conditions that had settled over the area. Out in the bay, just on the edge of the ice, a sailing vessel lay at anchor. She was the *Dagmar Aaen*, a Danish fishing vessel crewed by young people from Denmark, Norway and Germany. The previous year they had tried to sail through the North-East Passage but had been turned back by the ice. Now they were trying the North-West Passage, but it looked as if they were to meet the same fate as, trapped by drifting ice floes in Resolute Bay, the sea was now on the point of its winter freeze.

Single-masted, and with the obligatory crow's nest for ice navigation, the red and white vessel provided me with a sight of great and rare interest. I had not expected to see such a sailing ship with Griffith Island as a backdrop, something that had not been seen since the 1850s. It was also a matter of great pleasure that there were still young people in the world who were prepared to undertake such a journey.

The rapidly dropping temperature forced Max and I to turn back towards the hamlet where we had an appointment to join Peter and Pia at the house of Elizabeth, an Inuit woman. She had met Pia and Peter when they were passing through Resolute Bay the previous year. Again, Elizabeth turned out to be one of the native women of high education and ability. A graduate, and active in local politics, she was particularly concerned about the plight of the families who had been removed to Resolute Bay in the early 1950s. Only recently had contact with their Pond Inlet relations been made, and Elizabeth and others felt that there should be some compensation for the people who had been separated from their families forty years earlier.

As we left the warmth of her house to return to our unwelcoming accommodation, we passed a family of sledge dogs that were distinguished by eyes that glowed with the bright blue of English cornflowers, a rare flash of nature providing a most unlikely colour against the drabbest of backgrounds.

Eager to leave the bay the next morning, we set off early to return to the airport where we were to pick up our aircraft for the flight to Back Bay. The aircraft was being provided by the Polar Continental Shelf Project (PCSP) which had kindly agreed to donate five hours of flying time towards the expedition. Some of the sting was taken out of our comments regarding the 'Bate's Motel' when we were handed a box of sandwiches and drinks to see us through lunchtime. As things turned out, this 'Red Cross parcel', was to stand us in good stead when we came up against an extremely rare example of Canadian bureaucracy. We stood in a large hanger filled with Arctic stores as our supplies and personal baggage was carefully looked over by an officious individual who declared that, in his opinion, we were over-weight for the aircraft. After several attempt to cut down the weight we tried to get the weight checked again. Now the official looked up under the peak of his PCSP baseball hat and told us that he was 'Too busy'. We were left kicking our heels in the chilly hanger for the remainder of the morning until the official re-appeared at the top of a flight of stairs. Was this to be the go-ahead we were looking for? No, he was off to lunch.

Two hours later our morale was at a fairly low ebb as Peter came to me to ask if I could have a word with the PCSP official. We had been assured that the PCSP would provide us with a weapon during our stay on Prince of Wales Island, but the official demanded proof of my experience with a rifle. I told him that I had been a qualified naval marksman for many years, I had captained the winning Queen Charlotte Trophy team in the Portsmouth Command Small Arms Championships, held several other shooting medals, and possessed a shotgun licence. 'Yeah,' he said with a triumphant grimace, 'but have you got a Canadian licence?' Of course, I had to admit that I had not. 'Then you don't get no weapon,' he replied smugly.

About ten minutes later I met him again as we approached each other down a long passage. Stepping in front of him I asked him pointedly 'Are you really happy to be sending these people out without a weapon?' 'Not my problem' he said pushing past me. He was such a rare example amongst Canadians that he ought to have been stuffed, mounted and displayed in a fairground.

When I returned to the floor of the hangar, I found a minor eruption taking place as the pilot of our aircraft at last appeared on the scene to see what the delay was. He turned out to be the same 'Happy Harry' who we had last seen snarling northwards from King William Island. Now he wanted to know what the hold-up was. Luckily for us he was accompanied by his flight engineer, a cheerful young man who had obviously been given the job to counter-balance Harry's lack of charm. The engineer took one look at our baggage and not only declared it to be well within the limits but also expressed surprise that we were not taking an ATV. At last, we could get moving and soon had the kit loaded onboard the Twin Otter. Our joy was only slightly marred when Peter was held up by yet more bureaucracy which resulted in Harry slamming the aircraft door and shouting 'Come and get me when you're goddam ready!'

Much of the flight southwards was spent above low clouds. There was serious concern that the conditions at our destination might prevent our landing and the threat of a return to Resolute Bay loomed large for an uncomfortably long time. But as we banked to circle the landing site, a gap in the surrounding fog banks was large enough to allow the aircraft to land safely on the sloping gravel of the western edge of Back Bay. The bay (like so many other bays and features, including a major river, named after Captain George Back RN) opened out in the east to Peel Sound and was surrounded on all its landward sides by low, steep-shouldered hills separated by deep valleys. From its southwest corner, the bay flooded into a long, wide, arm fed from the west by a small river. Another, shorter arm reached from the bay's northwest corner. About half a mile off the centre of the bay's southern shore lay the domed shape of Browne Island, named after Lieutenant W. H. J. Browne RN, who in 1851 had sledged down Peel sound off the east coast of Prince of Wales Island and was to play a significant part in our stay at Back Bay.

Between the landing strip and the edge of the bay, five huts could be seen, one at least in a state of disrepair. Close by, a pile of fuel drums rusted quietly whilst another, larger, waste dump, could be seen just to the south of the huts. At the bottom of the ladder, a welcoming party stood waiting to greet us. There was Wayne Davidson, looking even more gaunt than when we had last met, and Martha. They were accompanied by Martha's children, Sampson (12) and Angela (8), and their friend, Matthew Mitak (14). Wayne had good reason to look gaunt. He had taken the family on a hunting expedition out into Peel Sound only to be caught in appalling weather that had very nearly cost them their lives. After running out of food, dangerously low on fuel, and with a boat of doubtful seaworthiness, they had suffered two days of severe buffeting on a lee shore before limping back into the safety of Back Bay.

We were also greeted by another Inuit family which was sharing the hunting camp for a few days. They were led by Ludy Pudluk, Resolute Bay's representative on the Northwest Territories Legislative Assembly and a Climate Change activist. He had, nevertheless, purchased an impressive, fossil-fuel-driven powerboat in Yellowknife, brought it down the McKenzie River, turned eastward along the top of North America, through the Queen Maud Gulf, round King William Island and up Peel Sound to Back Bay. His aim was to reach Resolute before the advancing ice put a stop to any water-borne travel.

Our introductions were swiftly brought to a halt by 'Happy Harry' shouting from the doorway of the aircraft 'Save the hellos 'till later – let's get this goddam plane unloaded!'

A brief survey of the available accommodation led us to conclude that we would be better off erecting our tents than moving into one of the huts. Wayne, Martha and the children lived in one hut, another was used for storage. A third, separate from the others and closer to the beach, was in use by Ludy Pudluk and his family. That left one that had not been used for eight years and was in poor condition, and a second one that was on the verge of collapse.

Leaving Pierre in charge of erecting our tent, I made my way down the beach where Wayne seemed to be having some trouble with his boat. As I reached the shore, I quickly realised that it did not need an expert seaman (which I am not) to see what the problem was. With a heavy sea running, Wayne had managed to ground the bows of the boat on the shingle whilst every wave poured over the stern. Knee-deep in water, Wayne was desperately trying to bail the boat out, but it was an unequal contest, and the well of the boat continued to fill. Conscious that I was supposed to know something about the sea – and aware that a number of eyes were looking at me in anticipation – I shouted to Wayne above a stiffening wind to throw me a stern line. My plan was to haul the stern towards the beach whilst pushing the bows out.

There was, of course, a risk of broaching as the 22-foot-long boat turned beam-on to the waves but, with care, it could be done. In any case, it was no more risky than continuing to allow the water to swamp the boat in its present position. But it was not to be. Wayne shouted back that the outboard motor would get damaged if the boat was turned stern-on to the beach. Cutting off my obvious reply, he continued to shout that the engine was stuck in its position and could not be lifted clear of the water. In fact, the weight of water already in the boat had caused the propeller to begin striking the seabed with every lift of a wave, seriously putting it at risk.

Seeing that the situation was getting now extremely serious, Wayne reinforced my view that he was the nearest I was likely to see of a 'voyageur' by jumping over the side into the icy waters. Waist-deep, and with the boat bucking violently, he worked his way to the stern, grabbed a stern-line and began to haul the boat off the beach. I pushed the bows out as far as my boots would allow me, unclear as to what Wayne's intentions were. With the boat clear of the shingle, Wayne, after a huge effort, managed to clamber back onboard. Half a dozen attempts later the engine coughed into life and Wayne made his way precariously to the midships throttle lever and inched it forward. There was another balancing act to get back to the engine where a jury-rigged tiller turned the engine and thus the boat. At last under power, Wayne turned the boat and headed off to the northern arm of the bay where it would join Ludy Pudluck's boat in a tiny natural harbour. The boat was named *Qakudloo*, the Inuit name for the giant petrel whose wide-winged flights over the bay mocked the spluttering boat as it pitched and rolled painfully towards sanctuary from the threatening sea.

Returning from the beach, I made my way to the least decrepit of the huts where, as usual Max had worked wonders. Clearing away much of the debris he had built a shelf on which sat our two stoves. Already water was boiling ready for a warming drink. The hut was entered via a small windowless compartment littered with scrap iron, animal bones, and shreds of fur-bearing skin. The smell of rancid seal-blubber dominated.

Having made myself a warm drink, I went outside to have a closer look at our new surroundings. Looking eastward, out of the bay and across Peel Sound, the cliffs of Somerset Island could be seen on a moderately clear day. Through binoculars, a scatter of small rocky islets could be seen just outside the bay. An onlooker on the same spot in the summer of 1846 would have been amazed to have seen two stately sailing ships making their way slowly southwards through Peel Sound to an appointment with destiny in the ice of Victoria Strait.

To my right front, about two miles away, lay the humpback form of Browne Island surrounded by a skirt of rubble beaches. In certain light conditions, the red-brown island would disappear from view as it merged with the larger southern horn of the bay curving northwards immediately to its rear. The western end of the southern arm marked the mile-wide narrows that formed the entrance to the south-western arm of the bay.

Behind me the ground sloped gently upwards past the huts, past the airstrip, and then reared up to make a steep-sided hill the colour of the limestone on King William Island, but actually made of granite rubble. Beyond the lower northern slopes of this hill lay the northern arm of the bay. Across a small river inlet, the northern side of the bay was lined with a series of low brown hills separated by steep valleys.

A white line of ice sat at the entrance to the bay whilst large, fantastically shaped, multi-year ice floes sat grounded about fifty yards off our shore. These impressive structures ranged from flat, tilted blocks to towering, broad pinnacles that looked like ships under sail. One in particular caught the eye with wide ridged columns of a startling blue colour supporting a glistening white platform.

As I took in the view, I suddenly became aware of one of Ludy Pudluk's family running across my front whilst cocking a rifle. 'What's up?' I shouted, keen to know whether or not I should be running for cover. The man continued to run as he pointed out to sea and shouted back 'Beluga! Beluga!' I followed his directions and was amazed to see, between the shore and the grounded ice floes, certainly no more than thirty yards from the beach, a school of white whales surfacing and blowing as they passed from left to right.

With mixed feelings, I watched both adult and junior Inuit racing down to the edge of the beach to open fire with a variety of weapons on the beautiful creatures. To the natives they were just a resource to be harvested when the opportunity presented itself. The skin ('muk-tuk') would be eaten as a delicacy whilst the rest of the animal would be fed to dogs.

The hills rang to the sound of the shots as the Inuit fired round after round at the whales. There was a race northwards along the beach to get the *Qakudloo* out to sea, and soon we could hear the boat's engine racing as Wayne tried to cut off the whale's seaward escape. Bursts of firing echoed across the water as they closed in on a surfacing animal. Then there was silence. Incredibly, the hunters came back empty-handed. How they managed to miss such a large number of large targets whilst having fired off the equivalent of a small country's annual defence budget, I do not know. But I felt relieved to see the whales escape what I thought would be an inevitable slaughter.

That evening, after our evening meal, Wayne invited me over to his hut where he had something he wanted me to see. Despite a large window the hut was dark inside. Two bunks faced the inside door; a kitchen area faced the window beneath which was a table and chairs. A curtained partition led to a bedroom area. In a corner, a television and video player sat incongruously. The equipment was battery-driven but as no batteries were available, the children made eager use of one of Dan's portable generators to spend the whole night watching tapes of horror films.

Wayne produced a length of wood which he placed in front of me and invited my comment. The piece was about three feet long, round, with a diameter of about three inches. One end showed signs of having been snapped off; the other end had clearly been neatly sawn. Although the wood did not appear to have been machine-turned, its surface was too smooth and regular to have been natural. From simple observation it was difficult to tell what it might have been. Wayne felt that it could have been the end of an oar, but that did not feel quite right to me as it seemed too thick and of constant diameter over its entire length. There was certainly no wear at any point. Eventually, the best guess that I could come up with was that, if it had been any part of a boat, it might have been the bowsprit of a small sailing boat or possibly a yard used to carry a light sail. Wayne then pointed out a small patch of red paint near the ragged end, but I still could not come to any more sensible conclusion than I already had. With a shrug, Wayne threw the wood into a corner and told me that he had a larger item he wanted to show me that he kept in a bedraggled-looking outhouse. However, that would have to wait until tomorrow as it was getting late and we had a busy day ahead of us. Peter had arranged with Wayne that he would ferry us out on the *Qakudloo* to one of the small islets beyond the entrance to the bay. Before I left, Wayne loaned me what must have been the world's oldest .303 Lee-Enfield rifle. With its simple rear sight set halfway down the barrel it looked as if it could have been used during the Boer War. If it had been, it had not been cleaned since, as oil and the sight of a '4 by 2' cleaning rag were little more than a fond memory to the weapon. Nevertheless, if it went 'bang' and ejected a lump of lead at high speed from the opposite end to me, I was more than satisfied.

As I took a final look at the scene before getting into my tent that night, I could see that my white ensign and 'sledge pennant', which I had hoisted on a large fish-drying frame, were flying horizontally in a stiff east wind. If it continued into the next morning, we would be heading directly into it, I wondered how the *Qakudloo* would cope with a boatload of passengers and a head wind.

Hauling myself out of the tent the following morning, the first thing I noticed was the bright fog which had surrounded the camp. Shivering in the cold air, I stamped my feet and rubbed my hands as I looked out to sea. Only very little of the water could be seen. The strong wind of the night before had flooded the bay with ice-floes. Where there had been the rolling blue-grey waves of yesterday, there was now the silver-grey of jagged ice crunching and hissing as each floe jockeyed for position.

By mid-morning, the fog had cleared, and the ice glinted sullenly beneath the Arctic sun. Through my binoculars, it appeared that the entrance to the bay was now solidly blocked but from higher up the hill to the rear of the camp clear lanes could be seen through the ice, which could lead out into Peel Sound. The people most affected by this turn of events were the Pudluk family who were desperate to get their boat up to Resolute Bay before the approaching, sea-enveloping, freeze. After some debate, they decided to take the risk and soon we could hear the deep-throated roar of their powerful engine as they threaded their way slowly through the ice barrier. With the Pudluks heading north, it now meant that their accommodation was available and Peter, Pia, Pierre and I moved in to sleep on the floor of the hut where the temperature remained around a cosy 2 degrees Centigrade. Dan was to join us the following day, but Max, Maurice and Francois stuck resolutely to their tents.

The departure of Ludy Pudluk and his family also meant that the entire population of Prince of Wales Island (a landmass the size of Holland) numbered thirteen: six team members, two television crew, and five locals.

By noon, my flags hung limply on their pole and a rising tide had loosened the jam of ice-floes so that several areas of clear water could be reached from the beach. At this point, Wayne came to me and asked me if I knew how to shoot. I assured him that I did and he responded by asking me if I would go on a seal hunt with Matthew and him on the boat. The hunting season had, so far, been a disaster. No caribou had been sighted, and even seals had proved difficult to catch. There was no winter stock in, a situation that just a generation ago would have meant starvation. Only Arctic char had been available, and that had barely covered their day-to- day existence as the lack of fish hanging from the drying frame demonstrated. The worst suffering must have been felt by Wayne's monstrous St Bernard dog 'Sidney'. All the dog had to live on was fish entrails until our party arrived and began to feed it chocolate bars. With the noble excuse that I would be helping to alleviate an animal's suffering, I agreed to go with Wayne.

It was an exhilarating feeling to be stood at the rear of the boat's canopy, gun cocked and ready in the hunt for seals. Wayne sat at the stern and operated the tiller whilst Matthew controlled the throttle on my right, the engine grumbling as we cautiously edged our way between the floes. About a quarter of a mile from the shore, Matthew suddenly jerked his arm out and pointed, 'There!' Sure enough, a curious seal had popped up to have a look at what the noise was all about. Using the cabin roof to support my arm, I brought the gun up and peered along the sights. With the foresight locked onto the target I squeezed the trigger, only to be mortified to see a splash about forty feet behind the seal indicate by how far I had missed. Matthew grunted. Minutes later, another seal broke the surface, this time even closer. Again I held it in my sights, fired, and watched the bullet strike the water twenty feet to the left. The pantomime continued to an accompaniment of Matthew losing control to a high-pitched giggle. No matter how I tried to 'aim off', the gun behaved as it had no intention of hitting the target. Even a huge bearded seal, who had been watching the performance from a grandstand position on an ice-floe, idly watched a bullet send slivers of ice into the air several feet from where he was sitting before unhurriedly rolling into the water. Apart from the humiliation of being the world's worst seal hunter (the children soon had me nicknamed 'Sharpshooter'), I also learned how close Wayne had come to the Inuit way of life.

When he, Martha and the children had taken the boat out into Peel Sound on the hunting trip that had very nearly turned to tragedy, he had killed a seal but had failed to reach it before it sank. Now he insisted that the sea could get 'very angry' if the life of a seal was wasted – an unusual view of meteorology for a 'Weather Technician'.

On our return, I learned that Dan and Peter had decided that as there was clearly could be no attempt to get out to the islet by boat, and that the ice looked as if it intended to stay, a helicopter should be ordered from PCSP at Resolute Bay for the following day.

Max and I walked about a hundred yards south of the camp to where there was a widespread rubbish dump. It was a deeply sad and sorry sight. Lying around the edge of a soggy patch of tundra were numerous polar bear skeletons. Most of the heads had been removed to be placed in the sea at the water's edge where small shrimp-like creatures stripped the flesh. The resulting skulls were then sold in Resolute Bay. Complete spines of beluga whales lay resting against the rib cages, skulls and broken antlers of caribou. The place was an Arctic charnel house, appalling to the eyes of comfortable western visitors, but reflecting the uncomfortable reality of a fragile food chain that includes mankind.

Dan had placed our luxurious toilet facilities behind a stockade of empty oil drums just a few yards west of the huts. To reach it, we had to pass by a tangled mass of majestic caribou antlers still attached to their skulls. More antlers were scattered around the hut area along with severed limbs (usually still covered with flesh and hair), broken polar bear skulls, and seal bones. The Inuit find this no problem at all as, for most of the year, snow covers the ground hiding the gruesome litter and in the summer children play games made up of poking each other with caribou legs (the delightful Angela was a particular exponent of this pastime).

It was, however, interesting to see how quickly we became used to living and eating in such surroundings. Francois, having discovered that the disreputable couch in our communal kitchen hut was actually a fold-down bed, deserted Maurice and moved in. On most mornings, we could guarantee having our breakfast sat on his bed whilst he quietly snored on regardless.

That evening Wayne called me over to have a look at another piece of wood. This time it was far too big to have been seen in his hut and was propped up outside. At first glance, I was not sure at all what I was looking at. Close on twelve feet in length, the timber had once had a diameter of twelve to fourteen inches, but a split along its length had caused almost half of its bulk to be missing. A very slight bend about five feet from the bottom spoiled an otherwise straightness along its length. Before I had chance to examine the object closely, Wayne pointed out a very faint triangle mark about six inches long on each side, halfway up the top section. This he believed would confirm to me that this was part of a Royal Naval vessel. In fact, it confirmed no such thing, whereas a 'broad-arrow' mark would have been very exciting. In his disappointment at my reaction, Wayne was about to abandon the object when I noticed two deeply gouged or chiselled marks one above the other, and about three feet apart on the upper section. The angles suggested that they had once been the positions of two cleats (fixtures for securing ropes or lines). I then looked further down and saw something of very great interest.

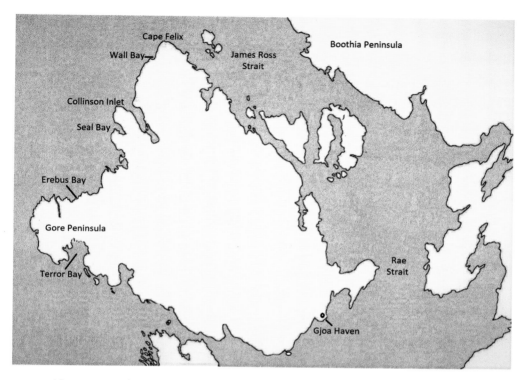

Above: Map of King William Island.

Right: The author, still dressed for town, getting a lift across a meltwater stream from Staff Sergeant Bob Martin of the 'Mounties'.

Below right: On the back of the trailer. (See page 20 for an explanation of this ungainly position.)

The return journey to Gjoa Haven halted by the muskeg. The blurred images are visiting mosquitoes massing for the attack.

Max's skidoo and sledge about to be used to tow the party eastwards along the Simpson Strait. The state of the ice is obvious.

Above: Max and his son on Todd Island reflecting on their return journey across the ice to Gjoa Haven.

Right: The White Ensign and the author's 'sledge pennant' flying on Todd Island. The land on the horizon behind the flags is the lip of North America (probably Richardson Point).

Below right: Examining 'Jack Todd's' skull.

Max towing the boat from Todd Island to Booth Point. Max is kneeling on the skidoo seat in case he needs to jump clear.

Amundsen's *Maude* in Cambridge Bay. She was later recovered and returned to Norway.

Looking forward to the next sixteen days based at Cape Maria Louisa.

Tent rings at the 'Peter Bayne Map' site. Franklin Point can be seen near the top right-hand corner.

The first sighting of the mounds. The photograph was taken from the east.

The larger Southern Mound. The photograph was taken from the west.

The 'grave' site. The depression, covered with slabs, and with smaller slabs remaining vertically inserted at the top and sides as if the site was once completely edged with small vertical slabs.

The ice off Cape Felix.

The Drift Calculator at Cape Felix. The ruined cairn closest to the camera is the most easterly sighting cairn. The other two can be seen in a straight line towards the west.

'Ross's Pillar' with Back Bay and the probable Victory Point behind, with Franklin Point curving westwards in the far background.

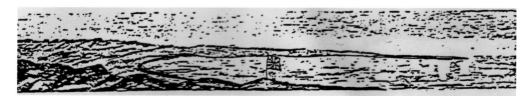

Above: Sketch made by James Ross with the Union flag flying from the probable Victory Point with Back Bay behind and Franklin Point in the far distance.

Left: The 'Franklin Project' box of emergency supplies close by Crozier's Landing.

Above: Just to the south of Crozier's Landing. The base of the Franklin Project cairn can be seen at the centre top of the esker. The cairn in the foreground is the supposed grave of Lieutenant Irving. His bones were returned and interred at Edinburgh.

Right: Captain (later Admiral, Sir) Francis Leopold McClintock RN, discoverer of the fate of Franklin.

Below: Looking east from the tent, a cairn and second anchor on King William Island. The site was encountered almost three decades later by an 'Arctic Science' expedition.

The author patrolling the estate whilst hoping for an aircraft.

Above: Geologist Art Dyke lifting the slabs at the 'grave site' on the Southern Mound. He declared the depositing and the arrangement of slabs as 'natural'.

Left: Despite the disappointment at the mounds, Pia entertains the party.

Above left: HMS *Resolute* and *Intrepid* winter quarters, Melville Island, 1852-3.

Above right: On the ice of Prince of Wales Island's Back Bay. The long pole was used to fend off ice at the bows of the *Qakudloo*.

Above *left*: The keel and forward upright (bow) of the boat surrounded by associated debris.

Above right: The rock oval at the 'Sledge Camp'. The 'gravestone' can be seen to the left rear. Peel Sound is behind with Sherard Head on the right horizon.

Top: Ice in Victoria Strait at ground level.

Above: Crossing Collinson's Inlet with pack and cart.

Left: The tent on the eastern edge of 'Fort Crozier'. At least a day went past before the author discovered the site on the rocky platform to the rear.

At the Dundalk conference talking to Professor Potter and Councillor Jim d'Arcy, Chairman of Dundalk Council.

Above left: Being greeted at a Mayoral reception at Point Barrow by an Indian Princess. The moment was followed by raw whale blubber hors d'oeuvres.

Above right: A warning noticeboard at Point Barrow.

Top: On South Georgia, taking the salute at a march past of King Penguins.

Above: The author, seated in the Admiralty Board Room, being bombarded with indignation by an Inuit politician. At his side sits Franklin's greatest friend, Sir John Richardson (actor Alistair Findlay).

Left: Lieutenant H. L. V. Le Vesconte on the deck of the *Erebus*. To the rear is the ship's double steering wheel.

Lt-Commander Gould's 1927 chart showing that he calculated that at least one of Franklin's ships would be found in Eastern Queen Maude Gulf.

Above left: Sir
John Franklin as
Lieutenant-Governor.

Above right: Jane,
Lady Franklin.

Left: Sir John Franklin
monument, Royal
Naval College Chapel,
Greenwich.

The lower part of the timber, beginning almost at the slight bend, had been flattened off as far as the bottom edge. This removal of the natural rounded shape which remained throughout the top part had been achieved by a process of chipping away slices of wood until the required width of levelling had been reached. I recognised the marks immediately, not least from my time as Officer of the Day in HMS *Victory*. The task had been done by using an adze, a dockyard or shipwright's tool. And a tool that – as far as I was aware – had no home in the native culture and could only have arrived in the area on a ship.

So now what was I looking at? My limited knowledge of wooden ship construction told me that the bottoms of masts had to be squared-off to fit into a mast 'step'. If the mast was not keyed into the base of the ship and was allowed to remain round in shape, once the yards and sails had been fitted the construction would tend to act like a weathervane and swivel round under the pressure of the wind.

Was this then a mast? If it was, it had to belong to a larger vessel than a small sailing boat. What was such a vessel doing in that part of the world? Wayne was somewhat vague as to where the 'mast' had come from but thought that there might be other parts of a vessel remaining. I felt that I might have stumbled across the cultural barrier that had been raised a fraction to let us see the area in the first place. Any further lifting of the veil would only be done with the approval of the local Inuit.

I awoke the next morning to the sound of a large bird walking across the roof of our hut. Not the usual, gentle flopping sound of a gull's webbed feet, but something that sounded hard, almost menacing. Looking out of the hut window, I found I could see little beyond the shoreline. Thick fog obliterated the view on all sides. Once outside the hut, I noticed that the air had a sharp edge to it indicating a considerable fall in temperature. None of this boded well for our plans to fly out to the islet by helicopter, and the ice blocking the bay could put paid to any ideas of going out by boat. Turning to go up the beach towards the kitchen hut, something caught my eye on the roof of the hut I had just left. It was a huge black raven staring directly at me. A bird of ill omen the expedition could well do without.

As I joined the others, I learned that Dan had cancelled the helicopter. There was no chance of the machine landing safely in conditions where even the hill behind us could not be seen. When Wayne appeared, he was adamant that the *Qakudloo* could not be risked, especially as the boat did not belong to him and was only on loan. It meant that the day was lost as far as visiting the important sites was concerned.

Throughout the morning, the fog slowly thinned and then lifted enough to see the low hills to the south of the bay. It also allowed us to see a welcome visitor. Around the northern shoulder of our hill, with a slow stately step, a bull musk ox came to have a look at us. At about fifty yards, he paused to take in the scene, then moved across the base of the hill to disappear over the southern slopes. The team members (apart from Pierre who had gone back to bed) decided to follow the musk ox's example and stretch their legs by going for a walk towards the interior of the island.

Keeping our hill to our left, we passed along its northern flanks and climbed up to the top of a granite rubble spur that tailed westwards from the hill itself. To our left, large clouds formed behind a far of range of dark hills. From their feet, a wide swathe of green and purple tundra led down to the southern arm of Back Bay, its waters packed with ice floes. Even from the elevation at which we stood, it seemed as if it was possible to walk from one shore to the other. Turning eastwards once again, more hills ranged on into the interior of the island. Since man had first walked on the planet there can have been very few people, even amongst the natives, who had roamed across those receding slopes and valleys. A tough Canadian fur trapper, William Kennedy, and Lieutenant Joseph-Rene Bellot of the French Navy had travelled by dog sled across the south of the island to Ommanney Bay and returned to Peel Sound before making their way up the Sound to Cape Walker during their 1852 search for Franklin. But their journey had been many miles to the south of us and, to the best of our knowledge, no-one else had bothered with the island until a zoological expedition visited the area in 1958.

To our right front a large, rounded hill, scarred with rocky outcrops, lost its eastern slopes to the edge of a fog bank which blocked any further view to the north. But this was no ordinary view of a fog bank. Soaring through its dark centre was an arch of bright light like a white rainbow. It was a 'fog bow' – a spectacular Arctic phenomenon that was a delight to encounter.

As we reached the end of the long granite spur, I raised my binoculars. There was our friend the musk ox who had obviously worked his way along the northern side of the ice, passed through the deep, wide valley of tundra that blunted the western tip of the spur, and now stood munching on tundra grass on the lower slopes of the hill to our right.

Max and Francois, both armed with expensive cameras, decided that they would like to get a close-up photograph of the animal. For most of the next hour they entertained the rest of us as they crept closer and closer to the unsuspecting animal. By now no more than two specks to the naked eye (although Francois's white boots flashed with his stride like a distant lighthouse), I could see them clearly through my binoculars as the musk ox ignored their approach until they were within twenty-five yards.

Then the lumbering animal turned its bulk and faced them with its fearsome horned forehead. Even though the two intrepid photographers were well over a mile away from where I stood, I could detect their inching retreat as they thought better of getting any closer. The musk ox then returned to its grazing as Max and Francois once again returned to the hunt. Suddenly, the musk ox turned, faced them, and began to paw the ground. The expected retreat had hardly begun before the animal started to charge. Our heroes' appreciation of wildlife took a back seat as they scattered down the hillside. Beside me I heard Dan gasp 'Oh, my God!' But he need not have been concerned. The musk ox had made its point and, faced only with the rapidly retreating backs of Max and Francois, it snorted to a halt and continued on its way around the middle slopes of the hill.

As we regained the northern edge of the hill behind our camp, we saw two tiny figures walking along a ridge near to its top. They turned out to be Martha and her daughter, Angela. Martha was carrying a rifle in the hope that they would come across a caribou. We told her that we had seen nothing apart from the musk ox, but, with their winter stocks so low, she decided to carry on looking in the hope that fortune might smile upon them. Sadly for them, it did not, and caribou were not to be seen again that hunting season.

We arrived back at the camp by late afternoon and having collected ice from the shore to provide our fresh water, I happened to be passing Wayne's hut. There on the ground lay the length of wood he had shown me on the first evening of our arrival. I idly glanced in its direction, and then stopped dead in my tracks. I had seen something that I had not seen in the dim light around Wayne's table, and something that could cast new light on our investigations. Wayne had pointed out to me the patch of red paint towards one end of the piece of timber. What I had missed was that there were several other remnants of paint, but these were of a yellow-ochre colour – exactly the colour used by the nineteenth-century Royal Navy for the upper works of its ships. I knew the colour well from my time in HMS *Victory* and, even today, the Maritime Auxiliary Service still used the shade for its tugs and harbour ferries. I took the wood along to Peter and Dan who were crossing the beach on their way to the kitchen hut and explained the possibilities that arose from the fragments of paint. I also pointed out that it was possible that a Hudson Bay Company vessel might have used the same colours for their upper works, and even that it was possible (but unlikely) that whaling ships could have followed the Royal Navy's example. However, as there was no shred of evidence that either type of ship had been in the area, the options were very limited indeed. We then talked the matter over with

Wayne who, having already thrown the length of wood out, insisted that we take it back to England to have the paint analysed.

That evening we all kept an anxious eye on the weather. A helicopter had been booked for the following morning and we had a sense that time was running out for us. The morning had dawned with a clear blue sky and soon we could hear the clatter of the approaching helicopter as it closed in from the north. Max had given everybody a briefing on the correct way to conduct themselves when boarding and disembarking from helicopters and Peter had drawn up a 'batting order' for passage out to the islet. I was to go on the second trip in order that the film crew could get a shot of Max, Pierre and me emerging as it landed. We would then be on the second trip from the islet to the boat site and, if fuel, time, and weather, held out, we would be taken on to the 'marker' site on the far (Peel Sound) side of the bay's southern arm.

Our pilot turned out to be a pleasant Swiss woman who was in that part of the world to gain experience in Arctic flying. She had brought out with her one of Martha's sisters who had been elevated to the important role of 'Elder'. This lady would provide an authoritative view of the sites we were to see. Anything that was Inuit would be declared so by her, based upon a lifetime of experience in that part of the Arctic. This was of vital importance in order to prevent us wasting time on features and, possibly even artefacts, which were commonplace to the natives, yet novel to us.

As we flew out to join the first party, I could see that the ice lining our shore was much less than had been for the past two days and, once the beach was left behind us, we had clear waters below us. A few minutes flying brought us level with Browne Island and then the red hill that formed the southern horn of the bay. Ahead of us, I could make out the small clutter of black rocks that had pushed upwards through the waters of Peel Sound and was to be our target. Dan's voice came over the intercom as we got closer warning everyone to wear headgear as the islet had proved to be home to a colony of Arctic terns who were enraged at our intrusion into their territory.

Nature had proved to be considerate in providing the helicopter with a minute landing site at the western extremity of the tiny rock-bound islet. It could have landed nowhere else. The islet was about eighty yards long from east to west and slightly less on the other axis. The rocks were piled higher to the south than to the north and were, for the most part, piled up loosely against each other. Only in the southwest corner was there any geological order where a succession of solid ridges made their way down to the sea and could be seen disappearing well beneath its clear waters.

Clambering over to where Dan and the Inuit Elder stood, I joined them as they were examining a small cairn of black rocks. It was a very simple structure with each rock being about the size that an average man could handle comfortably. No more than four feet tall, it grew from a base about three feet wide and was capped by a single rock that showed evidence of being used as a snowy owl's perch. The instant and unqualified opinion of the Elder was that no Inuit would have built such a cairn. It would have been useless as a cache and if they had wanted to build an 'inukshuk', there was unlimited building material on the islet to have built one in the native manner.

A few yards to the east, Peter and Pia were measuring something else which had been declared 'non-native'. Built of similar rocks to the cairn, it looked as if at one time it had been a circular enclosure about six feet in diameter. Whilst much of the lower part of the wall remained, only about a quarter of the wall was left standing to give some idea of the original height of the structure. The rest had collapsed leaving the floor covered by a jumble of rocks. Peter felt that the feature once had a roof and that the roof had also collapsed into the structure. I did not agree.

Matters were complicated by other members of the party finding another circle of stones, barely detectable against other rocks and more evidence of stone wall building, but with so little remaining that it was impossible to guess what its purpose had originally been.

What, however, was to be made of the enclosure and the cairn? If they were non-native, and old enough (which they certainly looked), our current knowledge would suggest that they could only have been made by the Franklin Expedition in 1846, Lieutenant Browne's party in 1851, or by Kennedy and Bellot in 1852. Franklin, I felt, would not have dawdled his way south through Peel Sound but with open water would have pressed on to seize whatever advantage he could from his good fortune. Even if he had stopped to build a marker cairn, it would almost certainly have been a much more imposing structure than that we had found, and in a more prominent position on one of the surrounding hills, all easily accessible from the Sound. In addition, it was well known that the ice was likely to thaw on western shores first and Franklin would have kept to the east of Peel Sound (off the western coast of North Somerset) rather than cross over to the eastern shore of Prince of Wales Island.

As for Lieutenant Browne, his report mentions that he passed 'a small rocky islet' and later camped 'about two miles to the westward of the islet'. With such detail in his report, I believed that he would have mentioned visiting the islet to build a cairn and a shelter. That only left Kennedy and Bellot. They would have come up the western side of Peel Sound to take advantage of the thicker ice and would have had difficulty in avoiding the rocky islet. If they chose to rest there, or were forced to by inclement weather, it is quite within the practice of the day to have built a stone enclosure to be roofed over with canvas. A similar structure from the Franklin Expedition had been found north of Cape Spencer on Devon Island. It also reminded me of the rock enclosure built by Wilson, Bowers and Cherry-Garrard during their Antarctic winter journey of 1911 – they had also roofed their structure with canvas.

What then of the cairn? It was a little unusual in that its height was almost exactly at the level of the highest rock piles that had been formed on the islet. It could therefore not be detected from an eye-level position out on the ice or water. This, I concluded, could only mean one thing. It was not a navigational marker for anyone passing through the channel. It was, instead, intended solely as a marker for someone who already knew about the islet and probably marked the site of a message or buried provisions (another common practice of the time).

Perhaps then, this was the situation. After travelling across Prince of Wales Island and returning to the shores of Peel Sound south of Back Bay the fourteen men of the Kennedy-Bellot sledge party stopped at the islet, either to rest, or because of bad weather. With scurvy having appeared amongst them, they would have known that to delay too long could have resulted in deaths. They knew that fifty miles to the north, at Cape Walker (their next destination) and beyond, on Griffith Island, there were almost certainly caches of provisions left by the Austin Expedition. Failing that, they knew that James Ross had left a cache at Port Leopold on the northeast corner of North Somerset. Speed was of the essence, but those suffering from scurvy would have weighed down the sledges. Far better, once the sledge dog's rations had been accounted for, to lighten the load by reducing their provisions or stores by enough to allow them to get to Port Leopold. These could then have been buried under the cairn. On the other hand, they might simply have recorded their presence by leaving a message under the stones. Whatever the reason for the erection of the cairn, we were unable to dismantle it without an authorised archaeologist present.

The helicopter, having brought out the final party, took Peter, Dan, and Maurice the cameraman to have a brief look at the 'marker' site and then on to the boat site. As the remainder of us roamed over the islet looking for any other signs of occupation, Martha and her sister picked up a couple of tern chicks to take back for the children. It seemed that they make excellent house pets. Max, meanwhile, probably took a different less charitable view as the terns, copying their King William Island cousins, swooped and dived on his blue headgear.

On its return, I climbed aboard the helicopter with Pia and Francois to join the first party. Instead of heading south, though, we turned westwards in the direction of the camp. Seeing my consternation, Francois shouted above the engine that we were going to refuel before joining the others. But after we had touched down by the huts, the Swiss pilot indicated that we should all disembark. Once we were clear of the aircraft, she took off, leaving us looking bewildered at this turn of events. It was not long before we found out what the problem was. The helicopter had only enough fuel to bring the other two parties back from the islet and the boat place, and then to return to Resolute Bay. I was fuming at this turn of events. The sites on the islet had been bottom of the priorities, well below the importance of the boat and marker sites. But, in reality, it was nobody's fault. Safety in such a region was paramount, and nobody could blame the pilot for refusing to take risks on our behalf. Nevertheless, I ate a rather disconsolate lunch as Peter, full of excitement over what he had seen at the other sites, insisted that I had to see them.

There then followed a frenzy of negotiation with Wayne. We wanted him to take us over to the boat place in the *Qakudloo*, but he was concerned about the amount of ice still piling up against the shore. Eventually, our argument prevailed. Max remained at the camp, and Pierre retreated to his sleeping bag (his excuse was that he wanted to get up at about one-o-clock in the morning to photograph the eastern sun low over the ice). Peter, Dan, and Pia boarded the boat with Martha as 'ice look-out' in the bows, Wayne on the tiller, Matthew as cabin boy, and with the throttle manned by me. Winding our way through the ice floes, Dan used a long pole to fend off any threatening ice as Martha called out directions to take us between the glistening obstacles.

The journey took about twenty minutes before we reached the southern rim of the bay, landing beneath the eastern slopes of rising shore. We were still some way from the boat site, but Peter, who for some reason never seemed quite at ease in the boat, requested that we get ashore as quickly as possible. As it turned out, it was probably just as well that we did. The beach to the east of our landing shallowed rapidly and we would have found ourselves with the inconvenience of having to wade to dry land. The two-mile walk required to reach our destination, however, proved to be a pleasant, easy stroll, for the most part over granite chip ridges.

We had not gone far when Matthew pointed out two large white objects, obscured for the most part by moss. They proved to be, according to Wayne, two ancient Bowhead Whale skulls. On one of the skulls, a series of small scratches was capable of being read as '1856'. A curious find at such a remote spot in the Arctic, but one that possibly indicated non-native whaling activity in the area.

After about twenty minutes, we breasted a small rise beyond which a valley, backed by a steep hill, levelled out into a bay about half a mile wide.

A small stream flowed across our path as we began to wade through the tundra that edged the bay. Then, almost in the centre of the curve of the bay, and about a hundred yards inland lying on a bed of very soggy tundra was the bow and keel of a boat. I approached the object almost with a feeling of awe. I did not need an Elder to tell me that this craft was not of native origin, it was clearly of western design. As I came nearer, I could see that a number of planks or strakes (hull planks) lay scattered around the keel and bow whilst less definable strips of wood lay in a jumble in all directions. To one side, a warped replica of the bows lay in the mud, its holes exactly matching those of its opposite number. Other regularly spaced holes appeared to have been drilled in some of the major planks that sprawled beneath the bow (I assumed that the end still attached to the keel was the bow, as if the boat had been hauled up the beach bows first).

Two holes near the top of the bow and stern posts suggested the point at which the upper strakes were secured to the uprights. A smaller hole further down could have been for fixing a rubbing strake (a thicker plank meant to protect the hull), or as a securing point for a bow-line. At the top of the bow (but rotted away from the separated stern post) the thickness of the wood had been reduced to produce a comb-like tenon probably intended to fit into a mortised apron or canopy.

The boat remains were not the only surprise in that remote inlet at Back Bay. A hundred yards further inland from the boat site was a typical early ('historic') Inuit or Thule 'sod' house which could have been several hundred years old. 'Thule' is correctly pronounced as 'Thyooley' but is referred throughout northern Canada as 'Tooley' and refers to a race who inhabited the Arctic before the arrival of the Inuit people. They were regarded as great whale hunters, and the collapsed dwelling we saw before us gave every evidence of such skills. Large, bleached vertebrae lay scattered around, mixed with white curving ribs and huge shattered jawbones.

Despite being extremely interesting in its own right, it was not the Thule dwelling that caused excitement to ripple through our ranks. With its edge only a few feet from the sod house, a large circle of rocks stood out on the surface of the ground. About fifteen feet in diameter, the circle displayed none of the Inuit attributes as suggested by Jim Savelle, and had been declared as non-native by the Inuit Elder. Furthermore, inside the circle, a hearth made of three rocks survived in which evidence of wood burning remained. This feature, as far as the Elder was concerned, firmly underlined her opinion that this was not an Inuit site.

Tent rings are not particularly unusual at Thule settlement sites but, even to the untrained eye, the age of the tent ring we were looking at was considerably less than the accompanying, moss-covered rocks of the sod house. As for the hearth, such features are not unknown in the Arctic, but usually originate in the Pre-Dorset (*c.* 2,000 years old) era and tend to be four-sided 'box hearths' (to confuse the issue, nineteenth-century Inuit were known to have used these ancient hearths during hunting forays). Again, the hearth we saw exhibited none of the signs of great age.

It did not take a great leap of the imagination to link the remains of the boat and the tent ring together. Someone could have brought the boat into the bay and erected a tent for themselves a short distance further inland. But who had access to a 27-foot, double-ended, whaler? One obvious source was the whaling ships. They, however, were generally restricted to the Lancaster Sound 'Polynya' (an area of clear water) or south into Prince Regent Inlet where by 1874 they had reached into the Gulf of Boothia. Even if they managed to push further west, few were recorded as ever getting beyond the ninetieth meridian – 170 miles from Back Bay. Nevertheless, whaling ships cannot be entirely excluded, for if a whaling captain found a good hunting ground there was always the possibility that he would not spread the word for fear of losing his advantage.

With a dramatic reduction in the numbers of whaling ships in the Eastern Arctic by the early 1840s, the emphasis moved to the Greenland Fisheries. Along with this development, Baffin Island Inuit were employed by the whaling men and may well have become skilled in the use of the whaler. But the idea that the Inuit (or anyone else) would have brought an open boat hundreds of miles from the eastern edge of the Canadian Arctic all the way to Back Bay on Prince of Wales Island is scarcely tenable. And even if they had, would they have abandoned it, unburned and un-salvaged, in an area where wood was a scarce and precious commodity? There still remains, however, the date scratched on the whale skull – could it have been a European whaler? Nevertheless, the most likely explanation of all was that the boat had come from the Franklin Expedition.

One of the great Arctic enigmas is the question why did the members of the Franklin Expedition, when they abandoned their ships, head south instead of north? They knew that the southern route to the nearest Hudson Bay Company Post meant a journey of well over a thousand miles through an entirely hostile terrain. Captain George Back's account of his difficult travels down the Great Fish River were well known, and their own recently deceased leader had written a graphic account of the suffering he had undergone in the barren lands. A voyage of less than three hundred miles to the north, on the other hand, would have brought them out into Barrow Strait, an obvious destination for any rescue attempts. Another journey, including some overland travel, of about the same distance would have got them to Fury Beach, an equally obvious point for searchers to head for with the added bonus of ample provisions left by the 1829-33 Ross Expedition. It is unlikely that there was anyone onboard the *Erebus* or *Terror* who was not aware of the piles of stores at Fury Beach (despite a rumour that the stores had been pillaged by whaling ships). In fact, Thomas Blanky, the Ice-Master of HMS *Terror,* had served under John Ross when Ross's party had made its way north from the abandoned *Victory* to Fury Beach in 1832 and (assuming he was still alive) could have acted as a guide.

There is also evidence (albeit rather thin) that some of the Franklin Expedition did try to return to their ships for one reason or another. A boat discovered on the west coast of King William Island by Captain McClintock was found to be facing north, and the gravesite I had seen in 1992 just to the south of Crozier's Landing was said to be that of Lieutenant Irving. Irving had been alive the day prior to the departure from Crozier's Landing (he had been sent by Captain Crozier to collect the message left in a cairn by Lieutenant Gore). A medal found by the grave by Schwatka had been awarded to Irving and gave rise to the reasonable assumption that the grave was, in fact, that of Irving. What weakens the support for the Irving supposition is that he could have died on the day the march to the south began and been immediately buried at the site near Crozier's Landing. Nevertheless, no less an authority than that of Richard J. Cyriax (the doyen of Franklin research) has argued that this is almost certain to mean that Irving had returned northwards. As Irving is unlikely to have returned alone, it is reasonable to suggest that he would have been part of a party trying to return to the ships. If, on their arrival, they found that open water was within boat-hauling distance, it is likely that they would have set off northwards through Franklin Strait and up Peel Sound. Upon reaching the southern edge of Back Bay, they may have been forced to seek shelter or have been prevented by ice from continuing (the ships of James Ross's search expedition were stopped by ice at Port Leopold in the autumn of that year, 150 miles from Back Bay – he sent sledging parties down the east side of Peel Sound and to Fury Beach). If they had turned into Back Bay around the headland that formed the southern horn of the bay they would have seen the small sheltered bay – where the boat remains now lie – just ahead of them. They would have been no more than fifty miles from Cape Walker and the main Arctic highway for any rescue ships.

If that had been the case, what had happened next? I felt that they would have stripped the boat and made sledges from its planks. As they worked, the sail of the whaler would have been made into a round tent to provide shelter. With the sledge, or sledges, constructed, I believe that they would have tried to continue northwards, either having waited for the ice to thicken enough to bear their weight, or starting off around Back Bay to travel up the coast. If they attempted the ice, their remains probably lie at the bottom of Peel Sound, if overland, their bleached bones may rest on northwest Prince of Wales Island.

Wayne, however, under the influence of Inuit mythology, firmly believed in resurrecting an old and long discounted tale first told by Adam Beck, an Inuit interpreter in the service of John Ross who had sailed in close company with the Austin Search Expedition of 1850. Beck claimed that he had been told that Greenland Inuit at Cape York had massacred Franklin and his men. The story had been backed by Ross (who was thanked by Lady Franklin 'for murdering' her husband) but ignored by the rest of the naval officers. It was, of course, extremely unlikely that a hunting party of Inuit could have wiped out 129 fit men, armed with percussion muskets. Wayne, however, supported by Inuit tales, was of the opinion that a massacre had taken place, but that it had been carried out by an Inuit 'Shaman' (medicine man) at the boat place on Back Bay. Fearful of revenge, the native families living under the power of the Shaman fled eastward towards Greenland where Adam Beck picked up a distorted version of the story two years later. It was known that Central Arctic Inuit had reached Greenland at about the appropriate time (the kayak was re-introduced to the Greenland natives as a result). But they may have been part of a normal movement of peoples throughout the area.

Wayne's idea did, however, provide an explanation for one of the most puzzling aspects of the decayed boat. Throughout much of the Arctic, wood was considered to be a valuable resource. The further northwards wood was taken, the more value it acquired, and some of the more southern natives actually made forays into the barren lands to obtain timber for their own use and to trade. Why then, had so much of the boat survived in situ? As far as Wayne was concerned, the answer was simple. The area's connection with death had made it a place to fear, and the natives had avoided it ever since. This attitude had faded and when I asked Martha what had happened to a long piece of wood, the impression of which remained still clearly etched into the mud, she told me that her brother had taken it to make a fishing spear.

Just like another idea that Wayne had (that some of Franklin's men actually sailed one of his ships up to Back Bay – hence his excitement at my possible identification of the 'mast') it all sounded just a little implausible. On the other hand, we were actually standing at a site which might add a new, and singular, twist to the Franklin story, thanks to Wayne Davidson and his Inuit family and friends.

It was a tired but happy band that made its way back to the *Qakudloo*. The day had been full of surprises and thought-provoking discoveries, yet the surface had only been scratched. Speculation about the sites might be fun, but there was much more to do before the ghosts at Back Bay could be laid to rest. As we walked along the gravel slopes, the bay provided a spectacular backdrop to our thoughts. Large, grounded ice floes fronted the still, grey-green, waters of the narrow strait between the south shore of the bay and the high, curving rise of Browne Island. To the west, a late summer sun brushed the tops of the dark blue hills and flooded the ice-choked western edge of the bay with a burnished bronze light.

Back at the hunting camp, and after one of Max's splendid meals, I took a final walk along the beach before retiring to my sleeping bag. The following day seemed full of promise as Wayne had agreed to take us around the southern horn of the bay to the 'marker site'. I was somewhat concerned, however, by a drop in temperature, and the strange sight of thick fog pouring through the valleys between the hills surrounding the bay. The fog was advancing from the north, south and east and was already rolling down the ridges that lined the shore to the south. A pincer movement closed on us down each of the arms of the bay, but the most breathtaking sight was to the north where the fog oozed through the gaps in the hills like porridge overflowing the rim of a badly chipped breakfast bowl. I also noted that there were now two ravens perched on the roof of our hut.

As I peeked out of my sleeping bag the next morning, I soon realised that the temperature had continued to fall during the night and a decided chill now hung in the air. This was doubly confirmed when I stepped out of the hut and saw the wall of fog that circled the camp. There was also another development. Not only was the bay choked with ice that had drifted in during the night, but the sea itself had begun to freeze.

After breakfast, and as we were going nowhere until the fog had lifted, those who were interested were entertained to a mini-lecture by Peter Wadhams (wearing his glaciologist's hat) on the formation of sea-ice. Stamping our feet, and with our noses as cold as iron, we heard Peter talk about the growth of 'spicules' and saw demonstrations of ice flexibility as shock waves flashed away from a point of impact. It was all fascinating stuff. Even as the fog slowly lifted during the morning, it was clear that none of us would be able to leave the camp that day. Wayne refused to risk the *Qakudloo* amongst the wastes of ice that jammed up against the shore even though I felt that a cautious passage between the ice-floes could have got us out of the bay and into Peel Sound.

There was nothing for it but to catch up with our notes and attend to any domestic details. By midday I decided that I would like a warm drink so strolled over to the kitchen. Upon opening the door I was met by an astonishing scene. A civil war had broken out amongst our French companions over covert leaks to the French press – and both sides seemed to be guilty. Peter was shocked into a fury by the news, Pia made some robust suggestions, Dan meditated and calmed things down, and Max made everybody a warm drink.

Eventually, Pierre agreed that he should not have acted in the way he had and that it would not happen again. Maurice, on the other hand, would henceforth be treated like a journalist and only be informed about progress on a 'need to know' basis.

As we all sat together in the gloom of the kitchen hut, it was agreed that the outbreak of the French conflict and its causes had pointed up a strange omission from our expedition. Probably due to the fact that we seemed to get on so well together, we had never appointed an official Expedition Leader. Most of us had taken the lead at some point depending upon our abilities or knowledge, and the need for a de facto leader had never arisen. Now we felt that the time had come to put up a single name to act as a central authority who would act as our spokesman. It turned out to be the easiest election imaginable. Peter Wadhams was currently on his twenty-seventh polar visit. He had been a visiting Professor with the United States Navy and had travelled beneath the Arctic ice on Royal Navy submarines. When I had first met him he was the Director of the Scott Polar Research Institute. He now held the low-key sounding title of 'Reader in Polar studies' at Cambridge University. His leadership skills proved to be those of example and persuasion. Rather than lead a charge at the enemy guns, he would convince the enemy gunners that statistics demonstrated their chances of success were less than an acceptable percentage. I was to give him his first test.

I had not liked the tenor of the day. There had been the combination of the elements to prevent us from getting to the 'marker site', and our French allies had taken to the barricades. Having seen the boat site, I was now desperately keen to get to the 'marker site' but, with only two full days left there was the strong possibility that the weather would prevent me from getting there. I told Peter that if it appeared that I would not be able to get there by boat the following day, I was going to do it on foot. At this, Max, who was feeling the confinement to camp even worse than I was, perked up and demanded to come with me. Peter, however, did not seem terribly keen on the idea and pointed out that it would mean a sixty-mile round trip. With Max to stiffen my backbone, I pressed on to say that I had not come thousands of miles to miss something that could be of considerable interest a mere thirty miles away. Peter, faced with this act of insubordination so early in his leadership, suggested that we could discuss the matter in the morning.

I arrived at breakfast the following wearing my walking boots. The ice had closed in, once again reducing any chance of getting to the south horn of the bay by boat. I felt good and was eager to get underway, especially now that Max would be with me. However, I had not got far in my preparations when I was cornered by Dan who told me that he and Peter were 'not happy' with my proposal. Peter then joined in and suggested that, in his opinion (he was, after all, a glaciologist) the ice would break up enough for the boat to be got through later in the day. There was clearly no mileage in pressing the issue at this stage, but I did get Peter to agree that if it proved impossible to get the boat out, Max and I could start out later in the day.

As it was clear that there would be no movement of the ice for some time, Max and I decided to walk along the western spur. We set out to walk straight up the hill (approximately 300 feet) behind the camp rather than walk around its shoulders. Having got no more than a few paces out of camp, we were hailed by Maurice who wanted to know what we were doing. When we told him, he decided that he would join us, dragging a somewhat reluctant Francois with him. Whatever anyone thought of Maurice, there was no denying that he was a large, powerful individual who thought nothing of hoisting a heavy camera onto his shoulder and ploughing on to his destination regardless of conditions or terrain. Francois was quite happy to plod along bowed under the weight of a huge tripod and his recording gear. During his time in camp, he had begun to look more and more villainous as his chin stubble grew darker by the hour and as he whittled away at a piece of wood with a wicked looking knife. Most of the women that I knew would have surrendered to the merest flash of his dark eyes.

We were away for about three hours, returning by the northern shore of the southern arm of the bay. As we rounded the southern slopes of the hill, I could see that the ice had not moved. If anything, the high water at midday had brought more ice in with it. Determined now to tell Peter that I insisted on being allowed to walk around the bay, I had the wind completely taken out of my sails when he informed me that Dan and he had decided to order a helicopter for the following day.

That night, having scoured the beach for scrap wood (under the strict guidance of Martha) we followed Dan's lead in building a bonfire so that he and Max could have the pleasure of teaching the children how to toast marshmallows over an open fire. The event was a great success marred only by my getting into a fight with eight-year old Angela over the possession of a wooden toasting 'fork' (she won). Max took great delight in demonstrating how to get many toastings from a single marshmallow, but Francois told me in a low voice, having tasted one of the sickly sweetmeats for the first time, 'In France we not 'ave zis marshmallow – now I know why.'

The following morning, with the fog pouring over the rims of the surrounding hills, it was with great relief that I heard the clatter of a helicopter after breakfast the following day. The machine banked and swooped in a way that suggested that the pilot was of a different stamp to our cautious Swiss woman of a few days earlier. Whereas she had flown her aircraft with the precision of one of her country's watches and landed it elegantly to the west of the camp, our new pilot rocketed in, reared up, and plonked the machine down in the middle of the camp. He turned out to be a cheerful, broadly smiling young American whose greeting of 'Hi! How ya doin' was a tonic in itself.

The plan for the day was that the helicopter would take Peter, Pia and Pierre to the boat site where they could carry out tests around the boat with the Proton Magnetometer. When that had been done Max, Dan, and I would be joined by Wayne and Martha and be taken over to the eastern side of the southern horn of the bay. Our orders were to 'survey' the site. I was to look for any evidence that could prove – or even suggest – that it had been the sledge camp of Lieutenant Browne.

Our pilot chose the most direct route to get us to our destination, flying over the prominent hill that formed the southern horn of the bay. Under Wayne's guidance, we made our descent and landed just south of the most easterly bulge of the beach. As we came down, I could easily make out a bizarre collection of rings and piles of stones no more than a couple of dozen yards from the water's edge.

With the helicopter gone, I was able to look around at the place. I could not remember a more dramatic spot during thirty-three years of travelling with the Royal Navy. To the north, the rock-strewn beach rose up gradually to form a skyline about half a mile away. To my left (the west) the view was dominated by the huge bulk of a bare, rounded hill, its base about two hundred yards from the sea. Both the beach and the hill were composed of rust-red coarse granite flecked with blue grains. Another large dark hill, Mount Matthias (a misspelt memorial named after an Assistant Surgeon Henry Mathias who died on board HMS *Enterprise* in 1849 whilst serving under Sir James Ross) lay to the south with a fringe of low cloud or fog rolling over its brow. Behind it, the bluff cape of Whitehead Point (named after Edward Whitehead, a Clerk-in-Charge, who died in 1851, also under Ross) jutted out into Peel Sound. Out of sight, but just to the south of Whitehead Point Sherard Head (named after Lieutenant Sherard Osborn who had crossed Prince of Wales Island in 1851, and who died as a Rear Admiral) jutted out into the waters of Peel Sound. As I looked eastwards, those same waters, sprinkled with drifting ice-floes, lay flat calm with the only ripples coming from the heads of seals as they broke surface just offshore to see what was happening. But we were not there to take in the view. We knew that we did not have long to work before the helicopter returned to lift us off.

From where we stood, the most prominent features were the 'gravestone' I had seen on the slides at the Scott Polar Research Institute earlier in the summer, and an oval-shaped ring of stones. Whilst Max and I took measurements, Dan used his GPS equipment to fix their exact positions. The 'gravestone' continued to mean absolutely nothing to me. Its black upright with two roughly globe-shaped rocks on either side was at the head of a larger, heavy looking black rock which measured about two feet in length. Although it stood out to us standing on the beach, it would never have served as an indicator from any distance.

The stone oval was equally mysterious. At four points on its circumference four or five rocks had been placed on top of each other to form a small pillar of stones. These odd features were not spaced regularly but were at approximately 12-o-clock (the eastern extremity), 2.30, 4-o-clock, and 8.30. Other rocks seemed to have been placed in the centre of the oval but apart from a possible small circle, little else could be defined.

A few yards to the north, a small, hollow cairn had been erected. As I examined it, I noticed that one of the stones had fallen from the top leaving a patch of colour that differed from its immediate surroundings. That gave me an idea. Walking a few yards up the beach, I stopped and bent down to lift up random rocks that were unconnected with the features we were interested in. Beneath each lifted rock, the lower stones glowed bright rose pink. I then returned and lifted one of the stones from the oval circle. The same, lighter colour appeared once again. The simple test suggested that the oval had been built a long time ago – long enough for the upper surface to have weathered to the same colour as the rest of the beach, yet protecting the stones underneath. There was even an opportunity to look at the test from a different angle. A few paces from the oval shape, two depressions could clearly be seen which suggested that they had once held stones used to make up the design. In both cases the hollows were weathered to the same dull surface red. It was probably many years since the rocks had been lifted.

Just to the south, four more small rock pillars had been built marking out a large, rough, rectangle – perhaps a 'convex quadrilateral' or trapezium. A possible fifth pile lay just to the south. This construction seemed to have collapsed, but two ball-shaped stones gave an echo of the 'gravestone'. Twenty-five yards inland, and higher up the beach we came across two stone circles, one larger than the other. Beside them I spotted a much more exciting sight. In the absence of Professor Savelle, I was prepared to state that the rock outline in front of me was the right shape and size of a Royal Naval Pattern sledge tent. Its straight sides were strangely out of tune with the curves and circles I had seen elsewhere on the beach.

Further to the south along the beach were a variety of other features that had been declared as 'native' by the Inuit Elder. They consisted of stone circles and 'fireplaces' (my description).

I made my way back along the beach to where Max was getting a hot drink and a snack prepared. Just as I had left the last of the tent rings, I came across a sight that caused my heart to pump rapidly. I probably have more experience than anyone in the world in the art of marking out the shape of an anchor with Arctic rocks. There was an anchor design by me somewhere in the middle of King William Island and another on the eskers near Cape Maria Louise. In both cases, the laying out of the stock, ring, shank and arms had been straight forward, but I had difficulty in finding flat, square or triangular, slabs to represent the flukes. Only when such shapes had been found did I consider my anchors properly finished.

At my feet a row of rocks had been laid in a straight line of about five feet in length. Eight feet away, and in line with the first row of stones, there was another, longer, curved line. At each extremity of this second line a flat, angular rock had been deliberately placed. The ground between the two lines looked as if it had been artificially levelled and contained a number of larger rocks scattered randomly along its length. Had someone else laid down an anchor pattern over 140 years earlier? The stock, the arms, and the flukes remained as their designer had left them. The scattered rocks between the two main elements could have easily fitted together to form a connecting shank. I brought Dan over to take a look. Unprompted by me, he looked at the arrangement of rocks for a while and then declared that they meant nothing to him. I suggested that they might be an anchor. Dan looked again, but he was not impressed and was of the opinion that I was suffering from 'wishful thinking'. I had to be careful, the mounds near Cape Felix had been claimed to be an example of nature imitating man. But, in that instance, the feature was totally random and – apart from the nearby 'Franklin camp' – unconnected to any other obviously man-made features. That was not the case on the shores of Peel Sound where we were surrounded by man-made (or man-assembled) shapes. And one of them looked like the outline of an Admiralty Pattern tent.

Could such a thing be possible? To find the answer recourse had to be made to the Admiralty Blue Books, the resting-place of all the Franklin search expedition reports. In them can be found the report made by Lieutenant William Henry James Browne (we had a copy with us at Back Bay).

Browne was the son of the Dublin Harbour Master. He had first gone to sea with the Mercantile Marine but whilst at Fiji he transferred as a Master's Assistant to HMS *Sulphur*, then under the command of Commander Edward Belcher who was to lead the 1852-54 Franklin search expedition. Following Belcher into HMS *Samarang* and at the same time transferring from the 'civil' branch of the navy to the 'military' branch, he served as Mate. His promotion to Lieutenant was followed by an appointment to HMS *Enterprise* in which, under the command of Captain James Ross, he took part in the first of the Franklin searches. With the ship iced-in at Port Leopold, Browne, whilst earning a reputation as an Arctic artist, led a sledge party down the east coast of Prince Regent Inlet, returning after eight days. On his return to England, he was re-appointed to HMS *Resolute* to take part in Austin's Franklin search expedition.

Ordered by Austin to search the eastern shore of Prince of Wales Island, Browne travelled south down Peel Sound and reached 'the point' at twenty minutes to one on the morning of 2 May. He noted that his men 'were constantly getting frost-bitten, and also complained of numbness in their arms. I determined on stopping for some improvement in the weather, so, at 1am, encamped close to the point.' Could, therefore, the 'marker-site' on which we stood, with its array of mini-cairns and rings, be Browne's campsite?

Wayne Davidson believed that this was not the case. He felt that to have reached 'the point' (which he considered to be the northern tip of the southern horn) and then, in the face of high winds and driving snow, and towing a sledge, to have reached a position 2 miles to the south in twenty minutes would have meant a six mile-an-hour dash. Something his long experience in the Arctic told him would be impossible. However, I felt that his logic only made sense if 'the point' actually *was* the northern tip of the promontory on which we stood. And I believed it was not.

During his journey down the coast, Browne used a number of capes or 'points' to act as aiming marks. In each case the position has been the next easternmost point of land. The tip of the southern horn is not an eastern projection, nor is a subsidiary extension (known on modern maps as 'Whitehead Point'). The most easterly spot is just below the highest point of the 545-foot hill that dominates the entire promontory. And that 'point' was less than half-a-mile from the 'marker-site'. That would mean a 'dash' of one-and-a-half miles an hour. I felt, therefore, that the site on which we stood not only could quite easily be Browne's site but in all probability actually was.

Wayne then produced yet another objection. Both he and the Elders felt that the selection and arranging of rocks in a snow-covered landscape is an extremely difficult business. No-one, they believed, would have gone to the trouble of digging down through the snow to ensure a solid foundation for their rocks. I could not agree. Firstly, I felt that the stone tent rings – and the 'anchor' – could simply have been laid on a covering of snow and maintained their position as the snow melted. But perhaps that was not even necessary. Browne repeatedly comments in his report that much of the ground was bare. The day after he left 'the point' he found a low beach with a 'quantity of small vegetation, such as mosses, short grass, and dwarf willow'. The snow, if any, cannot have been very deep at that point. Later he saw 'a great quantity of moss and other small vegetation'. Even as far north as Cape Walker, he had been able to report that on 'all the slopes, especially those with an eastern or S.E. aspect, a quantity of moss and short grass was seen.' In his 'Remarks', Browne notes that all 'the flat land was covered with moss and short grass and occasional tufts of dwarf willow; there are also many pools of fresh water.' And eight days after he had left the southern part of Back Bay, the temperature was high enough for him and his men 'to dry our furs'. Perhaps the spring of 1851 was an exceptionally warm season and they had no difficulty in arranging rocks during the sixty-eight hours they spent on the site – they certainly found time to bury a message.

What of that message? Browne wrote that at 7.00am on Sunday 4th May, in thick fog and heavy snow, he deposited 'a paper on the first slope of the beach' (the snow failing to prevent the burying of the message and its metal container). The most obvious place was, of course, beneath the large, heavy black rock at the base of the 'gravestone'. I felt that this was too obvious. Long experience had taught the early naval explorers that the natives would soon pull down cairns and destroy graves in the search for anything that they may contain. To try and avoid this loss, they resorted to building small cairns which supported boarding pikes to which was attached a 'pointing finger' sign with a Roman numeral beneath. It was believed that the Inuit had no understanding of symbology, so would not understand the import of the direction indicator. They certainly had no knowledge of Roman numerals. Consequently, they would be extremely unlikely to have worked out that the prize they sought was buried so many paces in the direction indicated by the sign. The weakness in the plan was, however, that the natives would have been just as delighted to have found a boarding pike as anything else, and promptly removed it, directions and all.

If I were to seek for Browne's message on that beach, I would look more closely at the trapezium-shaped four small piles of stone. At first Peter (who had only been able to pay a flying visit to the site), thought that they might be an arrangement laid out in a diamond or lozenge shape to reflect the central design on Browne's sledge flag. Wayne was convinced that they were part of some type of long-distance navigational aid which pointed out the direction to Beechey Island, 150 miles to the northeast. The stones were arranged so that the two piles closest (and parallel) to the shore were approximately 65 feet apart. Also parallel to the shore, but about 40 feet further inland, the other two piles were approximately 40 feet apart, the centre of their interconnecting line being about the same position as the centre of the longer line. As a result, if lines were drawn between the two southern piles and the two northern piles, and the lines extended inland, they would cross at a point approximately another 40 feet up the beach. This, I felt, could be the burial place of Browne's message, there being (for me) no other reason for such an arrangement of stones. But wherever the message might be (assuming that we were actually on Browne's site), we could not dig for it as we did not have an authorised archaeologist with us.

If the site is studied more closely, and if it proves to be Browne's sledging camp of the 2nd to the 4th of May 1851, it would appear that the man who helped to send the Franklin searches in the wrong direction had camped less than four miles from what may prove to be the most northerly achievement of Franklin's men after they had abandoned their ships. A cruel irony that does not reflect the courage of Browne and his sledge crew as they fought their way through a strange and hostile land in search of their lost comrades.

By mid-afternoon, heavy fog was pouring over the hills to the south and it was clear that we could not have much longer on the beach. Then the sound of the helicopter reached our ears as it swung around from the north. Within minutes, it was over our position and we crouched down as it first hovered and then settled down on the red granite beach. Keeping the engine running, the pilot signalled us to get onboard. Once we were all securely strapped in, the helicopter rose into the air with the sheer red cliff of the promontory hill to our left. The nose of the machine dipped slightly, and we began to head north to swing around the tip of the southern horn of the bay. We had not been in the air more than thirty seconds before we saw, heading towards the spot we had just left, two polar bears, their muscles rippling beneath the pristine white of their fur as they loped along in an effort to escape the noisy object that swooped above them. We had, it seemed, made our departure at just about the right time.

Back at the camp we waited for the helicopter to return from the boat site with Peter, Pia, and Pierre. Their magnetometer survey had not produced any surprises, such as metal objects, around the wreckage of the boat, but it was clear from the readings they had obtained that there was a considerable quantity of wood just below the surface of the mud.

Early that evening, as the sun began to dip behind our hill, I was in the sleeping hut writing up my notes for Peter on the 'marker-site' when Pierre called me outside. There, once again, amongst the crowded ice-floes just offshore, was an enormous school of Beluga whales, their white arching backs flashing silver in the rays of the dying sun as fountains of spray from their blow-holes glittered with rainbow colours against the ice. This time there was to be no blasting away at them with rifles, just the uninterrupted delight of their passage along our beach.

That night, with the entire population of Prince of Wales Island crammed into a single hut, we held a final cheery gathering (even the French seemed to have declared a truce). Songs were sung, and exaggerated stories told whilst the children giggled at the ridiculous sight, the whole being floodlit by television lighting.

The following day we were lifted out by a Twin Otter, leaving Wayne, Martha and the children to follow in another aircraft. At Resolute Bay, light snow was falling as we said goodbye to Maurice and Francois who would be making their own way back to France. Dan suggested that we flew back via Yellowknife, a proposal agreed by everyone as it would give us a chance to make a report to the Prince of Wales Northern Heritage Centre.

There was thick cloud below us as we flew down Peel Sound except, that is, that by sheer good fortune, a hole appeared in the grey beneath us just as we passed over Back Bay. There it all was, the whole of the western part of the bay, including the two arms, white with ice. The only sound to disturb the lonely vigil of the musk ox would come from the grinding of the ice as it pressed against the barren shore. Soon that would give way to the moaning of the wind as it brought the winter snow to hide the remnants of the boat, the whalebones and the stone rings of the red beach.

We spent that evening yet again in Back Bay, but this time off the city of Yellowknife. We were being hosted by the City Clerk, who had spent time studying at the Scott Polar Research Institute and who was not merely keen to make our stay in Yellowknife as pleasant as possible but also to learn of our experiences further north. It was my first dark night in that part of the world, and I was delighted to see the slender, green threads of an early showing of the Northern Lights as they hung suspended high above us.

In the morning I led the team across the city to the Heritage Centre where we met Margaret Bertulli. She listened with interest to our tale and explored the ramifications arising if our early conclusions proved to be correct. I lacked the courage to tell her that in my baggage there was a tiny fragment of wood from the Back Bay boat which I intended to give to Peter for him to have examined by experts at Cambridge University (this had actually been suggested by Professor Savelle – but I knew that Margaret had taken scalps for much less). She then took us – an extremely rare privilege – to see naval buttons, musket percussion caps, parts of snow goggles and clay pipes, along with the crumpled remains of a copper cooking pan – all valuable fruits of her archaeological work at Erebus Bay during the Arctic summer. Included among the artefacts were fragments of wood the same colour and patina as the wood we had found at the boat site.

As for our work during the Arctic summer of 1993 only time will tell of its achievements. The cairn and shelter on the islet, the remains of the boat, and the strange man-made formations on the red beach all have their story to tell. The trick is to get the correct means of listening, the will to use it, and the imagination to paint in the background to a story of other men's courage and endurance before it fades beyond recapture.

The remains of the enclosure on the rocky islet. The structure was declared to be 'Non-native' by the Inuit Elder.

12

One More Heave?

I turned the car off the Plymouth road and followed the signs to the tiny Somerset village of Pitminster. It was a bright day during the spring of 1994, and I was keen to find the grave of one of the Arctic's forgotten heroes, Captain W. R. Hobson RN. As a Lieutenant, Hobson had served for three winters in HMS *Plover* and HMS *Rattlesnake* off the coasts of Alaska in the search for Franklin. On his return to England, the threat of half-pay was averted when he was appointed second-in-command to the redoubtable Arctic explorer Captain F. L. McClintock RN. With the Admiralty having withdrawn from further Franklin searches, Lady Franklin had purchased a steam yacht – the *Fox* – and had gained Admiralty approval for McClintock to lead an expedition in a last desperate search for evidence of her husband's disappearance. McClintock, in turn, had asked for Hobson's services as his deputy and again the Admiralty had given their approval.

The *Fox* sailed in 1857 but spent the first winter in a hair-raising journey travelling the wrong way down Baffin Bay trapped in southward-drifting ice. With a combination of sheer luck and brilliant seamanship, McClintock managed to escape, and, by the late summer of 1858 the little company had reached the eastern Prince Regent Inlet end of Bellot Strait. Their attempt to pass through the strait's narrow waters was foiled only by a plug of ice at its western end (if this barrier could have been surmounted it is almost certain the McClintock would have led the first expedition through the North-West Passage by sea). Retreating to Prince Regent Inlet, McClintock set up a winter base at Port Kennedy before setting off towards King William Island the following spring. Hobson and his team of seamen and sledge dogs accompanied McClintock as far as Cape Victoria on the west coast of the Boothia Peninsula before separating to explore the northwest coast of King William Island. His journey was hampered not only by poor weather conditions but also by the onset of scurvy in Hobson himself. Pressing on, he reached the cairn-mounted ridge on the eastern shore of Back Bay, the cairn surrounded by a huge amount of abandoned stores, clothing and equipment. Inside the rock pile, Hobson found the harrowing note telling of Franklin's death, the desertion of the expedition ships, and the survivors' proposed journey south down Back's Great Fish River. After crossing Collinson's Inlet and finding a further note from the Franklin Expedition, Hobson was forced to turn back towards the *Fox*, his scurvy rendering every footstep both exhausting and agonising. Eventually, he could no longer stand and had to be placed on the sledge. The party arrived at the ship after a journey of seventy-four days yet, within a month, Hobson had recovered enough to prepare the ship for sea.

Inevitably, as the commander of the expedition, McClintock's name has become firmly attached to the discovery of the fate of the Franklin Expedition. He went on to become a famous Admiral, surviving to see Commander Scott off on his first voyage towards the South Pole. Hobson was promoted to Commander on his return and eventually given command of two small warships, HMS *Pantaloon* and HMS *Vigilant*. Six years after his return from the Arctic he was promoted to Captain and placed on half-pay. Because of his experience of scurvy, he was called to give evidence to the Board of Inquiry into the 1875-76 Nares Arctic Expedition. He died two years later at Pitminster and was buried at the southern edge of the churchyard.

My interest in Captain Hobson extended beyond his pivotal involvement in the Franklin story. After he had limped northwards from the cairn-site on the ridge at Back Bay, King William Island, I was the next 'Lieutenant, Royal Navy' to visit the site, 134 years later.

The grave itself, once described as 'overgrown', had clearly been restored, but a young tree had taken root near the simple gravestone. It existed as an unpretentious memorial to a brave, almost forgotten, Arctic explorer.

There had been yet another memorial to Hobson. Shortly after his death, the parish magazine of July 1881, brought attention to the fact that the church's chief defect was the 'somewhat mean squares' of the old east window. The problem was remedied by Hobson's wife who paid to have a 'painted glass' window installed 'to the glory of God and in memory of her late husband'. Unfortunately, as a plaque in the church revealed, this window had decayed to such an extent that it had been replaced in 1989.

In an effort to track down a photograph of the whole Hobson window I wrote to the Incumbent to see if one was available. My letter was passed on to the parish historian who, in replying, gave me the information on the history and fate of the window (part of which I managed to obtain in exchange for a donation to the church). She also informed me that her father had been the bacteriologist at the Liverpool City Laboratories who, in April 1926, had examined the contents of tins retrieved from the Franklin Expedition. I began to wonder whether such an extraordinary coincidence was a heavy hint from some unknown quarter that I should not call a halt to my attempts to shed more light on the Franklin mystery.

In the spring of 1994 I learned that the Erebus Bay site had been revisited, this time by Barry Ranford who described himself as 'an amateur historian', with the Arctic archaeologist Margaret Bertulli and an anthropological expert in skeletal biology. Under the latter's direction, bones and artefacts had been removed for examination and with tiresome inevitability, were claimed to reveal – yet again – that Franklin's men had died of lead poisoning and cannibalism. Upon reading this, I promptly resolved to undertake an extremely un-scientific and non-expert investigation of my own into one of these corroding, yet proliferating, myths surrounding the Franklin Expedition.

An example of the myths raised by the article was that a boat stem (the forward vertical extension of the keel) found in Erebus Bay by McClintock in 1859 and returned to the National Maritime Museum by Schwatka after identifying the same stem in McClintock's book by its markings, was not, according to a magazine article, the same stem. Ranford, the author of the piece gleefully wrote '...the stem of the boat that Schwatka carted off does not match that described in McClintock's note and sketches.'

I rang the Museum's Polar Curator – the person I had challenged once before. She, from memory, confirmed what Ranford had written. According to the Curator, the Museum's boat stem carried no markings. There was nothing for it but to go down to Greenwich to have a look at the stem myself after obtaining permission to have a close look at the artefact, now in storage some distance from the museum itself.

Some weeks later, in the subdued lighting of the Museum store, I helped the Curator to lift the heavy wood and metal object down to the floor. Between us we peeled off the protective layers of wrapping to reveal the curved timber *and the carved lettering and numbers* that I had consistently believed in. After regaining her composure and recording the stem's markings, the Polar Curator accepted the error. My general view of the writer's deliberations was that if he could be so wrong on the question of the boat's stem, he could be equally wrong in his other conclusions.

Another aspect of my third visit to the Arctic was the still open question of the boat and the beach site on Prince of Wales Island, and the islet a short distance off, brought to our attention by Wayne Davidson. A talk to the National Maritime Museum's expert in sea craft produced little more than 'Yes, it looks like a whaler.' Evidently, the next step was to carry out the research myself.

The largest single, positively identifiable item remaining among the boat debris was the keel and the stem post. Assuming that the wreckage of a single boat was involved at the site, the distorted stern post lay on the surface of the mud, firmly indicating that the vessel had been 'double-ended'. The double-ended design of the boat (first introduced in 1790) had attractions for the Royal Navy (they were to last in the Service until the 1980s). Not only could it be used as a lifeboat, but it also made an excellent work boat capable of being fitted with main and mizzen masts in addition to being pulled in either direction by oarsmen. The design also recommended itself to anyone sailing for polar waters. In this the Royal Navy was no exception, and whalers were carried on ships undertaking exploration duties to the North and to the South. Franklin's ships may have had as many as four on each. Later, Captain Scott carried five on the *Discovery* and Shackleton's *James Caird* was a whaler.

There was unlikely to be a specifically 'naval' design of whaler in the mid-nineteenth century, especially for ships such as the *Terror* and the *Erebus,* which had undergone significant modifications from their original role as 'bomb vessels'. Commercial sources could easily provide such vessels. Metal fittings might be expected to bear the 'Broad Arrow' mark if such fitting came from official sources, but no such items were immediately obvious at the boat site.

There are at least three possible occasions when a party from the Franklin Expedition could have taken one of the ship's whalers to the north. Firstly, when Crozier wrote 'and on tomorrow 26th for Back's Fish River', he may not have meant the entire Back Bay party as is usually accepted – some could have chosen to try a northern, or an even eastern, escape route. Secondly (the most favoured by 'northern' advocates), some of the men, having tried the King William Island overland route and finding that they were incapable of continuing, returned to the ships and launched a boat northwards. Finally (and a novel idea that has yet to be considered), having dragged at least two boats as far as the southern shore of Simpson Strait, and having found that the ice to the southeast and east of the island had cleared away, a party could have decided to attempt a passage through a possible route at the south of the Boothia Peninsula which, if it existed, could have provided a route to Hudson Bay (Simpson's chart had the inviting note in the area, 'G(ulf)of Boothia (supposed)'). However, having found the passage did not exist (as confirmed by Rae six years later), the open waters of Rae Straits and the James Ross Strait may have suggested a route to the southern entrance of Peel Sound (they would not have been aware of Bellot Strait as it had yet to be discovered).

By whatever method or route it was done, it remains possible that some of Franklin's men, sailing or rowing a whaler that had been 'decked in' and prepared for a long open voyage, could have reached Back Bay and rounded its southern horn to find shelter in a small cove.

With the sea freezing behind them, they could have dragged the boat clear of the water and used the whaler's sails to make a shelter using the plentiful rocks to anchor the tent down. As the temperature dropped, they would have climbed the large red hill that dominates the southern horn of the bay to watch for help that never arrived. Finally, when the sea had completely frozen over, they would have broken up the boat and made a sledge from the strong thwarts, strakes, and stern and head sheets. Removing the nails to provided spikes for their boots and as gifts for any Inuit they might come across, they set off northwards up Peel Sound to oblivion.

There was yet one further tantalising fragment of information on the subject. In 1891, Captain A. H. Markham published a biography of Sir John Franklin. Markham knew McClintock (who checked his proofs) and also Allan Young who was with McClintock in the *Fox*. Markham wrote 'Allan Young left the ship ... in search of the ship supposed to have been wrecked on the coast of Prince of Wales' Land.' Had a rumour been picked up from the Inuit that an 'umiak' (an Inuit boat larger than a kayak) lay wrecked on Prince of Wales Island? Could that boat have been the whaler that had carried some of Franklin's men northwards up Peel Sound to be investigated by an expedition 145 years later?

By a bizarre coincidence, Professor Savelle – just before leaving Cape Felix, and having already demolished our hopes for the mounds – let it be known that he had previously visited the boat-site at Back Bay, and that there was nothing more than a wrecked umiak. Fortunately, this depressing news was ignored. Later, Wayne met someone at Resolute Bay who was able to confirm that Savelle *was* in Back Bay at the time he stated. Consequently, the only inference that could be made from this news was that the expert who destroyed the idea of the mounds being a burial site did not know the difference between a native umiak – made of driftwood, bones, and animal skins – and an Admiralty-pattern whaler. Was he trying to keep us away from the site?

What then of the site of the message hidden by Browne on the beach outside Back Bay on Prince of Wales Island? Whether a message is found or not, if it can be proved that the site is the location of Browne's sledging camp of 1851, it could turn out that he was within five miles of a ship's whaler that might have marked the furthest north position reached by the Franklin Expedition. Brown's 'small rocky islet' is found near the entrance to Back Bay. Its unprepossessing, man-made, non-native, pile of stones about four feet tall and about three feet wide at the base could, with proper investigation, also provide its own answer.

1994 was to present no opportunity for me to get back into the Arctic, but my wife and I did visit Calgary and Vancouver so I could carry out some research and meet the friends I had made in that part of the world.

Whilst at Vancouver, I became re-acquainted with an antiquarian book-dealer and Arctic enthusiast, Cameron Treleaven I had met briefly on one of my visits to the city. He was a human dynamo who seemed most happy when juggling with six problems at once. Although trained as an archaeologist, Cameron had taken to the world of antiquarian books like a dagger to a rib cage. His knowledge was spectacular, he was always on the lookout for anything with an Arctic connection and had collected a wide range of items including the actual copy of McClintock's *Voyage of the Fox* that Lady Franklin had given to Lieutenant Hobson. However, despite his wealth of knowledge and store-rooms full of Arctic history, Cameron had absorbed the Canadian academic disdain for the Royal Navy's role in Arctic exploration which had come, in the main, from lack of understanding or knowledge of the Service, its customs or its traditions, and had, almost inevitably, accepted the lead-poisoning and cannibalism stories. Academic Canadian disdain for a tragic yet gallant history was evidently gathering pace – and needed to be questioned.

In company with Cameron racing ahead, I visited Calgary's splendid Glen Bow Museum, his presence smoothing the way for a visit to the rear storage areas to have a look at the Museum's collection of Royal Naval Polar artefacts. This chiefly amounted to the rescued contents of a cart abandoned by McClintock on Melville Island in 1853, and purchases from the old London United Services Museum. The latter included a sadly neglected trophy awarded to Sir John Barrow by 'The Old Arctics' and a plate from Captain Scott's 1910 Terra Nova Expedition. The cart used by McClintock had survived but was at another store and not available for viewing. Nevertheless, looking at photographs of the cart gave me the germ of an idea.

I had always had a problem with my back since, as a 17-year old, someone I was helping to stow a crane's ponder ball pulled the wrong lever, and I found myself trying to pull a 10-ton crane off the *Ark Royal*. The crane had won and my back was easily damaged by carrying weights since then. But a towable cart offered a whole new possibility. Parry had used a cart during his visit to the western Arctic in 1819-20, and McClintock had used one in 1850. The ground over the island was, for the most part, gently undulating and much, especially along the tops of the gravel ridges, was actually level. If I could organise a light-weight cart that could carry in the region of 75-80lbs there was no reason why

I could not pursue one of ideas I had regarding the Franklin party's attempt to escape southwards from Crozier's Landing.

It had always been assumed by numerous Franklin scholars that when Crozier led the 104 men south from the landing site on Back Bay, he would have taken to the ice and hauled his boat-laden sledges around the coast until he reached the McClintock boat site at Erebus Bay. From there, it was believed that he had crossed the base of the Graham Gore Peninsula until he reached Terror Bay, upon reaching which the ice travel was once again resumed.

I had no difficulty in accepting the latter part of the theory, but the early part of the imagined journey seemed to me to make little sense. I had seen the state of the ice off the western shore of King William Island. It had accumulated as a result of the great ice-stream pouring down McClintock Strait and colliding with the island's shore. Even during the short summer seasons I had spent on the coast I had seen the ice heaped up in continuous piles each larger than a country church. Even the best areas had broken ice strewn around like rubble in a quarry. The suggestion that parties of around thirty seamen, some almost certainly suffering from scurvy, could drag 30-foot oak sledges on which were mounted 27-foot whalers, each topped to the gunwales with supplies, over miles of shattered ice seriously tested my credulity. Furthermore, if that had been the plan, why did they land on the shore of Back Bay in the first place? It would have made much more sense to have headed directly south from the ships to the tip of Franklin Point, thus saving many miles of agonising hauling. That they did not do so gives rise to the probability that the ships had not been deserted in haste.

On the other hand, with the low-lying land covered in snow, an overland journey would have made much more sense. Not only would it have been clearly seen that there was no high ground barring such a route, the expedition down the coast in 1847 led by Lieutenant Gore would have meant that the coastline would have been known to Crozier. Gore was dead, but his second-in-command, Lieutenant Le Vesconte was alive and could have acted as guide.

There was no reason why I could not celebrate the 150th anniversary of Franklin's leaving England by taking a cart down my proposed overland route from Crozier's Landing to Erebus Bay, and even on to Gjoa Haven.

An early supporter of my idea was one of the 'Calgary Light Horse', Ron Staughton. A practical man with an artistic bent, Ron's attempts at sculpture had led to his works being exhibited. His mind immediately switched to the practicalities of a collapsible trailer that had to be both rugged and light. He offered to send me his proposals and even undertook to organise the cart's construction in Calgary.

On my return to England, I contacted Margaret Bertulli, still the Senior Arctic Archaeologist at the Prince of Wales Northern Heritage Centre. Could she see any problem with my plan? In fact, she turned out to be most encouraging upon my guarantee that I would restrict my visit to walking over the ground, I would not tamper with any possible sites of interest and would report any such finds to the proper authorities.

The new year presented what I hoped would be an excellent opportunity to air my views on the myths surrounding the causes of failure of the Franklin Expedition. As a Fellow of the Royal Geographical Society, I received notice that Peter and Pia Wadhams were to give a talk at the Society's headquarters on our 1993 expedition. I decided to attend and sit quietly in the audience – that is, until I also found out that Ranford of the Erebus Bay expedition (which had 'proved' that Franklin's men had died of lead poisoning and resorted to cannibalism) would also be speaking.

I arrived early in order to get a seat within throwing range of the stage. (When Doctor John Rae addressed the RGS on his findings of 'cannibalism' amongst Franklin's men, the President of the Society was warned that there would be 'many jostling seamen' present – I intended that I would be at least one of them that night.)

Before a packed auditorium, the President (Lord Jellicoe) led Peter and Pia on to the platform followed by the Erebus Bay expedition speaker. Pia started the proceedings with a tour through the background to the Franklin Expedition and Peter followed an account of our visit to King William and Prince of Wales Islands. In Peter's account, I appeared as the world's 'greatest discoverer of lemming holes' (Peter's opinion of the cause of the readings from the Proton Magnetometer at the 'grave' site).

Then it was the turn of the Erebus Bay speaker. He confirmed his views that cut-marks on the bones could only mean cannibalism and announced that he had 'just received' information that the examination of the bones had provided proof of lead poisoning, even that the lead intake had been 'recent' (i.e. not long before the men had died). My mind was spinning with questions. Some I had prepared earlier; others sprang immediately into my mind. Where, for example, had the 'recent' lead intake come from?

My feelings might be imagined when the President stood up on the conclusion of the speaker's talk and closed the meeting without the chance to ask questions. I promptly made my way down to the front where I was met by Pia who introduced me to Ranford. His greeting was cheery enough, but I had not come to exchange pleasantries and promptly asked him about the lead found in the bones. Before he could answer, his lady companion broke in with 'It was recent.' I then asked what did 'recent' mean? At this he backed away waving his arms and saying, 'I don't want to get into this.' Before I could remind him that he was already 'in it', the President took his arm and led him away through a door. I was left feeling like the guest who had spilled wine on the new carpet. One thing was certain – I was even more determined to get to the bottom of the lead poisoning and cannibalism myths and expose them for the nonsense I believed them to be (it was later discovered that Ranford had appropriated some archaeological finds for his own private collection – a brave deed with Margaret Bertulli within range).

Peter and Pia's contribution was followed by an article in the *Daily Telegraph*:

> The mounds at the north end of King William Island were found to be natural formations, and a rectangular shape once interpreted as Franklin's last resting place turned out – despite exciting reading from a proton magnetometer – to be a natural pile of stone slabs and lemming's burrow.

Actually, no-one (certainly not me) had claimed the depression on the western flank of the southern mound had been the site of Franklin's grave. If he was there, he would have been at the centre of the mound. The report continued 'According to Dr Wadhams: "It was an extraordinary example of nature imitating man."'

It was also an early academic shot across my bows that was to become a fusillade across the North Atlantic.

An entirely different meeting took place at the National Maritime Museum later in the year. Organised by Ralph Lloyd-Jones, a Franklin enthusiast, to commemorate Franklin's sailing in 1845. Descendants of both McClintock and Parry attended it, with descendants of James Ross and Crozier sending their best wishes. Peter and Pia were there with members of the Scott Polar Research Institute staff, and the museum's Polar Curator organised a small display of Franklin search expedition artefacts. Ralph Lloyd-Jones gave a talk on the fiftieth anniversary meeting of Franklin's sailing ('attended by several distinguished admirals including McClintock') and Nicholas McClintock, the senior McClintock descendant, talked of his illustrious ancestor's unflappability at times of danger. I then spoke of my plans for the summer after reminding everyone that far from getting 'several distinguished admirals' as they would in McClintock's day, they would have to settle for a clapped-out recruiting Lieutenant who, because he was wearing a blazer and tie, had just been asked for the way to the rest rooms.

I had been in considerable correspondence with Ron Staughton and he had sent me a number of sketches of his proposed cart for me to drag down King William Island. He then changed tack and suggested that we might be in the business of re-inventing the wheel. Ron had spotted a newspaper article covering the work and products of a Calgary company named Chariot Carriers Inc. The company was in the business of manufacturing lightweight, collapsible, two and three-wheeled carriages for towing behind bicycles or for pushing children whilst hiking or jogging. Ron had approached the owners of the company and they were interested enough to have a close look at my ideas. Would I mind if Ron gave them to go-ahead? I agreed readily, hoping that the cost would not prove crippling.

I was becoming concerned about the lack of funding when I had a most surprising telephone call from Calgary. It was Cameron Treleaven to tell me that he wanted to go the King William Island with me. Of course, being Cameron, it was not quite that simple. The telephone earpiece almost glowed red as Cameron told me that I needed him (he was, after all, a trained archaeologist), that he was perfect for the mission with experience of living rough on the plains, that he could get much-reduced airfares within Canada, and he could take care of the logistics of the expedition, including tent and sleeping bags. I told him I would think about what he had said and would write to him in reply.

This new turn of events was not what I had envisaged. It was true that there were a number of advantages to having Cameron onboard (they were all tattooed on my starboard eardrum) but I could also see some clear disadvantages. Over the next day I weighed up the problem and decided to agree to his accompanying me. It had not been his archaeological expertise that had swayed me, there was unlikely to be any time (or permission) to spend on the rigorous and detailed search of a site, it was his offer to handle the logistics that swayed matters. I decided that, no matter what, if he was to have the responsibility for supplying the expedition he should not have any interference from me. I had seen teams fall apart when even well intentioned suggestions were seen as intrusions. It was not to prove a wise decision.

I wrote to Cameron pointing out the possible difficulties that I could foresee. He, for example, was a highly fit, volatile Celt whereas I was an unfit, lumpen Saxon. He would have to get used to travelling at 'convoy speed', the speed of the slowest ship (me). I suggested that he got in touch with Ron Staughton and Margaret Bertulli as quickly as possible.

It was not long before I had another telephone call with an incandescent Cameron on the other end. Once Margaret learned that he was a qualified archaeologist, she decided that the affair had become an 'archaeological expedition' and applications for permits for such a visit should have been in months earlier. I telephoned Margaret for clarification of the situation and found that she as always was determined to stick closely to the regulations (Canada can produce red tape that makes the Indian Civil

Service look positively sloppy). It was one thing for a non-archaeologist to tramp over the ground with a look-but-do-not-touch permit, but quite another to allow an uncontrolled specialist free rein, like putting a fox in charge of the hen-coop. Nevertheless, despite the very late application, Margaret was happy to send an appropriate application form to Cameron with the promise to try and steer it through its Byzantine convolutions in sufficient time.

I had no difficulty in obtaining permission from the Royal Navy to head north once again. The Commander, indicating approval with his 'Go for it my dear boy, go for it.'

The ice of Victoria Strait (west of King William Island) from about 1,500 feet. For a long time, this was believed to be the route taken by Crozier's men after the ships were abandoned. Clearly the level surface of the west coast of King William Island – already traversed by members of the expedition – would have been preferable.

13

Some you win...

I was met at Calgary airport by cheery John Howse who, along with his wife Inga, had offered me accommodation during my time at Calgary. The following day, I made my way over to Cameron's bookshop. Cameron showed me the progress he had made on the logistics front. He had collected the cart from Chariot Carriers and had done sterling work towards bringing it up to the standard required for our enterprise. A 'mountain bike' type of wheel that carried deeply treaded tyres had replaced the normal wheels. From the front, two long shafts extended, the ends of which were coupled to a padded towing belt. A weatherproof cloth cover sheltered the main body of the cart, the bottom of which Cameron had strengthened. To the rear was fitted a handlebar that could be used to push the cart from behind. It also supported a rigid gun case. The fully collapsible trailer clearly matched my requirements.

From the gun case, Cameron pulled a highly polished and clearly expensive single-barrelled shotgun. The weapon had been loaned to the expedition by one of Cameron's customers. There was no doubting that a solid ball charge from its barrel would cause concern to even the largest polar bear, but my concern was that it was a little too grand for our purposes. I would have preferred the type of weapon I had carried in the past, battered but useable.

I suggested to Cameron that we should try and obtain from the Canadian military the same type of Arctic rations that had served me so well in 1992. If my contacts failed to get them directly, we could always try the company which had produced them. Cameron, however, had been talking to a friend of his who had spent some time in the north and had persuaded him that he should take commercially available dehydrated food and lots of dried fruit, nuts, etc. I felt that the problem was that we would have to take cooking pots and plates, adding significantly to the weight to be hauled.

I had brought out with me a small water-proof rucksack which I felt would be ideal for any 'branch' expeditions; for example, if we wanted to visit a particular part of the coast which was off our track, we could take just enough to cover both supply and safety requirements whilst leaving the cart on the route. It was therefore with some trepidation that I saw Cameron produce two huge rucksacks. I reminded Cameron of my stipulation that I could carry nothing on my back (my small rucksack excepted), but he countered with the entirely sensible point that if the cart failed we would have no choice but to put as much as possible on our backs to reach our destination.

He also had plans to take a few other items ranging from the indulgent to the bizarre. One of his Arctic treasures was a set of silver cutlery which had belonged to one of Franklin's officers on Buchan's 1818 Spitzberg Expedition. He intended to use these whilst travelling down King William Island. Rather than enjoy the fun and games provided by the use of an erratic compass, Cameron had taken what I felt to be a rather wimpish step, and had purchased a hand-held Global Positioning System instrument – a more advanced version than the one that had failed on my solo visit to King William Island. This was a clever piece of kit which not only told you where you were, but where you were going, which direction to go, and how long it would take to get there. I felt that such items took away the pleasure of finding your own way. It also seemed to take about three hours to set up.

Finally, Cameron was keen to take an emergency radio locator beacon. I did not like the idea as, again, it seemed to be against the spirit of the enterprise but decided to agree to one being taken along on the grounds that we would be without a radio.

The tent upon which Cameron had decided was set up in his backyard. Compared to anything I had used in the Arctic it was a rather Roll-Royce affair in which I could almost stand upright. It had, for me, two setbacks. Firstly, it seemed a little heavy when compared to the tiny tent that, even with my bulk, I had shared quite amicably with Pierre Souvedet in 1994. Secondly, it was a large ridge tent with copious flysheet which I felt would be at severe risk from some of the winds I had experienced at Cape Marie Louise. Nevertheless, even if the flysheet was lost in a gale, the tent itself was sound enough for continued use. It also had a similar shape and size (but without the curved apse-shaped extension) to the tents used by the Royal Navy in the Arctic in the nineteenth century. If it was good enough for them it was good enough for me.

All that remained was the purchase and packing of the food. This began with a raid upon a huge retail warehouse where Cameron marched down the aisles filling his supermarket trolley with packet after packet of dried fruits, raisins, nuts, sunflower seeds and a strange mixture that looked as if it had been swept up from the bottom of a parrot's cage. This was followed by a visit to a camping shop where the dehydrated food was purchased.

I spent the next day sorting out the supplies into a variety of canvas bags. In an effort to cut the weight down, I constantly nibbled at the mountain of food and fed Cameron's daughter every time she rocketed past in pursuit of a dog. My comeuppance came when I attempted to eat some of the sunflower seeds; my mouth suddenly filled with splinters. Apparently, in Canada, the seeds were sold in their shells and years of experience enabled the suitably skilled to chew them in a manner that separates seeds from shells, the latter being spat out by experts in a nonchalant, manner. Cameron had acquired the art and could spit with ease. I, on the other hand, had neither the time, nor the inclination, to become acquainted with such strange behaviour, yet was expected to drag the unappetizing junk around the Arctic.

The combined weight of food, stove, and cooking implements seemed extraordinary. Although almost everything could be got into one of Cameron's giant rucksacks, the resulting mass could hardly be lifted clear of the ground. I could only hope that Cameron was soon to show the same stowage skills as Max Wenden had done in 1993, especially as Cameron had suggested that the cart should take no more than 100-120lbs and the load already accumulated did not include fuel.

On our way to the airport, I spotted a sign outside a church that read 'Failure to prepare means preparing to fail.' I felt in good spirits, moderately fit, and eager to put my route theory to the test. The sign, however, reminded me of my greatest concern, the weight we were carrying. The full horror was revealed via the airport scales. Excluding the cart, the digital counter ran up to 280lbs. Cameron was clearly placing a lot of faith in the cart.

We left Calgary on 30 June and reached Yellowknife via Edmonton, Fort Smith and Hay River.

Our next port of call was the Prince of Wales Northern Heritage Centre. The Director Dr Charles 'Chuck' Arnold and Margaret Bertulli were away (the latter on King William Island with a number of Franklin searchers). We were, nevertheless, warmly welcomed by the remaining staff who took us into the storerooms to see the items recovered by Margaret the previous year. I could never see such artefacts without experiencing a pang of sadness. Before me, laid out in display trays, were the few remaining fragments of personal property of men who had set out from England with high hopes, the cheers of the nation, and the heartfelt wishes of loved ones. As before, all that remained were corroded bits of metal, broken pipe stems, and a few buttons.

Leaving the stores, we went to the library where Cameron drooled over a number of books including an album of photographs from the Royal Navy's 1875 North Pole Expedition under Captain Nares. I, in turn, came across a gem of my own in the shape of an Inuit woman who worked at the centre. She was a member of the Netsilik tribe – the same tribe who occupied King William Island in 1848. Approaching me with a smile, she asked 'Are you Royal Navy?' When I answered in the positive, she floored me by saying, 'Oh, it was my people that killed your people.'

Ignoring my dropped jaw and fully widened eyes, she then continued to tell me that it was well-known among her people that their forebears had been responsible for the murder of Franklin's people as they tried to make their way south. When I pressed her to see if I was being subjected to a jest laid on for the benefit of the local Inuit population, she remained quite matter-of-fact stressing that the information was genuine and well-known among her people, yet something that previous generations had not broadcast for fear of some imagined revenge by the 'kabloonas'. In fact, the tribal Elders still strictly demanded that the subject should never be discussed outside the tribe. If this was the case, I was being told something that fitted in exactly with my developing theory about what had happened during that dreadful April and May 147 years earlier.

We could both have spent much more time at the centre, but we had to catch an aircraft to Cambridge Bay where we would stay until passage became available on one of Willy Laserich's 'Adlair' Twin Otters.

Cambridge Bay lay sullen beneath a grey sky as we bounced along the airport road to our accommodation at the Arctic Island Lodge. The temperature seemed quite high, despite the fact that the bay was still locked in ice. The high temperature did not bode too well as it was important that the ice of Collinson's Inlet – the first barrier on our route south – remained thick enough to walk across.

Having loaded our baggage, we were soon strapped into the Twin Otter. I had asked the pilot to drop us off on the eastern edge of Cape Jane Franklin, north of Collinson's Inlet. It might have made more sense to have been dropped off on the southern shore of the inlet but, not only did Cameron and I want to visit Crozier's Landing and my 'Victory Point', but I wanted to cover the actual ground over which I believed that Franklin's men had hauled their sledges.

After the drone of our engines had caused a few musk ox to flee with a mini-stampede north of Cambridge Bay, we climbed above the low cloud that blanketed Victoria Strait. Soon, towards the eastern edge of Victoria Strait, the cloud began to thin and then to clear completely allowing us to see the blue-laced ice of the strait as it abutted the deeply uninspiring shores of King William Island. Rather like an extremely ugly child that is capable of being loved only by its mother, the predominantly brown-grey desolation below us would be welcome to those only who sought answers amongst its barren rocks.

We had reached the island at the coast of Franklin Point and, just to the north, the huge blue-white gash of Collinson's Inlet could be seen. It was an important moment for me because, as we flew over it at about 2,000 feet, I could see that it looked solid enough to allow passage across its surface.

Soon we were flying just a few feet above a sea of honey-coloured rocks. The slightest of bumps as the wheels touched down, a loud revving of the engines, and we had landed. A minute later I felt and heard once again the crunch of frost-shattered rocks beneath my boots.

No time was lost in unloading our baggage, which was dragged clear of the aircraft's turning circle. Then with a cheery 'Good luck' from the aircrew, the engines were re-started and the aircraft turned into the slight breeze. With a roar and a lurch forward, the Twin Otter rumbled across the rock-strewn gravel and after an extraordinarily short run was clear of the ground.

As I looked around it became immediately clear that we could do no better than to make camp at just the spot where the Adlair aircraft had dropped us. The level gravel beneath our feet was ideal for the tent and we were within two hundred yards of a small meltwater pond. We were situated at the eastern edge of a vast shingle plain that took up at least a third of the horizon and sloped gently up to the western skyline. To the northwest, after a brief glimpse of the ice of Victoria Strait, the land turned to tundra as far as the eye could see, its drab grey-green arcing around to the east. Just beyond the point where the rock plain petered out to the north, a large lake reflected the blue of the clear sky (this was Cooper Lake, named after Paul Fennimore Cooper, a descendant of the author of *The Last of the Mohicans*, who had visited the island twice in search of evidence of the Franklin Expedition). From the eastern horizon and running south of our position all the way to the west, the ice of Collinson's Inlet glittered beneath a hard sun. Beyond the ice, the inlet's southern shore etched a sharp, undulating, brown line to separate the shimmering white from the sky's metallic blue.

The tent was erected with little difficulty and we soon had our sleeping bags laid out and a small fireplace arrange for the stove to do its work on our return. I had intended that on the first day (it being by now late afternoon) we should carry out a reconnaissance of the shore of Collinson's Inlet. This would not only allow us to look for a suitable place to set out across the inlet, but I was also keen to see if there was an obvious place where Franklin's men would have headed for on the southern shore – a small headland perhaps, or a particularly level stretch.

Some you win...

We set off at an easy pace towards the southeast shoulder of the rocky plain. Beneath us the shingle varied from small gravel to fist-sized rocks, all of which promised a good road over which to haul the cart on the following day. Crunching down the slope to the shore we reached the northern edge of Collinson's Inlet within half-an-hour of setting out. The ice had lost its sparkle beneath a greying sky and the blue meltwater on the surface hid its colour from the clouds. It was, nevertheless, an exhilarating sight as, between the piled-up ice at its rim we could see the far shore across the water-laced surface as far to the west as Franklin Point. This, then, was to be for us (as a Victorian author wrote), a 'crystal pavement by the breath of heaven cemented firm'. Such flights of fancy would have been lost on Cameron who was more interested in getting his latest toy to work and muttering beneath his breath whenever satellites failed to show quickly in sufficient numbers to give an accurate reading.

Through my binoculars, I could see that the southern shore echoed the light brown rock on which we stood. There was no obvious place towards which we should be headed on the following day so the task ahead of us was no more than to simply get across as quickly as possible. It was equally clear what the problems would be. Not only was the shoreline edged by ice ridges of varying heights from a few feet to size of a house, but also great areas of open water prevented free access to the ice over much of the shore. Once on the level ice beyond these barriers, numerous leads zigzagged in all directions linking large meltwater ponds. The crossing was to have many diversions before the far coast was achieved.

A tramp of about two hours along the shore brought us to the southwest tip of Cape Jane Franklin. We then walked eastwards up the snow-draped slope of the ridge towards a dominating cairn. At a position about fifty feet from the skyline, on the edge of a snowfield we came across a man-made structure. Built of rock to produce a rectangular space about three feet by two feet at its centre, it stood about two feet tall. The centre cavity was lined with flat slabs and was clearly used at some time as a cache, probably for meat or fish, which, once filled, would have a flat slab placed on the top weighed down by other rocks. There seemed no reason to seek for a Franklin link in what was obviously a commonplace Inuit construction.

The cairn, which had been erected at the highest point of Cape Jane Franklin, was unusual in that it had a pyramidal base surmounted by a tall pillar of rocks making the whole pile about eleven feet tall. The choice of site would have required little thought as it was the most prominent on that part of the coast, and would have been seen from all directions at almost any point from the skyline.

The view, even on a grey late evening, was spectacular. To the south ran the long east-west opening of Collinson's Inlet with its southern shores leading westward to Franklin Point. From the southwest horizon to the northwest skyline lay the broken and heaped ice of Victoria Strait. Northwards lay the coastline towards Cape Felix with, clear in the middle ground, the frozen waters of Back Bay edged by the shingle beach of Crozier's Landing. A man with a naval telescope, standing on this spot in April 1848, would have been able to see the landing parties as they hauled their sledges and boats onto Crozier's Landing before turning his lens towards the south in search of a route to safety. Would he have suggested an overland route, or one over the ice?

The question of a return route now faced us. To the east of our position lay a wide plain of rocks that reached to the horizon. Beyond lay our campsite. Cameron, shotgun on his back, started pacing off in a south-east-by-east direction. When I asked him where he was going, he assured me that he was taking the most direct route back to our tent. When I suggested that he was heading in the wrong direction, and that we should be going directly eastwards, he responded with a shrug and offered to follow any route I suggested.

It took us about an hour of slogging over the brown rubble to reach the eastern crest. As the view beyond revealed itself we suddenly caught sight of our green tent directly ahead. Cameron had been given a good lesson in the problem of navigation on King William Island.

We turned in at about 1.30am after a snack washed down by hot chocolate. The temperature had dropped slightly to just below freezing with the clouds blocking out the light, but as I wriggled into my sleeping bag, the sun burst through and promised fine weather for when we awoke.

I awoke later that morning having a good night's sleep despite being woken in the early hours by a gang of Canada geese passing raucously overhead. Outside the tent, I was greeted by a gloriously bright day baking beneath a cloudless sky, which made the area around the tent look like an Arabian desert. There was no wind to speak of as I made my way over the short distance to the north to reach the edge of the tundra and the nearest meltwater pond. Returning with filled water bottles, I soon had the stove heating a billycan of water ready for a morning drink of hot chocolate to help wash down a cereal bar for breakfast. Whilst preparing this simple meal, I made as much noise as possible in order to wake my companion, but Cameron proved to be a heavy sleeper who was reluctant to vacate his sleeping bag. By the time his bleary eyes hit the sunlight, I was ready to get on the trail to Victory Point, our target for the day. Cameron insisted that before we left, he had to set up his hand-held GDP system. This meant a great many pressings of buttons, repeated mutterings of 'C'mon, c'mon', and walking about in triangular patterns. It meant that we did not get away until 11.30am.

It had been decided that we would return to the Cape Jane Franklin cairn and continue on to the beach at the point we had left it the day before. As we left the cairn and made our way down the snow-clad western slope, I re-emphasised my theory that Franklin's men, under Crozier's direction, would have dragged their boat-laden sledges along the flat land rather that taken them over the broken ice of Victoria Strait. Ahead and below me, I could clearly see the rationale for my idea. The ice, particularly by the shore edge, was heaped up in a chaos of random piles and beyond, the summer remnants of pressure ridges rippled out to the horizon. Wide leads scored their way across the jumbled surface; in winter the waters between the ice-edges would freeze to provide yet another obstructive hazard to anyone attempting to haul a sledge across such a rough surface. On the other hand, I could clearly see that the beach was, for the most part, level with several long stretches of gravel that would have presented a most welcome surface over which to drag sledges. Cameron felt my idea was wrong, he believed that the officers and men of the *Erebus* and *Terror* would have hauled their sledges over the ice.

Having run out of ways in which to persuade Cameron further, I turned down his suggestion that we should concentrate our search along the shoreline and suggested that if he wanted to look in that direction he was welcome to do so, but he would be doing it on his own. Consequently, he crunched his way across to the beach, about two hundred yards from where I was intending to begin my search on a parallel track. He had not been walking by the edge of the ice for long when he came to the attention of a squealing and swooping flock of Arctic terns which caused him to constantly duck his head as they dived in screeching attack. My sympathy knew no bounds.

I had not been searching along my chosen path for long when I came across a very shallow gully that may, at one time, have been a watercourse. Positioning myself somewhere near its centre I slowly paced northwards swinging my head from side to side as I searched for some evidence that Franklin's people might have passed that way. Then I saw it. Lying on the top, completely exposed to the elements, was a shape I recognised immediately. A wooden toggle such as was used by the Royal Navy in Franklin's time to secure a flag or ensign to a halyard (it was superseded a few decades later by the swivelling metal Inglefield clip). Excited as I was, I knew that such an identification was likely to be open to challenge. Hobson, McClintock and Schwatka had passed this way and, although they were unlikely to be carrying flags large enough to warrant a four-inch long waisted toggle, they could have used such a fastening as part of their sledge harness – but, then again, so could Franklin's people. I knew that the Canadian Army had been in the area in the mid-1960s and that the military still used such toggles on their flags but not only did it *look* far older than anything they would have been using, it appeared to have been – at least in part – hand-made.

Cupping my hands I shouted to Cameron and waved him over to join me. When he saw my prized new find there was re-appraisal of his view that an over-ice route had been chosen in 1848. Cameron used his GDP equipment to get the exact co-ordinates of the spot where the toggle had lain. He then returned to his self-imposed exile on the beach. I placed the toggle in a polythene bag whilst looking forward to handing it over to Margaret Bertulli on our return to Yellowknife. At long last I had found something that *could* have come from the Franklin Expedition, *and* which could support my contention that an overland route had been taken by Crozier.

Cameron was not allowed to make his way along the beach for long before I was shouting and waving at him once again. Through my binoculars I had seen, ahead of me and still on my predicted route, a most surprising sight. A cairn lay directly in my path. I arrived at the pile of rocks moments before Cameron and tried to make sense of it. It was almost a replica in size and shape of the one I had built at Cape Maria Louisa, about four feet high and two feet in diameter. Carefully constructed of layered rocks it bore one distinguishing feature that I had always associated with Wayne Davidson. He had pointed out to me the slow rate of lichen growth in the Arctic – a simple pointer to the age of any non-natural formation. The more lichen, the longer the feature had remained undisturbed. The cairn we looked upon was covered in bright orange jewel lichen. Moreover, it was not recorded on any list or map that I had seen. Whilst Cameron recorded its position with his GPS instrument, I concluded that the cairn could have been missed by anyone travelling along the beach (as I had done in 1992) as it would have blended perfectly with its background when seen from that direction. Did this mean that I could claim it as yet another small confirmation that my proposed route was perhaps the path taken south by Crozier's men? Possibly.

From the cairn's position to the ridge at Crozier's Landing (now clearly in view ahead of us), the beach and my route converged so Cameron decided to remain with me as we crunched our way over the rubble towards the Irving Grave site. Some fifty yards from the start of the shallow slope of the dominant ridge I pointed out to Cameron the small cairn that marked the site of Lieutenant Irving's grave. He showed his depth of reverence for the moment by draping his jacket over and leaning the shotgun against the monument. I shuddered inside, but said nothing, only removing the offending articles shortly afterwards with the excuse that I wanted to take a photograph. I also noticed that the small cairn had very little lichen on its surface. This could have been because it had been reconstructed as recently as the 1930s.

At the southwest corner of the ridge, I found the same 'Franklin Probe' wooden box I had seen three years earlier, still secure with its survival contents intact. What was new, however, was a plastic badge lying by the side of the box. Its design consisted of a star (polar?) supported by two narwhals over a red and white twist of rope. Who it belonged to, or represented, I had not the slightest idea, but someone had thought it important enough to leave as an 'offering' to the site.

A few yards to the north lay the circular base of the great cairn, by now almost levelled to the surrounding rocks that made up the top of the ridge. We toasted 'Brave men' with wine Cameron had kept from the flight and fired two shots out across the ice of Back Bay. This time there was no answering echo.

After a short search of the area to see if anything remained from the vast array of stores that had been landed from the ships in 1848 (nothing was found), we headed northwards towards the unnamed river that had caused me some interesting moments in crossing it in 1992. As we reached its southern bank, I could see that it still poured westwards with a stiffish current, its warmer waters carving a well-remembered crystal amphitheatre as it ate into the ice of Victoria Strait. I explained my crossing of the river to Cameron and my return via the ice. Cameron, was unimpressed. He snorted with derision at my puny efforts and assured me that he could have crossed it with the greatest of ease. Clearly, Cameron was having difficulty in getting into the right spirit for a tramp over such a historic area. I then led him up the nearby ridge to the small pillar of rocks that marked my 'James Ross's Furthest' but did not go too deeply into my reasoning to avoid yet another dismissive opinion. Nevertheless, I was pleased to find that a second visit had confirmed in my own mind that the site bore all the hallmarks of the spot where, in 1830, James Ross had turned back to return to the *Victory*, locked in the ice on the far side of the Boothia Peninsula. I was also able, with the use of Cameron's GPS instrument, to record exactly the position of *my* 'Victory Point' – 69 degrees 40′ 66′ N, 98 degrees 18′ 20′ W.

It was now time for us to turn back. This was as far north as I had intended to go, and we had to get under way the next day. We chose to go back along the ridges that lined the shore rather that return to the beach itself. There was always the possibility of coming across something of interest, but nothing turned up until after a couple of hours of hard slog across frost-shattered rocks we reached the northern slopes of Cape

Jane Franklin. At this point Cameron began to head straight up the rocky scree, I stuck to the tundra (much easier on the feet) to reach a position level with our camp from where it would be a gentle slope up to the tent. I continued eastwards, keeping the honey-coloured ridge on my right-hand. I had not been walking for more than a quarter of an hour before I found yet another cairn. This structure did not have the look of my earlier find. It was smaller, certainly under three feet in height, roughly constructed, and although the possibility that it might contain a message could not be ignored, seemed more likely to be a route marker. Again, there was a modest amount of lichen present, so I guessed that it could have come from one of the Fennimore Cooper expeditions; Cooper lake was less than half a mile to the east.

Cameron not being a rum drinker he was unable to join me in a nightcap to round off a meal of hot dogs and soup. The night passed without incident or alarm, and I awoke to steel-blue sky and still air that was already being warmed by a blazing sun. After a visit to the nearby pond to freshen up and top up the water bottles, I returned to the tent and soon had water hot enough for a morning drink. With no sign of life from the tent's interior, I began to whistle loudly and rattle the cooking implements, but it took some time before Cameron's head appeared at the tent flap. As soon as he had left the tent, I hauled everything out and began the packing needed to get us on our way. I was, however, quite limited in what I could get on with, apart from forcing my sleeping bag into its bag and rolling up and securing the self-inflating ground mat. The rest would have to be packed and stowed by Cameron, and I was feeling great concern over how it was to be achieved. The best idea that I could come up with was that the cart would be topped up first and the large rucksacks secured on the top. But I could not really see this working, as the weight would be far more than the cart was designed to carry. Perhaps the small rucksack that I had brought along for any 'branch' expeditions would have to be filled with lighter items to help relieve the pressure on the cart.

My worst fears were realised when, having packed the cart to bulging point and having rammed what was left into the large rucksacks, Cameron pointed to one of the monstrous packs and said, 'That one's yours.' He estimated that the weight I would have to put on my back would be in the region of 30-40 lbs, but I guessed that it was closer to 50lb. With his pack being of a similar weight, it meant that the cart (intended to carry no more than 120lb) would be carrying 180lb, plus the fuel.

Having earlier lectured Cameron on dealing with problems 'as they were, not as you wished they were' I felt in no position to complain about what I felt to be his failure to ensure that my weak back would not be put under unnecessary strain. I said nothing but bore an undying hatred of all those dried fruits and other 'nibbles' that Cameron had insisted upon bringing along. We then had to wait until Cameron had set up his GPS equipment. Yet again, this consisted of walking in triangles cursing those satellites. As a result we could not set off until it was past eleven-o-clock.

The moment I hitched myself to the cart, I knew that I was going to have a difficult time. The rucksack's lower straps would not fit comfortably around my waist, with the result that the entire weight was pulling on my shoulders. To this was added the problem caused by the waist-belt of the cart's harness being constantly forced down by the base of the rucksack. Nevertheless, the compacted gravel underfoot was ideal, and we were soon striding out towards the northern shore of Collinson Inlet. For the most part, the cart behaved excellently with my hauling and Cameron's pushing. However, by the time we reached the ice, Cameron had grown concerned about the way the bulging canvas on either side of the cart was being rubbed by the wheels. Not only was this having a braking effect on our progress, but also the canvas was beginning to be worn way. He called a halt to investigate the problem, whilst I decided to use the time to press on ahead to look for a way onto the ice.

The inlet's expanse of glittering white was, for the most part, level. Obvious obstructions existed only at the shoreline where the ice had piled itself up in a jumble of random blocks existing for about fifty yards out from the land. Much of the ice at its edge had melted, leaving few solid avenues onto the inlet – and these were frequently scored with cracks and open-water leads that appeared to bar progress across their surface. After about half-an-hour of forays onto different parts of the ice, I managed to find a circuitous path through the piled-up ice to the shimmering plain beyond. Just to stand there had in itself been worth all the effort to reach that spot.

Behind me the ice ridges blocked off the land, to the right and left the inlet streamed in a dazzling blaze of blue and white from east to west, and ahead of me, some two and a half miles away, lay the southern shore. The scene lay locked in a vault of absolute silence. Not that wildlife was entirely absent. Through my binoculars I could see a number of seals basking in the sunlight unaware that they were being observed. I could also see the problems that were to lie ahead. Even though I had reached the level ice, a flowing stream of clear meltwater just to my front barred the way. Beyond, stretched scores of meltwater ponds and long, brilliant blue, leads. The distance to the southern shore would have to be multiplied many times by the continuous diversions these obstacles would enforce.

A trail of footprints on a thin surface layer of snow helped me to find my way back to the shore. I marked the spot with a small pile of rocks before setting off to re-join Cameron.

My companion had applied a measure of practicality to the problem of the wear on the bulging sides of the cart. He had taken the foil windshield of our stove and had taped sections of it to the points of contact with the wheels. It had done little to reduce the rubbing of the wheels, but would at least protect the canvas, and thus the watertight integrity of the cart. Once again, I took up my position between the shafts of the cart and led the way to my little marker, about half a mile westwards along the beach.

The cart behaved extremely well on the ice, and it soon became clear that our best plan would be for Cameron to scout ahead and try and find a way around the leads and ponds that crossed out path. This worked well until he decided that some of the leads could be jumped rather gone around. It might have been fine for Cameron with his light frame and athletic background, but my reaction left him in no doubt of what I thought of the idea. Not only did the leads seem far too wide for me to leap, but the idea that the ice might give way on the other side as it felt the full impact of my weight, or that I might end up in the extremely chilly water, held not the slightest attraction. Then two things happened at more or less the same time. We had come up against a lead that seemed to extend from horizon to horizon and would have taken a very long time to have skirted. The crack, however, was really quite narrow, indeed, little more than a normal pace in width. As it seemed absurd to try and walk around its unknown limit, I stepped out from the shafts of the cart, took off my rucksack, and, after a short run up, took a flying leap at the obstacle, probably the first time I had attempted a long jump since I had left school. Not only did I clear the lead with ease; I landed so far away from the edge that even Cameron broke out into (possibly derisory) applause. Before long I was leaping such 'leads' with all the grace of a Himalayan hog, returning to haul the cart across as Cameron pushed from the other side.

The second, entirely unforeseen event, was the onset of a constant drying of my lips, and the onslaught of a savage thirst. What caused this I could only guess at, but I found myself repeatedly scooping up water from the meltwater ponds in an attempt to ease my condition. However, no matter how much water I drank, I could not stop my lips from cracking. Possibly allied to this was the uncomfortable effect of my rucksack causing the harness of the cart to repeatedly pound against the area of my kidneys. This made me feel continuously nauseous, a sensation which could only be prevented by stepping further back along the shafts, connecting the harness across the shaft ends in front of me, and pushing against the harness itself.

Although we had made a number of stops on the way across, by the time the far shore had come into immediate view, I was beginning to feel very tired and was glad of the opportunity to rest at the base of the ridged ice lining the southern rim of the inlet whilst Cameron went in search of a way through. Once again, a circuitous route over broken ice had to be taken before we eventually scrambled up the rubble of the beach. Astonishingly, the two-and-a-half mile crossing had taken us five and half hours. All I wanted to do was to curl up and sleep.

I unhitched the trailer and lay full length on the shingle wheezing like a winded walrus. My mouth remained constantly dust-dry despite frequent gulps of water from my water bottle, and my lips burned where they had cracked. There was also an ominous dull ache at the base of my spine. Cameron, little effected by the crossing, after a few minutes rest made his way up a ridge running parallel to the beach and disappeared over its rim in search of a route along the path dictated by his GPS.

As I rested, looking back across the inlet to the thin brown line that marked our point of departure, I was distracted by something no larger than a pinhead making its way across a frost-shattered rock just inches from my face. During earlier visits, I had seen butterflies and caterpillars but now, through my pocket magnifier, I watched a tiny spider scuttling over the rock surface. The insect had a hard-looking, shiny black body that glinted in the weak light. What sustenance such a creature could find in that part of the world baffled me. I could only hope that it dined on the many-times larger mosquitoes that had proved to be such an irritant in past visits (although they had failed to find either of us on this visit to King William Island so far).

Cameron returned to report that the way south looked straightforward. An esker showing on the map was easily found and looked as if it was heading in the right direction. Before departure, and over a mug of hot soup, I proudly pointed out to my companion that the last time a Royal Navy officer had crossed Collinson Inlet it had been Captain McClintock in 1859. Cameron replied 'Go figure.' Once back between the shafts, I found the going not at all difficult although I soon resorted to my previously adopted method of pushing against the harness to avoid feeling nauseous from the rucksack thumping against my kidneys. I also needed to keep taking great draughts of water.

All had gone well until, by early evening, the ridge along which we were travelling began to peter out and we were presented with our first real obstacle. A wide stream, virtually a mini-river, which fed a lake hidden behind ridges to our right, lay directly across our path. To our left, the stream ran into a wide swathe of tundra beyond which a ridge led away in the direction in which we intended to go. This same ridge, however, could be reached by crossing the stream ahead of us and Cameron was keen that we should tackle the problem by stripping off our lower garments, emptying the cart, and ferrying everything to the other side in a number of repeat crossings. I felt that this idea, although practical, contained too many risks. The terrain over which we were passing provided innumerable opportunities to damage our ankles: how much would the risk be increased when wading over an unseen riverbed? I had always been taught that it was of paramount importance to keep clothing dry, and yet here it was suggested that we should risk a stumble that would not only soak the clothing which we were wearing, but also whatever we were carrying. Such an idea was, I felt, a last – rather than a first – resort.

The stream-laced tundra to the east, on the other hand, when viewed through my binoculars looked a better bet. Several routes suggested themselves across the brown-green surface to the far ridge and, if the worst came to the worst, the entire area could be skirted at the cost of an extra mile or so. This, I decided, would be the way we would go.

Cameron was not at happy with my decision and continued to mutter under his breath as we made our way down the rocky slope to the edge of the tundra. From there to the far side it was about half a mile and I decided that we might as well strike out straight across, making our way around a small number of meltwater ponds. We had not been going for long when I felt I was able to congratulate myself on my decision. The ground was reasonably firm, supported by a surface layer of rocks that had obviously crept down from the esker to our rear, and the cart, although not running as well as it had on the shingle, did not increase the hauling effort too markedly.

After about five minutes of steady progress, my left foot disappeared in a patch of muskeg. Hauling the cart to the right in an effort to keep the cart clear of the slime, I felt my right foot begin to sink whilst, at the same time, the weight behind suddenly felt as if brakes had been applied. Cameron was having difficulty in pushing from the rear as his boots, and the cart's wheels, began to sink. This was no time to stop and calmly consider our options. We had to get clear of the muskeg as quickly as we could, but we soon found that our progress would be extremely difficult and wearying. As I heaved against the harness my feet sank deeper into the revolting soggy mass. Soon, every step I took drove my legs calf-deep and, such was the suction it threatened to remove my water-filled boots.

To my rear, Cameron was probably having an even worse time, for he not only had to heave against the rear of the cart in an effort to get it going, he also had frequently to lift it clear when its wheels sank to their axle. He was, nevertheless, about to get an enforced rest.

As I waded through the muskeg with a grotesque slow-motion rhythm, my left leg sank almost to the knee. My right foot, fortunately, found a clump of tundra grass against which I could push to release my other foot; but this time, it seemed that it simply would not break free. Without thinking, I threw my weight onto my right foot and pushed down hard. Then two things happened. My left foot, with a gurgling sound, pulled clear of the slime just as the grass beneath my right foot gave way. At that point the pack on my back took charge, and I toppled over like a felled oak tree into a layer of muskeg covered by six inches of water. My instant rage knew no bounds – especially as I had brought my situation on entirely through my own actions. I bellowed out words I had learned on the Stoker's messdeck many years earlier.

Cameron helped me to my feet, and I stood feeling extremely miserable with water cascading inside my clothing. So much for my desire to keep dry. There was no time to feel too sorry for myself. We were still in the middle of a field of muskeg, where even standing still for a few moments led to a sinking feeling. Although I might be wishing myself to be almost anywhere else in the world, there was genuinely no option but to keep moving.

We continued our heaving at a snail's pace with frequent stops to gather breath and for me to scoop up water to try and assuage the thirst that was still giving me severe problems. Eventually, inevitably, we reached the lower slopes of the ridge, and I crumpled gratefully to the ground delighting in the escape from the loathsome slime.

By now it was late evening, and we would soon have to think about finding a campsite for the night. After a short rest I resumed my position between the shafts. Although the ground was not the best we had come across, it was extraordinary how easy the passage seemed when compared to the patch of tundra we had just crossed. I was, however, feeling desperately tired and aching all over as we plodded along the line of the ridge. Soon, I had to call a halt with increasing frequency as my legs turned to jelly and my breath rasped in my throat. Still we pressed on, Cameron quite capable of continuing, and with me not wishing to stop whilst there was reasonable ground underfoot. Then nature applied its own brake.

We were just hauling up the slope of a ridge that barred our path when I heard Cameron shout 'Holy smoke!' Wondering what had caused this outburst I attempted to turn around, only find that I was incapable of such a movement. The reason being that I was sprawled face down on the rocks. How I had got there I did not have a clue. One second I was leaning into the harness, the next I was lying full length. There was no way of telling whether I had simply tripped or had fainted for a split second.

My first instinct was to stay where I was in the sincere hope that I might drift off into a deep sleep, but sharp pains in my knees and chest made me realise that it might be better if I tried to see if I could move. Having checked that my fingers and toes still worked, I then tried to stand up but found that the weight of my rucksack held me down. The only answer was to roll onto my side from where I could reach the straps that could free me. As I rolled over, I realised what had given me a pain in my lower chest. My binoculars had been suspended around my neck. They now lay battered – but fortunately not broken – on the shingle, the leather strap by which I wore them had ripped apart behind my neck. I had been fortunate in one respect. If I had been towing the cart as originally intended, with the harness around my waist, I would have taken the full force of the fall on my face. As it was, my pushing against the padded belt harness had meant that my face hit the rear of the padding rather than the rocks.

Despite the rocks beneath me, I felt strangely comfortable and experienced no desire to get up. As I lay there, Cameron wandered over the ridge and disappeared from my view. He returned within a quarter of an hour to say that there was an ideal campsite less than a couple of hundred yards away.

With great difficulty I hauled myself to my feet and, with even greater difficulty, restored my rucksack to its bed of pain on my shoulders. Cameron took charge at the front of the cart and began to pull it towards his proposed campsite whilst I pretended to push from behind, but actually having considerable difficulty just standing up. I ached in every joint from my hips downwards. By the time I reached the level stretch of shingle Cameron had selected, he had already started unloading the cart. Before long, we had the tent erected and I was able to get out of the wet clothes I had been wearing for the past few hours. It felt wonderful to get my boots off, put dry socks on and slip into the lightweight pair of shoes I had brought for use around the camps. I rigged up a line running from the top of the tent to a nearby rock and hung out my wet clothes to dry in a gentle breeze. With Cameron rattling around getting a meal under way, I decided to help by topping up the water bottles and, feeling extraordinarily light, walked over to a nearby pond. Reaching the edge of the water I balanced myself carefully on a pair of small rocks and bent over to fill up the containers. I had just reached the point of no return when the ground beneath me gave way and I sank yet again into a hidden patch of muskeg. The sudden descent caused me to topple forward and my pushed-out arms immediately sank almost up to my elbows leaving my nose inches from the surface and my body arched over the water. With infinite slowness, I pulled one arm from the mud and gradually 'walked' my upper body back into a position from where I could stand up. This was not too difficult, for whilst I was extricating myself from one end, my legs were sinking deeper into the muskeg and I could use my locked lower limbs to haul myself upright. That just left the problem of getting my feet clear. This could only be achieved at the expense of leaving my shoes deep in the grey slime. Once I had dragged myself free from the gurgling mess, there was no alternative but to stand in my stockinged feet and plunge my arms down the slowly filling holes to retrieve my shoes. Both refused to give way without a struggle. I cut a very sorry picture squelching around the edge of a meltwater pond in my socks whilst washing the muskeg off my battered shoes and lower arms. Having replaced my sodden footwear, I plodded back towards the tent, reaching beyond the halfway point before I realised that I still had not topped up the water bottles.

As I was having my private pantomime down by the pond, Cameron had cobbled together a meal and warm drink which, once I had added my newly wet clothing to the others on the drying line, I thoroughly enjoyed. It was a peculiar fact that although I am not known for my abstinence in matters concerning food, and I had been expending far more energy than was usual, I did not feel the need to eat beyond the odd nibble at dried fruit throughout the entire day. I had, however, suffered from that unquenchable thirst, and would frequently empty my entire water bottle with a single draught in an attempt to deal with it.

With the meal over, we both climbed into our sleeping bags, with me at least, immensely grateful for the opportunity to sleep. In the unlikely event that I would find difficulty in dozing off, I produced a bottle of rum and offered a swig to Cameron. To my deep gratitude he refused my offer and so, in true old Royal Navy fashion, I had to drink his tot in addition to mine.

I reviewed the day's events as I snuggled down into my bag. It seemed to me that my chief difficulty lay in not insisting on 'convoy speed' with me being, by far, the slowest ship. I had instead been trying and failing to keep up with the much younger and fitter Cameron.

Equally, I had been foolish in attempting to cut straight across the muskeg-riddled tundra when there was an obvious, if longer, way around. I also decided to stop any attempts at light-hearted banter with my companion.

I awoke the next day with my joints and bones in surprisingly good order. I ached all over, but no more than should have been expected after a day of hauling the cart. Again, Cameron was a reluctant riser which gave me the opportunity to get the stove going and water on the boil to greet him when he emerged into the grey light of the morning. I was delighted to find that my clothing had practically dried beneath the midnight sun and could be packed away into my rucksack. Cameron decided that he wanted to do some running repairs to the cart and gave me the job of cleaning the aluminium strips that protected the sides, I paddled barefoot in the pond rather than risk another farce like the night before.

With the cart packed and our rucksacks hoisted, we were ready to get under way by mid-morning. Through my binoculars (now secured by a length of bright orange nylon rope) I could see that our route ahead repeated the vast quarry-like fields of rocks, ponds, streams, and strips of dark green tundra that had existed all the way up to Cape Felix. Gentle waves of low brown esker ridges rippled away in a generally south-eastern direction. It was along the tops of these that we hoped to make good progress.

Our hopes were soon fulfilled as we began our day's journey over packed shingle that would have been good enough for most road vehicles. The cart rolled along at a cracking pace until, after almost two hours of good progress, the ridge on which we were walking began to tail off. Ahead lay a broad expense of tundra which, at least through the binoculars, lacked the sheen suggesting that shallow water lay hidden at the base of the tufted grass.

By now I had begun to feel the effects of the previous day's efforts and was still affected by a constant thirst. Once the tundra had been reached, and we began the far slower haul across its tussocks, I soon found that my energy, once again, began to drain rapidly away. Even before we were a quarter of the way across, I had to call a halt every few yards whilst I rested and took a drink. In an effort to speed things up, Cameron took up the position between the shafts whilst I pushed from the rear, but it had little effect on my performance which continued its rapid deterioration. Not only was every joint burning, but also an increasing pain in my lower back was beginning to cause me serious concern. It was proving difficult to stand upright with my rucksack, and the forward leaning posture I had adopted aggravated the pain, causing waves of nausea to surge through me.

Eventually, mainly through Cameron's efforts, we reached the rocky plain on the far side. I threw off my rucksack and sat with my back against it in an effort to provide some ease for my aches. I had just achieved a state of relative comfort when Cameron asked to borrow my binoculars. It was then that I noticed that they had vanished. When last seen they were resting on my chest, suspended by their orange cord, now they had gone. There was nothing for it, I would have to retrace our steps in an attempt to find them.

The tyre tracks across the tundra were, for the most part, easily followed. Compared to the slog with the pack on my back and either pushing or hauling the cart, the unburdened plod beneath a blue sky with a slight breeze was a pleasant experience. Having reached the base of the ridge we had so recently departed, and with no sign of the binoculars, I walked up its slope looking right and left in an attempt to find the cart tracks. But no marks of any sort remained to indicate our route and I continued on more in hope than expectation, desperately keen not to lose sight of the cart (by now a fraction below the line of the horizon, in front of a gently rising esker). Just as I was beginning to come to terms with the loss of the binoculars, and about to turn around to re-join the cart, I saw ahead of me a flash of bright orange – and the first smile for a long time broke out across my face.

On my return, with the errant binoculars safely tucked into one of my jacket pockets, I found that Cameron had been scouting ahead and had found a likely route towards the head of Seal Bay. After a brief snack on a bar of chocolate, I returned to the front of the cart and began hauling. It was not long before, yet again, I was having to stop to rest at increasingly frequent intervals, and our pace was reduced quite dramatically. There was, however, one cheering aspect. At first through my binoculars, and then without their aid, I could see the sparkling silver streak of Seal Bay.

The final haul was to be over a vast field of grey-green tundra. Fortunately, the area was for the most part dry and we were not troubled by muskeg, but I found simply moving getting more and more difficult. The halts I was having to make increased in both frequency and duration until, as we crested a low outcrop of broken rocks and shingle, I knew I was in no condition to go any further. I sank to my knees, threw off my pack, and sat with my back resting against it. Every bone I had ached abominably, and my spine felt as if it had been crushed from top to bottom. My continual thirst had dogged me all the way, and my spirits were woefully deflated.

Cameron wandered a slight way off and sat down on a rock looking over Seal Bay. I just lay there looking eastwards until, with a slight fluttering in the stomach region, I noticed something moving over the tundra towards us.

This visit to the island had been remarkable in both its good weather and its lack of wildlife. Arctic terns had made sport with Cameron, and we had seen a few seals as Collinson's Inlet had been crossed, but there had been no snowy owls or eider ducks. A few loons and Canada geese had passed overhead, but there had been no sign of the friendly knots which had crowded the ponds on earlier visits. Even the multitude of lemmings seemed to have vanished, and the hated mosquitoes.

I brought my binoculars up with a degree of trepidation and looked at the light-coloured shape that seemed to be heading directly towards me. As I watched the shape split into two and, before long, I realised that

I was looking at a pair of young caribou, not, as I had feared, a hungry polar bear. The welcome strangers trotted to within a hundred yards of us (certainly within a hunter's gun range) before stopping. They sniffed the air and paced backwards and forwards as if confused before they decided that discretion was the better part of valour and retraced their steps towards the eastern skyline. The impression they gave was that of an animal who had avoided extinction more by luck than judgement.

After half an hour, I rose very shakily to my feet and joined Cameron to make our way down to the shores of the narrows at the head of the bay. The water was less than a third of a mile away, but I found that I could only walk by throwing one side of my body forward, wobbling for a while, and then following with the other side. I must have looked like a drunkard picking his way home after a night on the town. Cameron had reached the water's edge long before me and was making his way across the hummocked tundra grass. This was necessary as the shore itself was an expanse of black mud with a stench that took me back to the miserable times in crossing Wall Bay three years earlier.

Instead of a stream linking the eastern inland lakes to the head of the bay, the waters had flooded the low-lying land to such an extent that the far side was now about six hundred yards away where a herd of caribou could be seen grazing on a patch of tundra. To the left of us the broad swathe of water reached the horizon whilst to the west the bay widened out to the Victoria Straits, its surface studded with drifting ice-floes before the mouth was sealed with ice that ran from shore to shore.

The waters directly in front of us were studded with rocks that breached the surface. This suggested that the depth was not very great, but I could also see, running across our front, a rippling that suggested a fast current over the route of the original stream. Cameron, nevertheless felt that the waters were fordable. I agreed, but with doubts over the fast-flowing water and the stinking mud which would swallow up a boot with the greatest of ease. As it was, however, I was in no state to cross a village pond, much less the obstacle in front of us, so it was decided to set up camp on the eastern edge of the rocky area where we had left the cart.

Our tent was set up on my part in a kind of slow-motion daze. Every movement, whether the raising of an arm or the bending of a knee, caused me to wince. I felt as if I had been run over by a combine harvester, and almost went without hoisting my white ensign and pennant at the top of the tent, but some standards had to be kept up. Once the sleeping bags had been placed inside, we both crawled into them without a meal or even a drink. Almost as soon as his head hit the pillow, Cameron was fast asleep. I, on the other hand, despite being desperately tired, found I could not get to sleep due to a severe onset of shivering. I had no idea what caused this reaction.

I was warm inside my sleeping bag, but my teeth were chattering and my whole body shook as if I was in the grip of some tropical fever. Eventually my clouded brain came up with the answer. With infinite slowness I fumbled around in my (still packed) rucksack and found my bottle of rum. A few mouthfuls of the splendid liquid were all it took, and I was soon gratefully drifting off in retreat from my pains.

I awoke fourteen hours later with Cameron still snoring in the depths of his bag. As I inched my way out of the tent, my spine felt as if it had been fused together and the tendons in my legs seemed to have shortened by a couple of inches. The collection of water from a nearby pond could only be achieved by my assuming a crouching position and lurching forward hoping my legs would act in some sort of co-ordination. Nevertheless, by the time Cameron emerged, I had managed to get hot water on the boil ready for a cup of hot chocolate. Over the steaming mugs, we decided that I should rest as much as I could in an attempt to see if the situation improved throughout the day. If it did, we would prepare to continue south – if not, there was little choice but to set off the despised Personal Locator Beacon.

Cameron was keen to get a closer look at the caribou that had approached us from the east the day before, so once breakfast had been finished he set off in that direction with the shotgun and binoculars. Whilst in that direction, he could also take a look at the lakes that fed the head of Seal Bay to see if there was a more practical way around its eastern edge – it was after all, a 'bay'. Even if we did not make it, such information might be of use to any subsequent travellers passing through this part of the country.

With Cameron gone, I settled down with my back arched over a convenient rock padded with a spare woolley-pulley. This, I found, made my aches more bearable as I wrote up my notes and listened to the call of a friendly snow bunting (sounding remarkably like the opening notes of that part of Offenbach's 'Orpheus in the Underworld' – the 'Can-Can'). I was banking on the assumption that whilst the bird continued its singing, it would indicate that there were no polar bears close by. Should, however, a bear come along my plan was to approach it waving the fibreglass gun case in the hope of scaring it away. If that failed, there were plenty of rocks at hand with which to pelt it. Not a bad plan considering I could hardly walk or raise my hand above my shoulders.

After about an hour, during which time I carried out some modest stretching exercises, I found that I could clamber to my feet and keeping to a gentle pace could move around without too much discomfort. Around me, in all directions, the land lay flat with just the slightest slope towards the north. To the south, Seal Bay lay speckled with ice floes that had either drifted in with the tide or had been blown in by the slight westerly breeze that ruffled the tundra-grass. Immediately to the west, directly behind the tent, a low ridge of rocks reared up like the outer ramparts of a bastion placed in that spot by nature to guard the narrows at the head of the bay. With plenty of time on my hands I decided to make my way up the rocks to see what lay beyond.

Once at the top, I could see that the site consisted of little more than a shingle platform with groups of larger rocks scattered over its surface and edges. There was also something else that I had not expected. Just a few feet in front of me, a prominent line of rocks curved around in a 'U' shape with the open end of the 'U' facing to the north. At some time in the past, a tent had been erected on that spot. Not a native skin tent that would have been anchored down by a circle of stones, but an Admiralty-pattern tent with its curved 'apse' such as were commonly illustrated in works on mid-nineteenth century naval polar exploration. I had seen the same pattern of rocks at my 'Peter Bayne map' site and on the red beach where Lieutenant Browne had set up his 1851 camp. And there was more to come.

A few paces directly northwards from the tent site, a large circle of rocks lay partially covered with a low-lying, scrubby brown shrub. The arrangement did not have the formality of the native tent ring, but reminded me instead of the broken-down cairns I had seen elsewhere on the island – not least at Crozier's Landing.

My mind did not have to work overtime to make the obvious links. It seemed highly possible, even probable, that someone with a naval connection had chosen this site to set up a camp and build a cairn. It was unlikely to have been McClintock in 1859 as he would have stuck to the coast, but both Gore in 1847 and Hobson in 1859 would have come south via the coast and could have returned using a short cut. On the other hand, Gore was eager to get back to Franklin with his news and might have preferred to retrace his footsteps. Hobson, suffering with scurvy, may not have been prepared to risk an untried route in his desperation to reach the safety of the *Fox*, locked in the ice far to the north at the eastern entrance to Bellot Strait.

Just one alternative remained. Crozier, in following the overland route I had long suggested, would have sent out a scouting party to reconnoitre ahead of his sledging teams. What better spot to erect a signpost could there have been than the site on which I now stood? Having reached the head of Seal Bay, the Advance Party would have set up camp on the only prominent feature in the area, marking the position with a cairn that could have been seen with ease from the northern skyline. Having long forgotten my aches and pains, I searched the immediate area as closely as I could for anything left behind by the last occupiers of the site. But nothing turned up, not even a single animal bone.

I then decided that I should take a closer look at the rest of the small plateau. With head down and eyes alert, I scoured the broken surface eager for even the smallest piece of evidence.

What I found was so large that it could have been spotted by satellite. Some fifty yards to the west of the tent site lay an area of roughly broken limestone rocks. Among the random scatter I saw, what appeared to be another possible circle of flat boulders. On approach, however, the 'circle' took on a more distinct shape – that of a double-ended boat.

It was the last thing I expected. But there was no denying that it looked as if someone had deliberately placed rocks to produce the outline of a Royal Navy whaler – the very type that was used on the *Erebus* and *Terror* as ship's boats, and exactly the same as the remains that had been found at Back Bay on Prince of Wales Island. I paced along the feature's length and width and worked out that it was about twenty-one feet long and nine feet wide. It was so large that it was difficult to photograph from ground level without retreating to a point where it almost merged with its surrounding rocks.

Assuming that the structure was man-made, what could it mean? It might have been nothing more than an attempt to leave behind a simple sign that someone had visited the site, in much the same manner as my stone anchors and the cairn at Cape Maria Louisa, just a personal mark on the landscape. It could, on the other hand, be a grave, albeit with particularly difficult ground to penetrate, especially in the late winter. Another more practical idea sprang to mind from its length. At twenty-one feet it was six feet shorter than the overall length of the standard whaler. If the boat had been removed from its sledge – for example, to effect a repair – it might have been convenient to keep the boat upright. This could be done by placing large stones around its base and would account for the shorter length of the feature.

Whatever the cause, I felt a disgraceful glow of pleasure that verged on smugness. *Someone* had been on the site many years before. *Someone* had erected an Admiralty-pattern tent and a cairn. *Someone* had arranged rocks in the shape of an Admiralty whaler. Whoever that *someone* had been, it seemed highly likely that he had a naval connection, and the chances were that he had belonged to a party dragging a whaler south over King William Island – exactly over the route I had suggested, and the very reason I was in that part of the world.

What was especially extraordinary was the fact that if I had not called a halt due to my inability to go further, we would have passed the site without a second glance.

I decided to unofficially name the outcrop of rocks with its reminders of the tragedy that was yet to come 'Fort Crozier', after the man on whose shoulders the mantle of leadership fell with the death of Franklin.

Cameron returned mid-afternoon. He had failed to find his caribou quarry and had seen no better crossing to the south than the narrows below us. My revelation of the discoveries I had made by simply scrambling up the wall of rocks behind our tent did little to make his disposition more sunny. The tent and cairn could not be ignored and I gained his agreement as to their possible origin, but he dismissed the boat-shaped formation as 'natural' (he was, after all, a trained archaeologist).

My companion then decided that he was going to attempt to wade across the waters to the far side where we had seen caribou the day before. I followed him down to the edge with the gun in order to keep a wary eye out for any unwelcome visitor. Cameron rolled his trouser legs up, donned a pair of 'wet boots', and stepped off. Twenty minutes later he had reached the far bank and spent the best part of an hour roaming around the nearby eskers before returning.

As was expected, the following morning revealed that my back had shown little, if any, improvement during the previous day and night. My problems with thirst had disappeared along with my general exhaustion. What remained was purely mechanical. Any sharp movement produced a grinding sensation at the base of my spine, a feeling accompanied by nausea when I tried to carry any weight for more than a few paces. The journey had clearly come to an end for me, and Cameron made no suggestion that he would like to continue on his own. The 'honour' of pressing the button on the personal locator beacon fell to me. A flashing light suggested that something was happening and, sure enough, six hours later the sound of an aircraft approaching could be heard from the north. It turned out to be the same RCMP Twin Otter that had first brought me up from Yellowknife five years earlier.

The aircraft made a few low passes whilst the pilot waved from the cockpit. It was clear that he was not going to attempt a landing on the tundra and, after a final swoop turned away to the southeast and was soon out of sight and sound. Now that we had been located, we collapsed the cart and packed our kit, with the exception of the tent, and sat down to wait for what would happen next. This turned out to be the unmistakeable rattle of a helicopter coming in from the east.

The noisy machine made a steep turn and landed just on the edge of the rocky outcrop. First out was a native RCMP Special Constable who smiled cheerfully when asking if there was any emergency medical aid required. When assured that there was not, he introduced us to his pilot, a woman with blond hair barely constrained by her baseball cap.

Both helped us to pack away the tent and to get our equipment stored aboard the helicopter.

It was with mixed emotions that I looked out of the aircraft as we flew low over the island on our way to Gjoa Haven. Beneath us the land rolled by with utter, unrelieved, bleakness, the dark tundra pockmarked by patches of grey-brown limestone and muskeg polygons all highlighted by streaks of snow lying in sheltered esker folds; and in all directions, sullen-looking meltwater ponds and still-frozen lakes laced together by innumerable streams and rivulets. In some respects my spirits reflected the gloom below. I had not achieved what I had set out to do, and events had clearly demonstrated that I did not have the physical abilities to keep up with my much younger and fitter companion. I did, however, feel that I had found evidence to support my theory that Crozier and his men had chosen an overland route rather than the coastal ice. I may have found my first-ever Franklin artefact in the wooden toggle (bearing in mind that the carved caribou bone I had picked up on Booth Point did not have a clear nautical link) and had been the first Royal Navy officer to cross over the ice of Collinson Inlet since Captain McClintock 136 years before. Equally, Cameron and I had been among the tiny handful of men who had walked across the un-named peninsula that surges north-westerly to Franklin Point between Collinson Inlet and Seal Bay since the sad procession of Franklin's men in the late spring of 1848. How feeble my efforts seemed when compared to their scurvy-ridden, hopeless predicament.

After a meal of caribou steak, I slept that night on the floor of the Gjoa Haven hotel, now vastly improved since my first stay in 1990. There was already an improvement with my back the following morning as I showed Cameron around the hamlet. Walking between houses on our way to the RCMP station we came across the local 'Honey Wagon' making its way around the sewage tanks. The native driver and I exchanged glances and recognised each other immediately. It was Max Kameemalik, the same Max who I had first seen as the cleaner at the local Medical Centre and who had volunteered to take us to Todd Island and Booth Point five years earlier. Max greeted me like a lost brother, and I was delighted to see his wide grin once again. Before long, Max had offered to let Cameron use his fishing cabin near Koka Lake as my companion had decided to stay at the hamlet for a few days in order to become more acquainted with the area and its people.

There was one further meeting of note before I left for the airfield to catch the aircraft for Cambridge Bay. At the hamlet Engineer's Office, I met a well-educated young native man who had an interest in history of his people. During our discussion, I asked him what, if anything, his ancestors (or he) would have done with the bodies of their enemies after an attack. Without hesitation he replied, 'Break their bones.'

'Mutilate them?'

Perhaps offended, he replied, 'No! But I would break their bones, in case they might come after me.'

Another link snapped shut in the chain of thoughts that was leading me to challenge the Franklin 'cannibalism' theory so loved by those keen to debase the role of the Royal Navy in the exploration of the Arctic.

Over two decades later, there was a strange – even eerie – echo of the possible Inuit involvement in the deaths of Franklin's men. In December 2018, news reached the Canadian press that a recent number of deaths at Gjoa Haven had caused great concern among the Inuit that Franklin's men were taking their revenge. Of course, no-one addressed the problem of the spirits of men long-dead seeking revenge for being massacred, but an Inuit blessing had taken place at the site of the *Erebus* soon after the wreck was located, another, delayed, version took place at Terror Bay to stem the further activities of spirits of Franklin's men. Again, no-one asked why the Inuit were reacting in such a manner? Could it have been guilt?

Having said goodbye to Cameron at Gjoa Haven, I reached Cambridge Bay by late afternoon and booked into the Arctic Island Lodge where I spent as long as I could stretched out on my bedroom floor to help my gradually improving back.

My aircraft for Yellowknife did not leave until after noon the following day so I was able to pay a call on Willy and Mrs Laserich before I left to thank them for all the help they had given me, not just on this visit, but for every visit I had made to the region. If the Arctic can still be called a 'frontier', they had never failed to demonstrate the 'frontier spirit' when the opportunity presented itself.

That same Arctic was to provide one last reminder of its nature when on arrival at the airport I learned that the aircraft was stuck at Resolute Bay as a result of bad weather. This meant another night at the Arctic Island Lodge where I gratefully learned that the only other passenger – a 'Topographic Technician' – had equipped himself with a bottle of rum which we both did serious damage to before I returned to my bedroom floor.

I had intended to spend a night at Yellowknife in order to bring the Arctic archaeologist Margaret Bertulli and Chuck Arnold up-to-date with recent events, but the enforced stay at Cambridge Bay meant that I would have no more than half-an-hour at Yellowknife airport before my connecting flight left for Calgary. As I result I rang Margaret to ask her if she would meet me at the airport as I passed through. 'Sure,' she replied, 'how did it go?'

'Not too bad. I've got something for you. Something I found.'

The tenor of her voice changed immediately and took on a menacing, steely note. '*You removed an artefact?*'

I decided that, after all I had been through, I was not going to be intimidated and stood my ground, 'Don't start shouting at me, Margaret.'

The same deliberate, even-toned, voice came back. '*I'm not shouting.*'

At this I cut my cables and ran, leaving her with my time of arrival at Yellowknife.

When it eventually arrived, the aircraft's route was to be full of interest to me. On taking off from Cambridge Bay it turned westwards and flew along Dease Strait, named after one of two Hudson Bay Company men sent out to explore the coast of North America in 1837. Out of my port-side window I could soon see beyond floe-strewn waters the brown mass of the Kent Peninsula. In late August 1821, on the peninsula's north-west corner on a spot indistinguishable to me, Lieutenant John Franklin had decided to turn back from his attempt to meet up with Lieutenant William Parry's ships (unknown to Franklin, Parry was already home after wintering far to the north at Melville Island). From 'Point Turnagain' Franklin had headed south in company with Surgeon John Richardson, Midshipmen George Back and Robert Hood, and Able Seaman John Hepburn. It was to be a journey of heroic stature. Most of their accompanying French-Canadian 'voyageurs' sickened and died despite stealing food from the naval men. One turned to cannibalism and murdered Midshipman Hood and was, in turn, shot by Surgeon Richardson. After desperate trials they eventually reached Fort Enterprise only to find that the local Indians had failed to leave the supplies that had been previously arranged. The Indians were eventually tracked down by Midshipman Back and sent to the fort where Franklin and his party remained barely alive after surviving on old animal hides and charred bones.

On leaving the western end of the Kent Peninsula, I could see the wide sheen of Bathurst Bay as we entered Coronation Gulf. It was along this route that HMS *Enterprise,* under Captain Richard Collinson, sailed in search of Franklin, and Roald Amundsen's *Gjoa* made the first water-born transit of the North-West Passage. In 1821, the only non-native vessels on those waters were the frail birch bark canoes of Franklin's party heading eastwards from the sand-choked estuary where we were just about to land.

The actual hamlet of Coppermine cannot be seen from the primitive airstrip on which our pilot had landed us with consummate skill. The site was surrounded by a series of low green hills to the south with islands off the northern shore. To the east, the Coppermine River flowed sluggishly into the Arctic Sea over a series of mudflats.

My arrival at Yellowknife later that day was met by Margaret Bertulli, dressed for the occasion in baggy shorts and a fetching straw solar topee. During the few minutes we had together, in return for a cup of coffee, I handed over the toggle and the co-ordinates of the sites of interest that had been found. It was a fair exchange. Without Margaret's help from the very beginning of my wanderings 'up north', I doubt that I would have got beyond my first visit to Yellowknife five years earlier. I was very grateful some months later when, in producing a report of the events on King William Island during the summer of 1995, she included my very unofficial naming of the site on Seal Bay as 'Fort Crozier'.

After a few days of indulgent recuperation at Calgary and with Sarah Allen – an old acquaintance – and her family at Vancouver, I arrived home where the first letter amongst the waiting mail informed me that, after 36 years' service with the Royal Navy, I was about to be made redundant. Perhaps that was to mean not just the end of my time with the Service, but also the end of my time beyond the Arctic Circle. If so, I had no complaints. I had met some splendid people who had been unstinting in their help for a complete stranger, I had seen sights that dazzled the eyes and made the heart leap, and I had walked in the steps of brave and gallant men who had given their lives in extending mankind's knowledge and vision. All in all, I felt I had not done badly.

My travels down the shores of north west King William Island had convinced me that if the site of Sir John Franklin's grave is to be found, and it proves not to be my mounds near Cape Felix, it is just as likely to be found by accident as design. Although recent searches have had all the advantages of modern technology, the ground actually covered still remains a small fraction of the total possible area. If a clue exists, it could be as small as the stub of a boarding pike, or so large that its very size renders it unrecognisable. It could have merged with the environment and remain indistinguishable from its background. It could be subjected to misinterpretation by those keen to follow their own pre-set agenda and to redirect slim resources to their own areas of expertise. It could be ignored by others trying to establish their own theories into which the key clue does not fit. I could claim to be an expert in nothing; not even in service with the Royal Navy because everyone who has so served can produce a different impression of the wide range of events and people they served alongside. However, any theorising I did stemmed from the simple application of logic contained within a framework of experience, against a background of a study of naval history. This enabled me to view events from a peculiarly 'Royal Navy' standpoint.

It was this angle on events that made me think that there may have been more (and remains more) in connection with the two large mounds I found south of Cape Felix. Not only did the experts declare that the features were 'natural' – based upon the fact that they consisted of apparently undisturbed soil and rocks – but also that they were too large to have been man-made. Having dismissed the mounds, the experts then went on to reject any idea that the 'drift calculator' was anything more than a line of native caches despite that fact that they were in the right place and of the right design. Both Hobson and Schwatka had reported finding the remains of a naval camp at the site; a camp both substantial and permanent enough to have been supplied with wardroom crockery and wine. The outlines of the camp's tents with their straight sides still remain just yards from the 'drift calculator' but were equally dismissed by the experts as being free of any 'provable' Franklin links, that is artefacts, in an area scoured many times by the local people.

I do not know what to make of my upside-down 'Peter Bayne map' site, but the chill that ran down my spine when I came across a rock arrow-head pointing in the direction I had already committed myself to stays vividly in my mind. All the right ingredients were present: a bay with a projecting prominence, a campsite, suitable cairn-building material (on a site where any cairn could easily have been swept away by encroaching ice), and a series of grave-like mounds in exactly the spot indicated by the (upside-down) map.

The intriguing sites on Prince of Wales Island, known to the natives but unexamined in detail before 1994, remain as a tantalising mystery. My own researches into both the remnants of the boat, and of the site on the shores of Peel Sound, have led me to the conclusion that the site deserves a most careful and sympathetic survey. At the very least, the remains of the boat should be preserved, and the Peel Sound beach site searched for Lieutenant Browne's message.

And will my 'Fort Crozier' prove to be of significance? Again, all the elements are present for a positive analysis. Sited on a direct route south from Crozier's Landing, overlooking the crossing point at the head of Seal Bay, with appropriate tent outline and cairn, and with the shape of a ship's whaler clearly and carefully delineated on the rocky surface, the obvious conclusions lead nowhere but to the site having considerable potential for examination.

'Considerable potential' is, however, not enough on its own. These, and other sites, may help to shed more light on the story of Sir John Franklin and his men, but when approached with the blunt weapons of current, fashionable mores that ignore the views and experiences of mid-nineteenth century travellers, the outcome is guaranteed to be not merely distorted, but also to be contemptuous of brave men. Better by far to follow the straightforward words of a speech made in the presence of Jane, Lady Franklin during a Franklin search fund-raising dinner:

> Sir, they must not be abandoned. Even if they have perished we want to know where and how they sank under their sufferings. We want to track their pathway through the cheerless tracks, and to know and mark for all time, the spot where their humble spirits bowed to the will of God, and where they laid them down to their last sleep. We want to recover the memorials they left behind them, the records of their trials and sufferings, and their last tokens of affection to friends and kindred.

No condemnation born of lofty hindsight, no denunciation of men who gave their lives in the cause of others, and no attempt to degrade the achievements of those who cannot defend themselves – just the honest desire to seek out the truth behind the veil of time.

The rock whaler outline on 'Fort Crozier'. The foreshortening effect of the camera height disguises the 26-foot length of the feature.

14

Headwinds and Rough Waters

My return to England and looming unemployment was met with the realisation that no-one was looking to employ a broken-down naval lieutenant. Even if I reached the interview stage, the usual question was 'What can you do?' As it was, I was much more used to the sort of environment to be found in everything from a submarine to an aircraft carrier where I was *directed* what to do – even if I had never done it before. My usual answer to the prospective employer was 'What do you want me to do?' In the end, I decided to try my hand at becoming a writer.

However, I never found the key to world fame as an author, although I never had a manuscript turned down by a publisher and I was nominated for the Mountbatten Maritime Literature Award (which I did not win). I was asked to open an exhibition on the exploits of the World War II 'X-Craft' miniature submarines, I was frequently interviewed on the local BBC station on matters nautical, and took the salute at a huge Sea Cadet parade. I gave talks at Women's Institute meetings, Rotary Clubs, Forces charity events (I was a Committee member of the local 'King George's Fund for Sailors'), and, of course, talked to numerous schools (usually primary schools – including my own). It was whilst talking to one school that I received the most original piece of advice on polar matters. When the teacher asked if there were any questions, one small boy, aged about 8 or 9, raised his hand and asked if I had taken a step-ladder into the Arctic with me. When I asked him why I should do such a thing, his eyes rolled up before he answered, with some exasperation, 'Because, if a polar bear came along, you could climb up it and it could not get you (Duh!)'. I replied by telling him that it was such a good idea that I would tell it to all the polar explorers that I knew.

It was just shortly afterwards that I met my first 'real' polar explorer. I attended an Arctic Club dinner and found myself sitting next to Wally Herbert who had led the first expedition to cross the Arctic from

Alaska to Spitzbergen. The journey took 16 months and became the first expedition to reach the North Pole on foot, and the first to traverse the Arctic Ocean. Unfortunately the timing was bad, the marvellous achievement was overshadowed by the first moon landing, and Wally Herbert's feat went unrewarded until, at a subsequent Arctic Club dinner, I was delighted to meet Dan and Jonolyn Weinstein once again. Before the meal, Dan took me to one side and asked me if I would lend my support to an idea he was trying to get off the ground. Wally, he felt, deserved better recognition for his polar work – would I send a letter to the Prime Minister requesting that Wally be awarded a knighthood? Of course, I agreed immediately (as did many other people with much greater 'clout' than I had). It worked.

In 1997, probably at the instigation of the Franklin enthusiast and librarian, Ralph Lloyd-Jones, I was invited to read the prayers in St John the Evangelist Chapel at Westminster Abbey at a service to commemorate the 150th anniversary of Franklin's death. It was a very pleasant occasion, particularly seeing the beautifully elegant comment about Jane, Lady Franklin, on her husband's memorial. The inscription read:

> This monument was erected by Jane, his widow, who, after long waiting, and sending many in search of him, herself departed, to seek and find him in the Realms of Light, July 18, 1875, aged 83 years.

Appropriately, a plaque beneath Franklin's memorial bore the inscription 'Here also is commemorated Admiral Sir Leopold McClintock 1819-1907, discoverer of the fate of Franklin in 1859.'

At about the same time, I was invited to appear on the BBC's *Today* programme to talk about Franklin and my view on the 'cannibalism' question. Rather grandly, they sent a car to pick me up and take me to Broadcasting House. 'The Grand Inquisitor' was a broadcaster of many years' experience, noted for his aggressive style of questioning. On the same programme was General Sir Peter de la Billière, the most highly decorated living British soldier and former head of the Special Air Service (SAS). The General advanced on me with a smile and his hand outstretched, with a brisk 'Morning!'

Without hesitation, I stood up and replied, 'Good morning, Sir.'

After a few seconds establishing levels of status (the General very, very, high – me, just above pond-life), we fell to chatting like old mess-mates. He was particularly interested in the fact that we had more than just a Service connection. Some time in the 1970s I had attended a luncheon at the Mansion House in London. As I took my seat, I was astonished to find sitting next to me Lieutenant-Colonel David Stirling – the founder of the SAS.

Fifteen minutes later it was my turn. I was ushered into a darkish room and directed to a chair opposite the Grand Inquisitor. There was very little greeting or introduction, and I soon found myself laying out my ideas about the so-called 'cannibalism' amongst Franklin's men. Suddenly, I was stopped in my tracks to be told that an expert in such matters was 'on the line' from Canada. To my astonishment, it was Margaret Bertulli. I responded by saying 'Hello, Margaret,' and getting a 'Hi, Ernie' in return – much to the obvious annoyance of my interrogators, who had brought in an expert in the hopes that sparks would fly. After I had made my case, and Margaret had put the Party line, the Grand Inquisitor looked at the clock and said 'Well, who cares anyway? Now for the Sports.'

As a disembodied voice began to drone on about Accrington Stanley, the two men on the other side of the desk turned towards each other and began discussing the next part of the broadcast. I sat in silence until I managed to catch the Grand Inquisitor's eye. Then, in a slow-motion pantomime, I pointed at my chest, followed by pointing at the door. The Grand Inquisitor waved a dismissive hand, and I departed.

I was to take part in a most unusual New Year's celebration to usher in the Millennium. During an earlier visit to the Royal Geographical Society, I was introduced to George Meegan, a former Merchant Navy officer, and a man I had never heard of, but who had an extraordinary interest – walking over vast distances. Amongst his achievements were the crossing of South and Central America on foot; the first walk from the Tropic of Capricorn, across the Equator, to the Tropic of Cancer; and the first walk from the Equator to the Arctic Circle; he had walked across 125°08′ degrees of latitude, and, most astonishing of all, he had walked – in a single journey – from Ushuaia on the tip of South America to Prudhoe Bay (named by Lieutenant John Franklin in 1828) on the Beaufort Sea. It took him five years to cover the 19,019 miles. He was now intending to celebrate the Millennium by walking from the Arctic Circle to the Alaskan settlement at Barrow (named by Captain Frederick Beechey RN after Sir John Barrow, an Admiralty official, but now reverted to its native name – 'Utqiaġvik') and then on to the tip of Point Barrow, a headland nine miles in length, with the Chukchi Sea to the west and the Beaufort Sea to the east. The most northerly extremity of the headland was also the most northerly point of the United States. The walk was intended to take place on 23 January, the day when the New Year is celebrated at Point Barrow as the sun just peeps over the horizon before immediately sinking down again. George, however, also wanted to celebrate the work done by Royal Naval explorers who were the first to survey the coast of Alaska – a list which included Franklin. To help celebrate their achievement, George also wanted someone to represent the Royal Navy to walk with him from

Barrow to Point Barrow – and he decided, that 'someone' should be me. I was prepared to help, but in my opinion the Royal Navy should have been represented by, at the very least, a senior officer – better still, by a Flag Officer.

I had spoken on several occasions to an Admiral who had met during a talk I gave at the National Maritime Museum, and so, with George's approval, I sent the Admiral a letter outlining what George's plan was. The Admiral replied that he was amenable to the idea if the visit would be on the same terms as I had been offered (flights, etc). A check with George not only confirmed the arrangements, but he also insisted that I went along as well.

In Anchorage over breakfast, I heard that there was to be a display of flags at the end of the walk. Consequently, I hauled out a White Ensign from my baggage that had flown outside my recruiting office in Lincoln and arranged to have it added to the other flags.

Someone came to me with a message saying that 'Colonel Vaughan would like to meet you.' Who Colonel Vaughan was, I had no idea, but, following instructions, I made my way to the indicated cabin and knocked on the door. The door was opened by a very elderly gentleman with a grey beard underlining a still handsome face. He invited me in. I sat on the only chair in the cabin. The elderly man sat on his bunk and asked a few questions about me, and what I was doing in Barrow. I replied to his queries and brought him into the conversation to try and gently discover his story. And what a story it was. After Wally Herbert, Colonel Norman Vaughan was the second 'real' explorer I had ever met. In 1928-1930, he had been a sledge-dog driver on the Byrd Antarctic Expedition, and now, at the age of 95, was the sole survivor of that expedition. I later found out that just two days before his 89th birthday, he had climbed 'his' mountain – the 10,320 feet high 'Mount Vaughan' in the Antarctic's Queen Maude Mountains named in his honour by Byrd. In fact, his whole life was a catalogue of astonishing achievement.

The next morning was the day of the walk along Point Barrow. The shock at being outside in the early morning was astonishing. In the blackness, wearing good quality cold weather clothing topped by a Russian fur hat bearing a Royal Naval cap badge, I could feel my beard beginning to freeze, whilst my limbs moved only reluctantly as the body fought to accustom itself to the iron grip of the cold. When we arrived at the muster point, the Admiral and I found a small group of Japanese schoolchildren stamping their feet and hugging themselves closely (George's wife, Yoshiko, was Japanese). George was there, but none of the local people who had been expected. As we continued to stamp our feet, a message arrived informing us that the walk as planned had been cancelled as it was 'far too cold'. Apparently, even the sledge dogs refused to show up. However, a bus arrived which was intended to follow the walkers as a mobile place of refuge for anyone who needed its shelter.

Immediately, a conference (including the bus driver and the Japanese contingent) was held. We decided that, as we had got that far, we should 'give it a go'. And so, with the bus crawling along to our rear and George ahead, we set off towards the frozen Arctic Ocean.

We had not gone far when we found out that it was not just the actual temperature that was the problem. A stiff breeze appeared to be stiffening still further, adding a severe 'wind-chill factor' to the festivities.

I quickly found that to breathe I had to cover my nose and mouth with my woollen scarf. I also lifted the large hood of my padded jacket over my fur hat and used the draw-strings to fasten it tightly round my head. Then, it was a question of keeping the head down, and plodding on regardless. I had to reach the end of the walk, if only to rescue my White Ensign.

As our trudge slowed down at the end of our walk, I could see a clutch of flags rippling horizontally (which, I assumed, prevented the flags from freezing like washing hung out to dry). There, amongst them, was my White Ensign. With a clumsy struggle, I managed to rummage inside my padded jacket and pulled out my camera. I managed just one photograph – a split second after I had pressed the button, the battery froze and died.

I felt sorry for George, but, as we chatted, he showed no disappointment, and was already planning the same event for the following year (according to him, 'The proper Millennium'). An interesting aspect of the walk was that we were told that the temperature had been -83° Fahrenheit (-64° Celsius), but American climate specialists we met later that day disagreed and presented the Admiral and me with a chart of Point Barrow with an accompanying note that (including the chill-factor) claimed a temperature of -114°F (-81°C). Whichever may have been correct, I know that I found it really quite chilly.

After the bus shuttled us back to Barrow, we went to a reception with several speeches; repeated attempts to please our hosts by chewing on their proffered whale blubber; and introductions to a young Native American Princess and a very senior clergyman who was much the worse for drink.

At one stage, I was hauled off to meet someone 'very famous'. In a corner of the room sat a lady, clearly of considerable age. I was presented to her and told to bend down. Thinking that we were going to rub noses, I bent over. Much to my surprise however, she reached up, gripped the end of my nose and gave it a tweak. Instantly, the group around me burst out laughing, and broke into applause. It turned out that I had been given a rather special honour, for the woman in question had met Roald Amundsen in 1906 during the closing stages of his voyage through the North-West Passage. He had just bent down to say 'Hello' to the young girl, when she took hold of his nose and tweaked it. An honour indeed.

On my final day, spent in Anchorage, I waded through deep snow to reach a bar recommended by Colonel Vaughan. Once inside, I approached the man behind the bar, introduced myself, and asked if Colonel Vaughan was likely to be in that day. He replied, 'No. Did you say your name was Coleman?' I confirmed it was, and the man made his way along the bar to a shelf which held a large brown envelope. He picked it up and handed it to me. Inside was a book with Colonel Vaughan's photograph and name on the front cover. It was entitled *My Life of Adventure*. I opened the book to the flyleaf, on which was written:

'Trapper Creek, Alaska. 23 Jan 2000. For Lieut. E. C. Coleman RN who took over my bed quicker than he would have jumped in my grave. Dream Big and Dare to Fail. Norman D. Vaughan.' (I had taken Colonel Vaughn's room when he left the hotel because it was cooler than mine.) Not a bad gift from a man whose experiences included giving a sledge ride to Pope John Paul II and inventing a parachute for sledge-dogs so that they could help to save lives in the frozen north. Colonel Vaughan celebrated his 100th birthday in 2005 by having a sip of champagne – the first time he had tasted alcohol in his entire life. He died four days later. I have always admired and passed on his motto whenever appropriate: 'Dream Big and Dare to Fail'.

Colonel Vaughan's book was not the only one I received. Sometime later, this time at home in England, another package arrived, from George Meegan. His book was entitled *The Longest Walk. An Odyssey of the Human Spirit*. Again, the flyleaf was inscribed, this time with 'In Celebration – The Importance of being Earnest who is Ernest C. Coleman Royal Navy participant and member. Band of Brothers who made the last journey of the 20th Century. Historian, we touched history. Your Pal, Meegan.'

On a space to the side, George had written 'Bed cover' above an arrow pointing down to my White Ensign flying from a mast above the name 'Point Barrow'. 'Your White Ensign cracks in my imagination and forever. Friend.' He then signed off with a little sketch of himself waving.

I cannot bring George to mind without thinking of the Victorian General, Charles Gordon (Gordon of Khartoum), who marched into battle armed with a Bible and a swagger stick. In George's case, it was matter of principle that native peoples should not be pushed to the borders of existence, and he stepped out boldly to make his point.

Dan and Jonolyn continued to surprise with their overwhelming generosity. Twice they sent me air and sea tickets to get me to Buenos Aires, followed by a flight down to Punta Arenas on the southern tip of Argentina, where I joined a Russian ship that was going to South Georgia and the Falklands. Whilst on South Georgia I found a reason to appreciate Shackleton more than I did normally. At his grave, a custom had grown up of 'taking a drink with the Boss'. This consisted of a substantial amount of whisky which was first tasted by the visitor, followed by pouring the rest on the grave. Several of the visitors found difficulty drinking their share of the whisky and asked me to help them. With much reluctance, I gave them all the assistance I could.

The Franklin story continued to resurrect itself. 2008 was a particularly interesting year. In January I was amazed to be asked to chair the McClintock Winter School held in Dundalk, in the Republic of Ireland. In reality, the 'School' was an academic conference, the like of which I was wholly unqualified to take charge of. Consequently, I immediately said 'Yes!' The event itself went fairly well – apart from the time I decided to ask for a slot during which I could present my own version of Franklin's history. When I said that I was amazed that searches for Franklin's ship were frequently undertaken in the wrong place, a large group rose, quickly filed outside, where they could be seen on mobile phones. Apparently, they were about to take part in yet another expedition to try and find the ship (I never learned whether or not they actually went). On another occasion, I tore into the 'lead poisoning' nonsense, only to have a high-ranking consultant surgeon barking at me that I did not know what I was talking about. When I assured him that several of his colleagues in the medical world did not agree with him, he replied that he could not care less as they were all 'wrong'.

I was delighted, however, to meet Kari Herbert, the daughter of Sir Wally (who had, sadly, died the previous year). I had last met her and her parents at Dan and Jonolyn Weinstein's house when visiting with my wife. She was being escorted by Dr Huw Lewis-Jones, a young polar expert. I was equally delighted to meet Professor Russell Potter of Rhode Island College. A specialist in English and Media studies, Professor Potter had long had a deep and sustained interest in the Franklin story. Unlike other academics, however, who seem to be trained to dismiss any new idea immediately, Professor Potter always gave the impression of carefully thinking about the suggestion – and then dismissing it. It was, nevertheless, a pleasure to meet someone outside the entrenched polar world who would listen to someone who might – just might – have a new view on the story based upon their own, entirely different, experiences.

Having survived the McClintock Winter School, I was invited to present a paper to the Shackleton Autumn School at Athy, also in the

Republic of Ireland. When talking to the organisers I pointed out that I was a severe critic of Shackleton. I could find very little about him that was be positive. Every time I mentioned this difficulty, I was met by a chorus of charming Irish voices telling me that it would be good to hear a different opinion. I decided to present a paper entitled 'The Royal Navy and Polar Exploration' on the grounds that I had just published two volumes under that title, and that Shackleton could be included as he had been made an instant Lieutenant in the Royal Naval Reserve in order to take part in Scott's first Antarctic Expedition.

Unused to academic style conferences, I found that the beauty of presenting a paper lies in the fact that the speaker can sit in the audience like any other attendee until his (or her) time has come. I had not long sat down when the speaker (it may have been Huw) said (I paraphrase): 'What was particularly shocking was that the Admiralty refused to send a ship to the rescue of Shackleton's men until they had a vessel available – and then only if Shackleton did not try to take command.'

The large audience practically simmered with indignation. Nevertheless, I could not let the accusation pass without comment. When questions were allowed, I repeated what the speaker had said (to much nodding of heads and dark mutterings throughout the hall) before saying: 'The Royal Navy had suffered a disastrous defeat at Coronel which left the German squadron under Von Spee free to rampage around the South Atlantic causing untold damage. Which do you think is more important, rescuing twenty-three men from Elephant Island, or hunting down Von Spee?'

Around the room, heads turned slowly towards me like close-range radars locking on to an approaching missile. The speaker, however, after staring at me for a few seconds, said 'Good point, and not one generally made. Next question.'

Matters did not improve the next afternoon when I mounted the podium to speak. After working through Frobisher, Cook, the Ross's, and local heroes such as Crozier and McClintock, I arrived at Scott's *Discovery* expedition – and up came a slide of Shackleton dressed as a Naval Reserve lieutenant. All eyes were upon me as I said 'Frankly, I would not go on a day-trip to Skegness with Shackleton in charge.'

In a very frosty atmosphere, I continued to explain that Shackleton failed in every expedition he led, abandoned his men many times (including Elephant Island) by saying 'Don't worry chaps, I'll go and get help.' Which was, of course, Shackleton-speak for 'If anybody is going to get out – it's going to be me.' He also altered the expedition photographs to make himself look good and confiscated all the expedition journals and diaries with the instruction that no-one was to write an account of the disastrous expedition but himself. He ignored advice not to go into the Weddell Sea, and lost two men in the other part of his expedition who were laying out depots unaware that he had not even landed on the continent (Shackleton still claimed that he had 'never lost a man').

My paper ended with desultory applause. I remember being rather startled when I learned that several members of the Shackleton family – including his grand-daughter were sat in the audience. I did, however, speak to Jonathon Shackleton, a cousin of the explorer and the family historian. To his credit, he did not make a fuss, but asked me a few pertinent questions, all of which I was able to answer without any unintended aggression.

One expert for whom I had great respect was A. G . E. Jones, who had acted as secretary to the great Franklin researcher, R. J. Cyriax. I first met him when at the Royal Geographical Society as a guest of Mrs Diamond she introduced me to him. She had warned me that he would quickly take offence, and to be very careful in the way that I handled him. In fact, I was there to learn from Mr Jones (as I always addressed him) rather than to be bombastic about my opinions. When walking up some stairs, Mr Jones looked up and pointed to a portrait of Shackleton saying 'Now, there's a man!' My success at 'handling' him may be seen in the fact that he supported Mrs Diamond's application on my behalf to have me elected a Fellow of the Royal Geographical Society. I was proud of a considerable correspondence with Mr Jones before his death in 2002 (nevertheless, I still think he was wrong about Shackleton).

In 2008, I received a letter from a Canadian film director who asked me if I would escort him around the Franklin sites of Lincolnshire. Of course, I agreed, and took him to Spilsby, Franklin's birthplace, and to Lincoln Cathedral where Franklin appears (alongside other Lincolnshire explorers) in a stained-glass window. As a result, he asked me if I would help as a consultant on a film he was making based upon a prize-winning book that had been written about John Rae. I knew the book well. I found it, for the most part, yet another attack on the 'self-sufficient donkeys' of the Royal Navy, with the dark background of Rae's involvement with cannibalism and his pathological hatred of the Royal Navy totally ignored. Nevertheless, I jumped at the chance.

A short time later, I received a letter informing me that a room in a central London hotel had been booked for me to stay for several days. I was then to report to Admiralty House on the morning of the first day. Admiralty House, the residence of the First Lord of the Admiralty, was part of the Old Admiralty buildings on Whitehall, and had access to the Naval 'Holy of Holies' – the Admiralty Board Room. Prime Minister John Major took over Admiralty House and Admiralty Arch (the London residence of The First Sea Lord) as flats for politicians and despite Greenwich Hospital being granted for the benefit of seamen by Queen Mary in 1669, he also handed the buildings over to a the newly created Greenwich University. The building contains the Royal Navy's own Franklin Memorial, and the bones of the Franklin Expedition officer, Lieutenant Henry Le Vesconte. Originally, the bones were buried in the floor of the Painted Hall in front of a magnificent memorial. Eventually, the memorial and the remains were transferred to the hospital chapel. Also in the grounds of the hospital there is a memorial in the shape of a large granite obelisk in memory of Lieutenant Joseph René Bellot of the French Navy, who lost his life whilst helping in the search for Franklin. During the time of filming, the occupant of the building was the Deputy Prime Minister, John Prescott – a somewhat different figure to former occupants such as the great John Jervis, Earl St Vincent; William IV as Lord High Admiral; and, of course, to Winston Churchill as First Lord of the Admiralty.

I discovered that one of my fellow consultants was none other than Pia Casarini Wadhams. Her husband Dr Peter Wadhams was somewhere in a submarine under the polar ice.

My immediate target was to to see the Admiralty Board Room once more – I still approached the room with great reverence. We were taken upstairs to the room by one of the film crew. When we entered, inside the room, wearing full dress uniform, were a number of very famous Victorian Admirals. Sir James Clark Ross (David Acton) was there, so was Sir William Parry (Andrew Alston), and Sir Francis Beaufort (Colin George) – the Hydrographer of the Navy, and inventor of the 'Beaufort Scale'. There was also – again in uniform – Surgeon Sir John Richardson (Alistair Findlay), Franklin's great friend (whose grave at Grasmere in the Lake District, I had only recently visited). Of course, they were all actors – but to be in their company whilst actually in the Admiralty Board Room was a very strange experience.

Then came a request for everyone to sit at the table. I was seated between Pia and Sir John Richardson. At the head of the table – beneath a portrait of Nelson wearing his hat 'flat-a-back' to keep it from making contact with the wound he had received at the Battle of the Nile – sat Beaufort; to his left sat Parry and Ross; then Tagak Curley, an Inuit politician; two Orkney dwellers, one – a former museum employee – who had written a chapter about John Rae's life as a child in a book on the explorer, and another who was a local Orkney historian and traditional story teller. To their left, wearing a fringed buckskin jacket, sat the author of the book on which the filming was based. The director, John Walker, sat at the other end of the table beneath a portrait of William IV. On the opposite side (we were privileged to gaze upon the iconic wind-dial, operated by a wind-vane on the roof), the scriptwriter sat on Pia's right. When I managed to obtain a copy of the script, I found it was adequate for a casual glance at the story, but nowhere near sharp enough for a serious documentary. Interestingly enough, the scripted part of the film was labelled as 'fiction' – an odd thing for a documentary.

The director had decided that the film should contain both historical re-enactments and segments of consultant's contributions. Accordingly, hand-held cameras circled the table as, starting with Tagak Curley, we were ladled out buckets of Rae hero-worship. To this tide of idolatry was added a loathing of Charles Dickens (who had challenged the idea of cannibalism being involved), the stupidity of the Royal Navy's explorers in not adopting Inuit clothing, and all naval officers in general. When it was Pia's turn, she resorted to a sort of cautious diplomacy based upon known facts – which earned her little more than a few eyes rolling upward.

I, with Nelson looking down on me, could do no more than open fire with the best broadside I could. I started by saying that Rae was a charlatan who had a venomous hatred of the Royal Navy, worked for an employer who wanted the North-West Passage to fail, and was involved in his own incident of cannibalism. Eyes popped, jaws dropped, and a collective gasp seethed around the table. I decided (remembering the Royal Navy's favourite signal – 'Engage the enemy more closely') to illustrate the Admiralty's view of Rae by repeating their answer to his assertion that they should imitate the Inuit way of life. I read out:

The prime consideration of fox-hunting is not the killing of the fox, but the observance of good form during the pursuit and at the kill. The objective of polar exploration is to explore properly, and not to evade the hazards of the game through the vulgar subterfuge of going native.

To which I added 'And quite right too.' A storm of fury burst out on the other side of the table. The Inuit politician writhed in shock and started shouting, the Orkney delegation grew abusive and threatening, whilst the author of the Rae biography flounced out of the room.

Instead of drawing a line under the matter (after all, everyone had been given a chance to state their opinion) the director gave the politician Tagak Curley a chance to reply to what was little more than a minor fact (but, a fact, nevertheless). I have retained the actual words used. My immediate mental responses as he spoke are inserted in *italics*:

Surely, if I may address the Admiralty and the English public, I think it is shameful, shameful, arrogant, to label people that you don't know as conspirators who murdered and conspired to take advantage of the weak people. (*Many explorers, from Franklin in the early nineteenth century, through to Rae himself, were attacked, or threatened with attack, by the local natives – usually involved with theft.*) I think that is a very strong accusation, and I really, truly, believe you don't have any evidence. (*There is substantial evidence – cut-marks from edged weapons, the traditional mutilation of the dead, the discharge of firearms shown by the number of used percussion caps found on the site, the probability that women and boys were involved in the attack, the number of graves elsewhere pillaged by the Inuit, the death threats to any strangers – including Rae – who intended to visit the area to the west of Boothia, and the rapid spread of the Netsilik tribe after the abandonment of John Ross's ship* Victory *with its wealth of metal. The Netsilik were well known to have the highest murder rate in the Arctic.*) You are relying upon second-hand information, or third or fourth-hand information. (*This is an extraordinary comment. Especially bearing in mind that Rae wrote in his report 'None of the Esquimaux with whom I had communication saw the "white men" either when living or after death, nor had they ever been at the place where the corpses were found, but had their information from natives who had been there, and who had seen the party when travelling over the ice.' Second-hand information?*) You don't – Charles Dickens didn't have, any facts, he had never been up to that part of the area. (*Dickens never claimed to have been in that part of the world. He did, however, obtain his 'facts' from research which pointed out that incidents with a Royal Naval involvement such as Bligh's open boat journey, a similar example with Captain Edwards of HMS* Pandora, *and, of course, Franklin's return up the Coppermine River, never saw anyone resort to cannibalism. The only examples he could find possibly associated with the Royal Navy were the 1834 wreck of HMS* Nautilus, *manned by the Indian Navy, in which, after just four days, the survivors dined on the corpse of one of their shipmates two days before rescue; a Midshipman's dog was eaten on board HMS* Wager – *wrecked off Chile in 1741.*)

After merely responding to a barrage of absurd criticism of the Royal Navy, my involvement with the project rather fizzled out. I eventually had a note from the director that they were unable to use any of my contributions (although my involvement around the table was, in the end, included, mainly so that Tagak Curley's response could be replayed). I never saw the completed film which, apparently, ends in an unrehearsed confrontation between the Inuit politician and Gerald Dickens, an actor and the great-great-grandson of the great writer.

According to a review in *The Scottish Daily Record*, Tagak Curley promptly attacks with 'Your grandfather insulted my people. We have had to live with the pain of this for 150 years. This really harmed my people and is still harming them.'

Astonishingly, despite being named after his grandfather Admiral Sir Gerald Louis Charles Dickens KCVO CB CMG, the author's grandson, Gerald Dickens apparently caved in immediately, and apologised on behalf of the Dickens family.

The documentary's blurb described the incident as 'stunning', and as a 'witness to history in the making'. Professor Potter described it as 'both deeply moving and somewhat absurd'. One anonymous reviewer considered the ending 'a ridiculous and irrelevant conclusion', whilst another thought that the documentary '...lurched slowly to its cringe-worthy conclusion'. I probably would have thought it embarrassing. Another anonymous reviewer thought that my conversation with Tagak Curley was '...a weird combative section'. All in all, and with only having seen parts of the film, I thought it was rather good. I was disappointed, however, that I did not appear in the film's consultancy credits after pointing out that Rae would never have entered the Board Room wearing a hat – the hat, as a result, was removed.

All of which pales into insignificance compared to an account of the events in the Admiralty Board Room in a description of John Walker's film career. The writer was not present at the event. I am introduced as a 'retired naval officer and historian' who 'expresses the Royal Navy perspective as if nothing has changed in 150 years' (so far, so good). I am then heard 'describing with pride the navy as the biggest and best in the world' (an indisputable fact in the mid-nineteenth century). The writer continues that I consider that 'without question Franklin did discover the passage'. (Again, pleased with that.) Then, the great writer is introduced with 'the written record as authored by Dickens and others is a fabrication but that does not stop Coleman from citing Dickens as an authority.' (Actually, I based my comments on the opinion of a man who researched his subject before coming to an utterly reasonable conclusion). He then notes that 'Coleman is steadfast in his refusal of cannibalism.' (That's fine.)

Then, looking for new prey to pursue, the academic picks on an organisation much bigger than me – although one that can match him for political correctness. He notes: 'The British response to *Passage* was to choose not to broadcast it in England on the BBC, though they had the opportunity to do so and even though it is funded with the participation of BBC Scotland.' Apparently, this is a result of an 'absence of voices critical of empire in the mainstream media [which] allows Coleman to speak with imperial defiance'.

The writer of the book also brings a new character into the story – Admiral Sir William Edward Perry (I wonder if he ever met Admiral Sir William Edward Parry?). A clip showing the Admirals at the table is entitled, 'Re-enactment of British *generals* discussing the search for the Northwest Passage.' Far worse, however, is the label attached to another clip: 'Political leader of Nunavut and Inuit MP Emeritus Tagak Curley tells an English *military official* that his insinuation of possible Inuit attacks on Sir John Franklin's men by the Inuit is wrong and ignorant.' To refer to me as a 'military official' – having been 'naval' since the age of sixteen – is a gross affront to me and every soldier who put on the Queen's uniform. Furthermore, I cannot recall ever holding an 'office' – just 'rate' and 'rank'.

I soon recovered, however, after an article in the Canadian magazine *The Tyee*, which, on the 3 October, 2008, told its readers:

> The film's most remarkable scene comes at the end when Ernest Coleman, a retired member of the British navy, confronts Tagak Curley. The war of words is riveting, and makes explicitly clear that the wounds of history are not gone or forgotten but still bleeding brightly, as fresh and painful now as they were hundreds of years ago.

The next news to emerge from the Arctic sent my telephone into melt-down. In 2014, scores of people called me at home to tell me that one of Franklin's ships – probably the *Erebus* – had been found. My delight turned to bewilderment when I learned that the discovery had been made in the wrong place. Something was clearly afoot.

15

Hope On, Hope Ever

('Hope on, hope ever' was the motto on the sledge pennant of Captain F. Leopold McClintock RN.) The Authorised Version of the misfortune of the Franklin Expedition is set down as firmly as the sword encountered by King Arthur was embedded in stone. Few dared to test the widely known challenge of the sword, with the consequence that the robber barons were rampaging through the dark forests with impunity. In a similar manner, what follows is the uncontested, standard edition of the story.

After the disappearance of Sir John Franklin, 128 men, and two ships into the Arctic mists during the summer of 1845, a large number of search expeditions failed to find any clue regarding the fate of the missing expedition. Dr John Rae of the Hudson Bay Company, however, during an HBC-sponsored expedition reached Pelly Bay and met an Inuit who was in possession of a Royal Naval officer's gold lace cap band. More Inuit were met at Repulse Bay and found to have possession of a number of items that could have come from no other source than the Franklin Expedition. From the natives, Rae learned that some years earlier, a large body of men had been seen trying to make their way to the south down King William Island. That these men were from the missing expedition was clearly 'beyond a doubt'. But Rae's uncovering of a corner of the mystery brought light to bear upon an unthinkable and abhorrent shame contained within his report. Franklin's men had apparently 'been driven to the last resource – cannibalism – as a means of prolonging existence'.

Two subsequent expeditions, Charles Hall in 1869, and Lieutenant Frederick Schwatka ten years later, both heard accounts from the natives that underlined Rae's report. The evidence, it seemed, was unchallengeable. Cannibalism had been resorted to by the pride of the Royal Navy, heroes of Victorian England.

Just over a hundred years after Schwatka, a Canadian anthropologist, Dr Owen Beattie, found a right femur on the southern shore of King

William Island. The bone had knife marks. An associated skull had been 'forcibly broken' with the face 'including both jaws and teeth' missing. Now there was 'physical evidence to support Inuit tales of cannibalism among the dying crewmen'.

After a further twelve years, the 'physical evidence' was to receive an impressive boost. Anne Keenleyside, an anthropologist on the staff of McMaster University in Toronto, took part in an expedition to King William Island which resulted in almost 400 bones being made available for scientific analysis. Her researches revealed that 92 of the bones had cut marks, over half being multiple cut marks. Furthermore, about a quarter of the bones had been cut close to the joints ('intentional dismemberment') and three long bones had been fractured ('possibly for the purpose of extracting the marrow').

If the 1845 expedition had ended in cannibalism, it had started out with its fate already cast amongst the ships' stores. Dr Beattie, the same anthropologist who had first found knife marks on bones, disinterred three bodies known to have come from the Franklin Expedition. Tests on the bone and hair of the well preserved remains proved to have very high lead levels. Similar levels had been found in one of the skull bones found by Beattie on King William Island. Not only had the men been suffering from lead poisoning, but the source of such an ailment was readily at hand. The cans used in the preservation of the expedition's meat and soup had been crudely soldered together leaving great lumps of lead-rich solder in contact with the contents.

When Barry Ranford, a Canadian school teacher and amateur historian, addressed the Royal Geographical Society on the subject of his Erebus Bay expedition findings, he announced that studies had revealed that lead found in the bones collected from the site was 'recent'.

Clearly the effects of lead poisoning upon the minds and bodies of the ships' companies had led to fatal decisions leading ultimately to the cannibalism that marred the end of a brave and gallant enterprise. All that could be done was to lower the curtain, dim the lights, and leave the auditorium.

On the other hand, however, if Rae's account of his meetings with the Inuit is to be believed, can any reliability be placed upon the information he gained? Charles Hall, the American Franklin searcher, frequently heard of a large Inuit meeting at Pelly Bay which took place the year before Rae's arrival (and which does not appear in Rae's account). What was the purpose of such a meeting? Was it just coincidence that many of the natives who were to give Franklin searchers information about the missing expedition were present (and, indeed, related to each other)? Could the subsequent efforts to stop searches to the west have originated at this meeting? If so, why?

Rae met a group of natives whilst heading north who in between stealing food from his sledge tried to dissuade him from journeying westwards (towards King William Island). If Rae was not impressed, his interpreter Ooligbuck certainly was, and promptly fled. Rae was forced to chase him for more than four miles before he was brought back. It was later claimed that the natives had warned Ooligbuck that Inuit would kill Rae and the rest of the party if they headed towards to the west. On his return to England, Rae claimed that this and later attempts to stop him from heading westwards came from an Inuit determination to keep him away from caches of food that had been established in that direction.

On arrival at Pelly Bay, the explorer met the native In-nook-poo-zhe-jook. Not only was the native wearing a gold lace cap band (as worn by Royal Naval officers of the period) but he also had a tale to tell about white men. At a point some ten or twelve days distance to the west, about forty white men had died of starvation beyond a large river. However, In-nook-poo-zhe-jook had not been to the place himself and could not point it out on a chart. Neither he, nor any of the other Inuit, despite offers of lavish rewards, would act as guide to the spot, and all continued to try and prevent Rae from travelling to the west. Travel he did – but only to continue his Hudson's Bay Company work. The man famed for his endurance over long distances made no attempt to reach the site suggested by the Inuit despite it being just ten days travel. On his return journey, both at Pelly Bay and Repulse Bay, he not only purchased more artefacts from the missing expedition, but received

> ...further particulars ... which places the fate of a portion, if not all, of the then survivors of Sir John Franklin's long-lost party beyond a doubt – a fate as terrible as the imagination can conceive... From the mutilated state of many of the corpses and the contents of the kettles, it is evident that our wretched countrymen had been driven to the last resource – cannibalism – as a means of prolonging existence... Some of the corpses had been sadly mutilated, and had been stripped by those who had the misery to survive them, and who were found wrapped in two or three suits of clothes.

The information had come from a race of people known for their eagerness to tell their listeners what they believed they wanted to hear, a people amongst whom cannibalism was endemic, and who clearly had something to hide from enquiring white men. Their own experience could have even provided the foundation of their story – Hall was told of a party of natives whose boats were crushed by the ice. The survivors had held on to life by murdering the rest and eating them. It was almost a model in miniature of what the natives were claiming had happened to the Franklin Expedition.

What of Rae himself? A loyal Hudson's Bay Company employee, he could not help but be aware that a breakthrough from the Atlantic to the Pacific across the top of North America would present a serious threat to the Company's monopoly. Tales of cannibalism might just slow down the desire to push westwards through the Arctic.

He was certainly no friend of the Royal Navy. His dislike of the Service probably began when he was selected as second-in-command to Sir John Richardson in the 1848 overland search for Franklin. Rae regarded the 60-year-old naval surgeon as lacking in vigour and overweight ('an excellent appetite and has filled out amazingly'). He considered Richardson's men (seamen and soldiers) as the 'most awkward, lazy, and careless set I ever had anything to do with'. Despite being one of Franklin's oldest friends, Richardson left Rae to continue the search and returned home, something he would have been loath to do had his companion shown the team spirit and loyalty expected among genuine searchers.

Matters hardly improved the following year when Rae, now in command of Fort Simpson, a Hudson Bay Company post at the junction of the Mackenzie and Liard Rivers, found himself hosting a party of Naval officers and seamen from the Franklin search ship HMS *Plover*. During their stay, Rae learned that one of the officers, Acting Mate William Hooper, had been making a record of his journey down the Mackenzie – a record which he intended to publish on his return to England. Among the notes was a detailed reference to recent 'misfortunes at Pelly Bank' when three of the Company's employees under Rae's command lost their provisions in a fire, and resorted to cannibalism – a detail Rae intended to keep from his Company superiors. The account was eventually published, and Rae allowed the fear of its publication to grow to such an extent that, eaten up by suppressed rage, fate offered him a dark weapon to use against the Royal Navy.

Furthermore, a subsequent naval Franklin search prevented Rae from his planned expedition up the Mackenzie River and consequently delayed his much-sought-after attempt to win the Founder's Gold Medal of the Royal Geographical Society (he was not, however, averse to using his naval contacts to eventually gain the award, naming the two most important sites of his 1851 expedition after the Society's Vice-President, Captain Sir George Back RN). He even took the trouble to sneer at the Royal Navy's habit of taking luncheon whilst out sledging, ignoring the fact that naval sledges were mainly hauled by men rather than, as in Rae's case, pulled by dogs. He ridiculed the seamen's concern about wolves that gathered around their ice-bound ships and called them 'bunglers' in their attempts at hunting. Royal Naval officers were referred to as 'self-sufficient donkeys' (sic).

In addition, Rae took a possibly more lethal swipe at the Royal Navy when, starting out in April 1851, he arrived off a large, protected harbour (later Cambridge Bay) and travelled along the south and eastern shore of Victoria Island. As he made his way along the coast, he found part of a pine flagstaff studied with copper tacks bearing the 'Broad Arrow' mark of British Government property. Nearby, he also found part of an oak ship's stanchion. His response was to make his way to England after a casual journey across the northern USA, visiting Chicago, Detroit, and New York. He arrived in England in April 1852 with a report on his finds dated 27 September 1851. The site of his finds was to the south of the position where Franklin's ships had been deserted, and on the other side of Victoria Strait from where Franklin's men had made their way down the west coast of King William Island.

However, Rae's reluctance to aid the search for Franklin was nothing compared to his outrage at the Royal Navy's refusal to accept his 1847 survey of Pelly Bay. When the subsequent Admiralty chart was published in 1850, the area was bounded by a dotted line, suggesting that Pelly Bay still remained uncharted. On 2 November 1850, he wrote to George Simpson of the Hudson's Bay Company:

> As I cannot well swallow quietly the injustice that appears to have been done me at the Hydrographic Office in leaving part of my discoveries in 1847 a blank on the charts, I take the liberty of inclosing another letter on the subject, which I would wish to be sent to the Editor of one of the leading London periodicals (say the *Spectator* or the *Athenaeum*) if you think after perusing it, that there would be any use in doing so, for little confidence can be placed in my own judgement in such matters.

Matters were to grow even worse. The Hydrographer (Captain Beaufort RN), Captain W. E. Parry RN, and Sir John Richardson agreed with each other in 1852 that Rae's positioning of his 'Fort Hope' – his base at Repulse Bay – was incorrect. Beaufort was Chairman of the Arctic Council, and responsible for the Franklin search expeditions; Parry and Richardson were amongst Franklin's closest friends.

As a final shot after Rae's return with the 'cannibalism' news, Beaufort issued a chart awarding the primary survey of the east coast of Victoria Island to Captain Collinson. Rae 'very mildly and humbly stated (his) opinion of the injustice of the act' to Beaufort's successor Captain Washington who, in turn, treated him with 'scorn and contempt'.

At this, Rae resorted to browbeating Washington 'in a manner that showed I meant it'. The new Hydrographer of the Navy, weary of the sound of Rae's voice and with more important things to do, agreed to 'patch' over the references on the chart to Collinson. Nevertheless, the Royal Navy quietly responded when Captain F. L. McClintock (the naval officer who *actually* discovered the fate of Franklin) noted that 'Mathison Island' recorded by Rae during his later survey of West Boothia, was, in reality, 'a flat-topped hill' on the south-east corner of King William Island.

Then fate took a hand. Having gained permission to travel to Repulse Bay to explore an un-surveyed part of the west coast of Boothia Felix (now Boothia Peninsula, close to where James Ross had discovered the Magnetic North Pole) as far as Bellot Strait, Rae set off with an entirely different aim in mind. To reach his supposed target area, he had to pass by Pelly Bay where he had carried out the survey that was much criticised by the Admiralty. On that occasion, with his chronometer broken, Rae had completed his survey by 'dead-reckoning' and was determined to demonstrate that his readings were accurate enough to be included on an Admiralty Chart. As mentioned earlier, he arrived at Pelly Bay on the 18 June 1854, with Ooligbuck as interpreter. The meeting with local Inuit promptly turned confrontational when he told them that he intended to head westward (to reach the western shore of the Boothia Peninsula – opposite the eastern shore of King William Island. It was then that Ooligbuck promptly fled – only to be forced back by Rae.

Rae then, when he met In-nook-poo-zhe-jook, purchased a gold lace cap band as worn by Royal Naval officers. The man

> ...had never met whites before but said that a number of Kabloonas, at least thirty-five or forty, had starved to death west of a large river a long distance off. Perhaps about ten or twelve days journey. Could not tell the distance, never had been there, and could not accompany us so far. Dead bodies seen beyond two large rivers; did not know the place, could not or would not explain it on a chart.

Despite this extraordinarily valuable information, Rae continued towards Castor and Pollux River, thus avoiding the opportunity to visit Chantry Bay into which Back's Great Fish River flowed, and the most obvious objective for men trying to make their way to the south. Reaching Shepherd's Bay, and heading a short distance north up the west coast of Boothia Felix, Rae found that he could see the eastern shore of King William Island with the waters between the island and his position without any isthmus or island obstacles to the passage of vessels. It seemed to him, therefore, that he had closed the last gap in the North-West Passage. With a hefty amount of prize money waiting for the first man to complete the discovery of the North-West Passage, Rae promptly decided to return to Repulse Bay – again bypassing Chantry Inlet. On his arrival, he found yet more Inuit eager to trade watch cases, engraved silverware, buttons, coins, and – most unsettling of all – the neck star of the Guelphic Order of Hanover that had been worn by Franklin. The hoard, combined with the information from In-nook-poo-zhe-jook, clearly indicated that the disaster – of whatever cause – happened within easy marching distance, especially for someone like Rae, famed for his long-distance marches. However, instead of setting out immediately, he lingered at Repulse Bay for *two months* claiming to record Inuit accounts of cannibalism amongst Franklin's men. It would not have been difficult, with a single interpreter (who was clearly not keen to return to the northwest to search for anybody), to both interpret the information in any way that he wished, and to construct a story that would have found favour with his hosts and his employers – a story that would reflect badly upon the Royal Navy, and firmly deflect any other suspicion as to the cause of the deaths.

For Rae himself, several gems had dropped into his lap. He had clear evidence from the material objects he had received from the Inuit that Franklin's expedition had come to a chaotic end. The only account of the expedition's fate had ended with cannibalism, and he had almost connected up two navigable strands of a North-West Passage – with its prize of £10,000 for the first person to do so. However, there was a problem with his finally linking the North-West Passage. Beyond the point where Rae turned back, lay Cape Felix on King William Island. The distance between that point and Bellot Strait contained over 150 miles of uncharted coast – three times longer than the stretch being claimed by Rae. Even worse, the prize of £10,000 had already gone to Captain Robert McClure who was just about to arrive in England having passed through a North-West Passage in the same year.

It was also to become clear that in their attempt to reach safety to the south, some of Franklin's men crossed Simpson's Strait to land on the shore of North America. A boat and skeletons were found at 'Starvation Cove' and were probably an attempt to proceed to the east and Chantrey Inlet where the mouth of Back's Great Fish River could be reached. That being the case, it would be clear evidence that the Franklin Expedition *had* discovered the North-West Passage, and that the prize money should have been distributed amongst the families of all those who took part.

Could the story of cannibalism be substantiated on the suggestion of a local native? Rae, it turns out, had at hand a ready-made story of horror to lend support to the Inuit tales. During the winter of 1846 – the same winter that locked Franklin's ships into the ice off King William Island – a wagon train bound for California had found itself stranded in the Rocky Mountains. When the survivors were rescued, it was found that they had resorted to cannibalism in order to survive. The facts shook the American people, and the media spread the story throughout North America and Europe. Dr John Rae could not fail to have been aware of this tragedy and its effect upon mid-nineteenth century society and may well have decided to attach a similar stigma to the Franklin Expedition. Rae also claimed to be unaware that a reward of £10,000 had been offered by the Admiralty for information regarding the fate of the missing expedition, despite rewards having been in existence since 1847, the year before he travelled north with Sir John Richardson. The particular reward of £10,000 had been proclaimed on 7 March 1850, four years before his arrival at Pelly Bay.

Passing by the mouth of Back's Great Fish River, Rae reached to within forty-five miles (two days march) of the last known position of the Franklin party before beginning his return to England, picking up further Franklin artefacts from the natives on the way. In a letter to *The Times* to explain away this extraordinary behaviour, Rae claimed that he had failed to continue his journey westwards to reach the scene of the disaster (and possible survivors) in order 'to prevent the risk of more valuable lives being sacrificed in a useless search'. When, no doubt, the weakness of this argument was brought home to him, he wrote to the antiquarian, Sir Henry Dryden of Canons Ashby, in February 1856:

> The information obtained on my outward journey was not sufficiently clear to enable me to fix with any degree of certainty on the position at which the party were starved, and to have travelled westwards without sufficient knowledge on this point, when the land was covered a foot and a half or more deep with snow would, as I then thought, and still think, have been useless.

This from a man who claimed to have travelled 6,555 miles on foot through the Arctic, 1,100 of them that very spring.

As for his reason why the natives tried to stop him from going westwards from Pelly Bay (cached food) the same letter revealed yet another reassessment. He now believed that the mouth of the Great Fish River (the region to which the Inuit tales pointed) 'is one of the most unfavourable positions for procuring food, particularly during the spring. The Esquimaux never remain there during that season.'

On his death in 1893, *The Times* obituary noted that he was both 'sensitive' and 'intolerant'.

His memorial in Kirkwall's St Magnus Cathedral declares that he was 'The Intrepid Discoverer of the Fate of Sir John Franklin's Last Expedition' – even though he could have had no idea of the 'fate' of Franklin and his men, having never visited the site of the disaster, and depended, at best, upon an interpretation of information from Inuit who also had never been to the site. A statue of Rae was erected in 2013 at the Pierhead in Stromness complete with an inscription describing him as 'the discoverer of the final link in the first navigable Northwest Passage'.

Such an accolade would have come as a complete surprise to Rae. He never claimed to have made such a discovery. He did not attend, nor put forward any counter-claim to the June 1855 Parliamentary Select Committee called to judge McClure's claim. He did, however, write a letter to *The Times* which appeared on 25 July 1855 in which he stated '...there was a water communication between Victoria Strait and Peel Sound to the northward.' Such a comment goes a long way to supporting the deceased Franklin's claim to the award.

The tragedy of both historical and current activities intended to denigrate Franklin and his men in order to elevate Dr John Rae is actually offensive to both. Rae's story of exploration, his work amongst the Inuit, his undoubted endurance, initiative and enterprise, are all worthy of as many memorials and statues as his admirers wish to erect. His personal aversion to the Royal Navy, based on misunderstanding, his support for his employers, and a distrust of social structures and hierarchies, instead of being pointlessly inflated should be dismissed and forgotten. With that done, Rae's achievements could be seen in their true light. Nevertheless, the placing of a small stone commemorative slab beneath the Franklin memorial in Westminster Abbey (a memorial which also contains a tribute to McClintock and Jane, Lady Franklin) seems hopelessly inept in view of his dogmatic and recalcitrant attitude towards the Royal Navy – it also makes no mention of his 'discovering' the North-West Passage.

When the news reached England, a predictable outrage swept the land. Charles Dickens took to his pen and scoured the history of the Royal Navy for examples of cannibalism. Despite great trials such as Lieutenant Bligh's open boat journey, and a similar experience by Captain Edwards of HMS *Pandora*, Dickens could list but a single example. Franklin's men could have been depended upon not to have indulged in such a horrible practice through their 'own firmness, in their fortitude, their lofty sense of duty, their courage, and their religion'. It should also be noted that Dickens never attached any blame to Rae. Instead, he wrote:

> Before proceeding to the discussion, we will premise that we find no fault with Dr. Rae, and that we thoroughly acquit him of any trace of blame.

He has himself openly explained, that his duty demanded that he should make a faithful report, to the Hudson's Bay Company or the Admiralty, of every circumstance stated to him; that he did so, as he was bound to do, without any reservation; and that his report was made public by Admiralty: not by him. It is quite clear if it were an ill-considered proceeding to disseminate this painful idea on the worst evidence, Dr. Rae is not responsible for it… With these remarks we can release Dr. Rae from this inquiry, proud of him as an Englishman, and happy in his safe return home to well-earned rest.

It is of course possible that Dickens' description of Rae as an 'Englishman' is the real cause behind the dislike of his detractors, rather than the reaction to his news of cannibalism.

As for Dickens' remarks about the Inuit that triggered a somewhat synthetic outrage amongst their twenty-first-century descendants such as Tagak Curley, it must be remembered that the great writer was expressing mid-nineteenth century anger at a horrific suggestion layered upon a national tragedy. The mourning of an incident regarded by many as the equivalent of losses in battle, and had seen special church services throughout the land for payers of salvation, had been gravely disfigured by an accusation in which no-one could believe – and Dickens had expressed their feelings of resentment and offence. The researcher, Huw Lewis-Jones, noted that some schools held Arctic poetry competitions and newspapers published their own examples expressing national remorse. One example in the *New Monthly* by Nicholas Michell remembered the lost:

Sleep! Martyrs of discovery, sleep!
Your winding-sheets the Polar snows;
What though the cold winds o'er ye sweep,
And on your graves no flowret blows,
Your memories long shall flourish fair,
Your story to the world proclaim
What dauntless British hearts can dare;
Sleep! Lost ones, sleep! Embalmed in fame.

Many decades later, R. J. Cyriax wrote:

It is difficult to realise at the present day, more than ninety years after Sir John Franklin sailed on his last voyage, what a painful sensation was caused by his disappearance. His rescue became a national question of the first importance, and carried the good wishes of all.

In the meantime, the *Sun* thundered:

> The more we reflect upon the 'fate of the Franklin Expedition', the
> less we are inclined to believe that this noble band of adventurers
> resorted to cannibalism. No – they never resorted to such horrors...
> Cannibalism! – the gallant Sir John Franklin a cannibal – such men as
> Crozier, Fitzjames, Stanley, Goodsire, Cannibals!

Sir Roderick Murchison, the President of the Royal Geographical Society,
merely skated over the awful accusation by saying that Rae had only
stated that he had been given an account '...by the Esquimaux of a large
party of Englishmen having been seen struggling with difficulties on the
ice near the mouth of the Back or Great Fish River'.

McClintock restricted himself to 'The information obtained by Dr. Rae
was mainly derived second-hand from the Fish River Esquimaux and
should not be confounded with that received by us from the King William
Island Esquimaux.'

Lady Franklin, with the most genteel of disapproval, urged upon
the Government '...the necessity of following up, in a more effectual
manner, the traces accidently found by Dr. Rae, and, in fact, of rendering
the search complete by one more effort, involving but little hazard or
expense'. She also regarded Rae as 'hairy and disagreeable'.

The Admiralty, concerned about the effect of the cannibalism charges
on the relatives of the dead, and on the public in general, had allowed its
publication without comment. They nevertheless requested the Hudson's
Bay Company to send an expedition to find out 'whether the report made
to Dr Rae be true'.

The response by the Company was, at best, tardy ('true to the instincts
of monopoly'). Despite the Admiralty's promise of 'freely and liberally
supplied' public money, the best the Company could do was to send
'a paltry expedition' under Chief Factor James Anderson with a small party
down Back's Great Fish River in 'three birch-bark canoes of an inferior
description'. None of the party had any knowledge of the area to which
they were heading, no map was carried, no interpreter was taken along to
communicate with the Inuit, and with no-one 'capable of taking a common
astronomical observation'. More relics were found, but no bodies, bones,
or graves in the nine days spent in the region of Montreal Island (at the
mouth of the river). So cursory was the search that a cache left on the
island by the 1834 Back Expedition was neither looked for nor accidentally
found, despite being an obvious place for a message to have been deposited.

On the additional evidence of the Anderson Expedition, and against the
advice of senior naval officers, it was decided that Rae should be awarded

the £10,000 prize for succeeding 'in ascertaining the fate' of the Franklin Expedition. The outrage now rose to even higher proportions. One writer claimed that all Rae had done was to 'ascertain the fate of sundry spoons, forks, and other relics'. Great play was made of Rae's turning his back when no more than two days march from the scene of the tragedy. His own account was used against him when he revealed that, 'We had still on hand half of our three months' stock of pemmican and a sufficiency of ammunition to provide for the wants of another winter.' His information about cannibalism was pointed out as being 'thrice-diluted' via the Inuit who were 'notoriously addicted to falsehood and deception'.

The Anderson Expedition had, it was felt, done nothing to support the claim of Dr Rae to the prize:

> ...admitting that Franklin and his party perished by starvation, – for we entirely and indignantly disclaim the idea of cannibalism, – we still think that had this occurred on Montreal Island, traces of the bodies would have been found, as well of relics from the ships ... we hardly suppose that Dr Rae, with all his desire to fasten cannibalism on his countrymen, will contend that they consumed the bones as well as the flesh of their emaciated comrades.

Worse still, was the possibility that some of Franklin's men might have been alive as it was widely believed that, 'all Arctic explorers tell us that, where aborigines live, Englishmen can also exist.' This likelihood received even more attention when one of Rae's party, Thomas Misteagun, disclosed in confession to his priest that 'perhaps one or two of the men may still be alive.'

Whilst Rae's contemporaries – men who had served with and knew the Franklin Expedition – treated his conclusions with extreme doubt bordering on contempt, has any new evidence come to light to support his report? Possibly.

An American, Charles Francis Hall, went in search of men from the Franklin Expedition. He believed that survivors would be found living amongst the natives despite the fact that the McClintock Expedition of 1859 had found proof that the ships had been deserted in 1848. Hall began his search in 1860, but it was to be a slow affair, another nine years before he stepped on to King William Island. In the meantime he had learned to live like a native and had become proficient in their language. He was still, nevertheless, stunned to find that after all his efforts to reach his goal, his guides would spend no more than one week on the island. Whilst crossing Simpson's Strait, the party visited Todd Island and found a human thigh bone, which, Hall's guides assured him, was all that remained of five white men whose bodies had been fed to the Inuit's dogs. Once on King William Island, Hall was led to a place where he had been told that two white men had been buried. Hall recorded the information:

> The bodies buried by placing stones around & over them – the remains facing upwards & the hands had been folded in a very precise manner across the breasts of both. Clothes all on & flesh on all the bones. On the back of each a suspended knife found. The bodies perfect when found but the Inuits having left the remains unburied after unearthing them, the foxes have eaten most sinews off all of the bones.

His reward for his determination was to find a complete skeleton, later believed to be that of Lieutenant Le Vesconte of HMS *Erebus* (the bones now interred behind the altar of the chapel at Greenwich Royal Naval College). But he had unearthed more than just bones.

Hall's guide turned out to be none other than In-nook-poo-zhe-jook – the same native who had met Rae fifteen years earlier and (according to Rae) tried to prevent him from heading westwards. Now, having visited a site on Erebus Bay already discovered by McClintock, he was able to tell Hall that he had seen a 'big pile of skeleton bones near the fire place & skulls among them … broken up for the marrow in them … close to the cooking place … long boots (which) came up high as the knees & and that in some was cooked human flesh – that is human flesh that had been boiled.'

Hall also learned from an elderly native woman that 'one man's body when found by the Inuits flesh all on & not mutilated except the hands sawn off at the wrists – the rest a great many had their flesh cut off as if someone or other had cut it off to eat.' A second elderly woman told him of a tent being filled with 'frozen corpses – some entire and others mutilated by some of the starving companions, who had cut off much of the flesh with their knives and hatchets and eaten it.'

In-nook-poo-zhe-jook's story clearly ranged from the mundane to the fantastic. It would not be at all unlikely that dying men would gather around the 'fire place', if only for warmth. Later evidence would show that bones can be broken for reasons other than to obtain the marrow and, as for the boots full of cooked flesh, why had it survived? Having expended valuable fuel in boiling it, it is reasonable to assume that cannibals would have taken the trouble to eat it. Why had animals failed to eat it? At least the story had the claim of being first-hand – the two women's stories were both hearsay, neither had seen the sights they were describing.

Ten years after Hall, Lieutenant Frederick Schwatka of the US 7th Cavalry led a successful expedition during which he discovered the last known position believed to have been reached by Franklin's men, Starvation Cove on the northern shore of the North American mainland (thus completing the North-West Passage). He also interviewed an Inuit woman and her son who had been to the site. Colonel Gilder, one of the Schwatka party, recorded that the son claimed to have seen 'bones from legs and arms that appeared to have been sawed off... His reason for thinking that they had been eating each other was because the bones were cut with a knife or saw.'

One of Hall's elderly witnesses had talked of 'the hands sawn off at the wrists', another – who had actually visited the site she was describing – mentioned 'one skeleton that had been sawed.' Now Schwatka's witness revealed that 'bones were cut with a knife or saw.' Under such circumstances sawn bones are less likely to have been evidence of butchery than of surgery. It is not at all uncommon in cases of severe frostbite for hands and feet – and even whole limbs – to be amputated in an effort to save the victim (a well-known example is Sergeant Elison of Lieutenant Greely's 1883-84 attempt on the North Pole – 'both feet and the greater part of both hands are now gone.') The idea of surgically removing a colleague's hands in order to eat them makes no culinary or sustenance sense as Dr Keenleyside has noted in her attempt to confirm the tales of cannibalism, 'In documented cases of cannibalism ... hands, and feet are consumed last.'

So the total 'evidence' in favour of cannibalism from the three decades after the event is restricted to third-hand statements by unfriendly natives reported by a man personally hostile to the Royal Navy in general. The single first-hand account comes from one of those original unfriendly natives and resorts to fabrication. The remainder, both first-hand and hearsay, support a previously ignored view that the men had received medical care.

What then of the twentieth century? The first item to return the spotlight onto cannibalism was the right thighbone found at Booth Point by Dr Beattie in 1981. When studied, the bone revealed 'three roughly parallel grooves measuring 0.5-1 millimetre in width and up to 13 millimetres in length'. These grooves, it was felt, 'were most likely knife marks'. A sketch accompanying the account revealed that the bone had the 'top end chewed off by animals'. The fact that the marks were in exactly the position an animal could have gripped the bone in his claws whilst chewing on the end is ignored. The claws of polar bears, foxes, or wolves could have left the marks. Also ignored was the Inuit custom of breaking the bones of someone they had killed. In addition, someone hacking at their victim is likely to leave the tell-tale parallel marks.

Oddly enough, of the almost 400 bones examined by Dr Keenleyside, not a single one is reported as being marked in any way by animals. This is an extremely strange result from an event that would normally have attracted a wide variety of wildlife to feed on the remains. Even one of Hall's first-hand accounts had the natives believing that the bodies were disturbed 'as if done by foxes or wolves'. It suggests that 'proof' of cannibalism was a prime objective of the study – even Dr Beattie's 'most likely knife marks' have been elevated by Dr Keenleyside to the definite 'marks made by a knife used to remove tissue from the bone' (despite the fact that butchery, whether professional, or simply carving the Christmas turkey, generally fails to produce incidental, or deeply cut, marks on bones).

So if it was not animals that caused the marks on the bones, is there any other way in which such marks could have been made apart from cannibalism? The answer is not merely an emphatic 'yes', but also with conjecture that has much sounder footing than the massive percentage of guesswork that provides the foundations for the cannibalism theory.

Cut marks – particularly those that are made at right angles to the length of the bone – are made almost exclusively by edged weapons used in attack. The single, or even repeated, hacking in a manner likely to produce such marks does not remove flesh. The many marks on finger bones found by Dr Keenleyside, rather than suggesting a somewhat fussy cannibal consuming the hand with a knife, suggest the horror of someone attempting to defend himself against a knife attack. But what of the 'cuts in the vicinity of the joints, a pattern indicating intentional dismemberment'?

Throughout the natives of North America, including the Inuit tribes, it was the custom to mutilate the bodies of fallen enemies. There are many recorded examples. On reaching a series of falls near the mouth of the Coppermine, Samuel Hearn became involved in an attack by Indians on a small group of Inuit. A young woman ran to him for aid:

When the first spear was stuck into her side she fell down at my feet, and twisted round my legs, so that it was with difficulty that I could disengage myself from her dying grasps. As two Indian men pursued this unfortunate victim, I solicited very hard for her life; but the murderers made no reply till they had stuck both their spears through her body, and transfixed her to the ground. They then looked me sternly in the face, and began to ridicule me, by asking if I wanted an Esquimaux wife; and paid not the smallest regard to the shrieks and agony of the poor wretch who was twisting around their spears like an eel! Indeed, after receiving much abusive language from them on the occasion,

I was at length obliged to desire that they would be more expeditious in dispatching their victim out of her misery, otherwise I should be obliged, out of pity, to assist in the friendly office of putting an end the existence of a fellow-creature who was so cruelly wounded. On this request being made, one of the Indians hastily drew his spear from the first place it was lodged, and pierced it through her breast near her heart... An old man ... fell a sacrifice to their fury; I verily believe not less than twenty had a hand in his death, as his whole body was like a cullender ... an old woman (who) in vain attempted to fly ... my crew transfixed her to the ground in a few seconds, and butchered her in the most savage manner. There was scarcely a man among them who had not thrust at her with his spear.

It is extremely unlikely that the bones of those victims would escape 'cut marks'.

The following notes by Dr William Bell on the mutilated body of Sergeant Frederick Williams of the US 7th Cavalry are equally graphic:

The muscles of the right arm, hacked to the bone, speak of the Cheyenne. The nose slit denotes the Arapahoe's and the throat cut bears witness that the Sioux were also present. I have not yet discovered what tribe was indicated by the incisions down the thighs and the lacerations in the calves of the legs in *oblique parallel gashes*. Warriors from several tribes purposely left one arrow each in the dead man's body. (Author's italics)

The body was also mutilated by being slit up the stomach and had at least three long slashes across the chest that must have penetrated to the bone. The right arm appeared to have been severed at the elbow (ie, 'in the vicinity of a joint').

The broken bones mentioned by In-nook-poo-zhe-jook ('broken for the marrow in them') and Dr Keenleyside ('fractured, possibly for the purpose of extracting the marrow') could easily, and probably certainly, have resulted from attack or mutilation, as could the missing face of Dr Beattie's Booth Point skull.

Furthermore, is there any actual evidence of cannibalism – particularly amongst the 'cut marks' found on the bones? Such indicators should be compared to a study that examined the skeletons remaining after a medieval battle. Such a study of bones bearing skeletal trauma covered sites at Wisby and Uppsala in Sweden; Fishergate at York, and Towton; and the Church of St Mary, Oslo, Norway. In all, over a thousand skeletons bearing cut marks were examined, and with the exception of cut marks on the hands (such injuries occur on weak, or defenceless men without weapons to defend themselves), all turned out to be identical to the bones found at Erebus Bay. Nobody, however, is demanding that the battle-caused cut marks be accepted as evidence of cannibalism (exactly as no-one is claiming the that cut marks on the skeleton of King Richard III are evidence of flesh-eating car park attendants).

Nevertheless, is there reason to believe that the natives of King William Island and the surrounding area would have carried out such an attack?

Hall had heard from one of his elderly lady informants 'very reservedly – in a way of letting me know a matter that is a very great secret among the Inuits'. Two Shamans, or medicine men, had cast spells so that

> ...no animal, no game whatsoever would go near the locality of the two ships, which were in the ice near Neitchille many years ago. The Inuits wished to live near that place (but) could not kill anything for their food. They really believed that the presence of the Kabloonas (white men) in that part of the country was the cause of all their troubles.

However, as Neitchille was a name given by nineteenth-century explorers to the Boothia Peninsula, the story probably refers to the John Ross Expedition of 1829-33 with his two ships the *Victory* and the *Krusenstern*. If that is the case, it demonstrates an early ill-feeling towards 'Kabloonas' that could have grown from a perceived threat to the precarious resources of the natives, into an outright attack to defend those resources.

The main tribe of the area was the Netsilingmuit, or Netsilik. They were considered by neighbouring tribes to be barbaric and savage to the extent that they had driven the once-large Utkuhigjalingmuit tribes from the lower Back River, the Ilivermuit from the Adelaide Peninsula, the Avertôrmuit from the Bellot Strait, the Arviligjuarmuit from Pelly Bay, the Qeqertarmuit from King William Island, and reduced their numbers to little more than a few families. Their access to the wood and metal from the ships abandoned by Sir John Ross had done away with their need to trade with the other tribes and provided them with a ready supply of superior weaponry. (One of the Orkney delegation who was at the meeting in the Admiralty Board Room in 2008 had, two years earlier, told *The Scotsman* newspaper that the 'natives in these remote areas only had bone tools – no saws or knives.') Knut Rasmussen, after studying the Netsilik, found that tribes who were linked by intermarriage with the Netsilik preferred to 'isolate themselves from them' for fear of their violence. Rae had been informed in 1846 that the Netsilik had a 'very bad character', and his guide refused to hunt for seal at night out of fear of them. Hall was warned about the aggression of the Netsilik and met a native family who were fleeing to

Repulse Bay to take refuge with relatives with whom they had a serious disagreement, and from whom they could expect violence, rather than stay within range of the Netsilik. In William Gilder's account of the Schwatka Expedition, he describes the fear of meeting the Netsilik amongst the Inuit of his party. On 15 May 1879, as they approached a group of igloos, 'The Inuits of our party, especially Ishnark and Joe, were very much frightened, and said the people we were about to meet were as warlike as the Netchilliks, and always wanted to fight when they met strangers.'

As they approached, the women were founding hiding behind sledges, whilst the men stayed inside the snow huts.

> It seems that when they first saw us they thought we were Netchilliks, and were in consequence very much frightened, so that while some of our people were dreading an encounter, these poor creatures were shaking in their shoes and afraid to come out of their igloos.

It was Netsilik natives who claimed to have met Franklin's men as they retreated down King William Island.

When Dr Keenleyside first examined the bones she caused a considerable stir by revealing that some of the remains belonged to at least one individual aged between 12 and 15. This news was declared to be evidence that at least one of Franklin's 'cabin boys' (a non-Royal Navy term) had to have been 9 years old when he left England. The problem with this idea is that the youngest men with Franklin were three 'Boys 1st Class' all aged 18 years when the ships sailed. No-one seems to have been prepared to consider the idea that an Inuit boy of 12 to 15 years would have been regarded as old enough to have taken his part in a warlike band, and that the bones were those of a young Netsilik killed whilst attacking Franklin's men.

The original, unsubstantial, claim that that there had been 'cabin boys', aged about 15 or 16 at the time of their deaths at Erebus Bay highlights a classic example of misinformation becoming lodged in the minds of some people who prove reluctant to accept a change of information. Keenleyside mentions the 'cabin boys' in her paper for the *Arctic* journal (Vol. 50, No. 1 (March 1997) pp 36–46), but admits to her error by adding that an examination of the sources 'revealed that the three youngest members on the expedition, all of them cabin boys, were 18 years of age at the time the expedition set sail'.

Despite this admission, an internationally respected forensic artist published an article in *Above and Beyond – Canada's Arctic Journal*, explaining how she had created an image of one of the 'approximately

15 to 16 years of age' cabin boys on the Franklin Expedition, over twenty-one years after Keenleyside had pointed out the inaccuracy.

In 2017, another wholly unexpected discovery was made. DNA tests were carried out on bones already held by the authorities, and on others recently retrieved from King William Island. The DNA was pronounced as 'good quality' by the experts, and soon the bones were declared as having belonged to men from Western Europe. But a shock was waiting. Some of the results indicated a lack of 'Y' chromosomes. This, under normal circumstances, could only mean that some of the bones belonged to one or more females. At first, there was an attempt to make a case for females being part of the ship's companies. When it was realised that such a suggestion was absurd, there was an immediate back-pedalling and the once 'good quality' DNA was instantly declared 'degraded' (and, therefore, unreliable). In fact, the scramble to avoid concluding the obvious was wholly unnecessary. There is no reason that Inuit women could not have been among the attackers – probably in the scramble to plunder the boats and bodies (McClintock described Inuit women as 'arrant thieves', but who were, nevertheless, 'good-humoured and friendly').

During a 2019 archaeological excavation in Agaligmiut (Nunalleq), Alaska, a seventeenth-century grave was discovered containing the bodies of Yup'ik natives (closely related to Inuits). Men, women, and children had been massacred. The story of a local war between aboriginal villages had been passed down through the centuries – a story that also included the suggestion that the Yup'ik women had dressed in men's clothing and taken part in the fighting. Could, therefore, the original DNA findings at Erebus Bay really have been of 'good quality', and the original findings have been correct?

On the other hand, can it be seriously suggested that natives using primitive weapons could have defeated Europeans equipped with the latest arms? The obvious answer is 'yes' – as was demonstrated by the Zulus at Isandlwana in 1879. But the natives must have an advantage. At Isandhlwana it was sheer weight of numbers combined with their enemy's military incompetence – on King William Island it was probably scurvy.

The disease haunted almost every Arctic expedition. It was scurvy that had driven Parry home in 1823 and would bring the Nares 1876 Expedition home a year early. Several cures and preventatives offered hope, but none gave a clue to the cause, the lack of vitamin C. Fresh fruit and vegetables were the simple answer, but little was available to the expeditions. Fresh meat would have helped but although the naval expeditions killed birds by the thousand and generally shot at anything that moved, the subsequent cooking of the meat destroyed the vitamin C content. As the Admiralty Committee on Scurvy reported in 1877, the sufferer was left in no doubt of his ailment:

> The colour of the face changes, the skin grows sallow and assumes a leaden hue, and the countenance may afterwards become bloated, and the eye assume a heavy expression. A general debility prevails, and an apathy of manner is noticed; there is a feebleness of the knees and ankles, and pains – resembling the flying pains of rheumatism – attack various parts of the body. Swelling of the joints, with rigidity, accompany these symptoms. This rigidity is especially observed in the hams, for which site a predilection seems to exist in the case of men engaged on walking exercise... The low spirits become confirmed, and the unfortunate patient indulges in the gloomiest of ideas; the fetor of the breath is now intolerable; the gums protrude as spongy masses from the mouth; the teeth become loose in the sockets, and frequently fall out ... every slight scratch degenerates into an ulcer, old scars break out afresh, and haemorrhages are now frequent from different parts of the body ... anxiety and despondency gives way to apathy and indifference. The breathlessness ... is frequently attended with faintings, especially on any exertion, and is sometimes accompanied with sanguineous effusion into the substance of the lung, and into the pleurae and other cavities. Death occurs suddenly in many instances.

In addition, the effects of frostbite became increasingly pronounced, leading to the onset of gangrene and the need for amputation.

The Franklin Expedition had been away from England for almost three years when the ships were abandoned. The 'preserved provisions' they had depended upon had proved to be a failure and the lemon juice carried would have lost its anti-scorbutic qualities. In May 1847, Lieutenant Graham Gore could write 'All Well' on a note deposited in a cairn. Almost a year later Captain James Fitzjames used the same note to tell the world that nine officers and fifteen men had died, indicating that nine officers and twelve men had died since the previous May. If they had been afflicted by scurvy, the large number would suggest that the disease had gained a firm hold on the ships. It may also be worth noting that sorrel ('scurvy-grass'), a plant known for its anti-scorbutic qualities was generally available throughout the Arctic Archipelago. In Parry's 1819 attempt at the North-West Passage, it was recorded by Dr Alexander Fisher, HMS *Hecla*'s Assistant Surgeon, that, during the summer 'The men were sent out twice a week to collect the sorrel, and in a few minutes enough could be procured to make a salad for dinner. After being mixed with vinegar it was regularly served out to the men.'

However, although it was usual to gather sorrel in bunches and dry it for later use, in such circumstances, the benefit of the vitamin C in the plant was greatly reduced and even eradicated. The fact that Franklin's ships were deserted in April may, among other reasons, indicate that the benefits associated with the plant had failed over the winter, leading to an outbreak of scurvy.

Charles Hall's eyewitnesses told him of one man seen with 'gums of lower teeth in terrible sick state'. Another 'presented a terrible sight about his lower gums'. The same witnesses reported seeing 'one man with very sore bleeding gums'. Dr Beattie studied the bones he had found at Booth Point and noted

...the first physical evidence to support a long-held belief among historians that expedition members suffered from the debilitating effects of scurvy during their final months. Areas of shallow pitting and scaling on the outer surface of the bones were like those seen in described cases of Vitamin C deficiency, the cause of scurvy.

In the case of the bones researched by Dr Keenleyside, in her March 1997 paper she cites previous works by Beattie, Geiger, and Savelle ('evidence of scurvy, lead poisoning, and cannibalism'), and says that 'New skeletal evidence described in this paper corroborates their findings.' She does not, however, raise further mention of scurvy.

There is also the question of whether or not the Inuit are an entirely peaceful people. At least one female Inuit writer definitely thinks not (I shall not give her name for obvious reasons). In her research into Inuit traditions, she noted, 'There is no absolute truth to the idea that Inuit are strangers to violence, or even warfare.' She continued:

(Inuit) violence is always in origin a 'problem-solver'. Whether effective or not, it is always intended to right a wrong, to address a lack – whether deemed defensive (resisting assault, theft, or invasion), acquisitive (taking food, slaves, territory, etc.), retaliatory (avenging murder, rape, vandalism, or insult), or merely as a cathartic expression of frustrated rage... The Inuit actually engaged in a startling amount of violence, much of which was organized.

The Danish scientist, Dr Kaj Birket-Smith, who had travelled with Rasmussen and remained with the Inuit to study their way of life noted in his 1927 book *The Eskimos*:

Murders were numerous ... as they were only a few years ago among the Copper and Netsilik Eskimos in Canada. The lack of any feeling of responsibility can make them cold-blooded witnesses of murder... Death is measured by a standard other than ours in these regions, where it must be faced almost every day. It cannot be concealed that we can draw examples of thoughtless brutality from the life of the Eskimos.

In another example, Birket-Smith writes of 'marriage by capture. It still happens among the primitive Netsilik Eskimos that a man takes a wife from another who is weaker than himself.'

What chance would exhausted, scurvy-riven men, weakened beyond measure from hauling heavily laden boats on sledges, have against such a people?

R. J. Cyriax did not believe that the Inuit would have killed Franklin's men. His rationale was that McClintock having 'found that the numerous relics there had not been touched', there could not have been any Inuit at the site. What is assumed is that the relics found by McClintock were the *only* ones ever at the site. Many other items, however, could have been taken, of which no trace remains. Where, for example, did the items found in the possession of the Inuit by Rae come from? Cyriax's assistant, A. G. E. Jones, wrote of the Inuit:

> Among them, death was a part of their life. As Stefansson said 'to kill a man was about equal to killing an animal' and the killer was 'entitled to two tattoos lined on his face' Amundsen told how Unkktudliu, in a rage, killed his stepson who had accidently killed his own son. Rasmussen hears the story from an old Eskimo who had killed his own brother, who was a danger to the community... (Elisha) Kane's party were once endangered by the Arctic Highlanders when they were a small, sick, party, and the Eskimos were many. Dr Hayes solved that problem by putting some drug in the Eskimo food, and the party left while the Eskimo were sleeping.

There is another aspect that seems to be overlooked. Cannibalism does not usually happen where food is readily available. The Arctic cannot be considered as a barren waste where birds, fish, and animals do not exist. Musk oxen, caribou, polar bears, seals, fox, hares, loons, ptarmigan, and eider ducks all exist in abundance on north-west King William Island. Would Hall's elderly lady have told him of the 'very great secret' when shamans cast spells to prevent animals and birds from going near the ships if they did not believe that such food was plentiful? Seamen throughout centuries have frequently taken the opportunity to cast a fishing line over the side. McClure noted that – even further north than the ice that held *Erebus* and *Terror* – the ship's company of HMS *Investigator* had taken

'112 reindeer, 7 musk ox, 3 seal, 4 polar bears, 2 wolves, and numerous fox, hares, lemmings, mice and a variety of birds and fish'. At Erebus Bay, Franklin's men clearly had guns, cartridge-making equipment, gunpowder, percussion caps, musket balls and shot. Peter Bayne, a whaler from Nova Scotia, one of five men hired by Hall to help him, reported that an Inuit had told him that during the first summer of Franklin's entrapment in the ice, his men 'caught seals like the natives, and shot geese and ducks of which there was a great number; that there was one big tent and some small ones; and many men camped there'.

There is not the slightest reason to think that during the due seasons they had not topped up their stores of locally hunted food. The most astonishing thing of all, however, is if they were starving, why did they leave '4 cakes of navy chocolate'? The ten-pound blocks of chocolate were used to provide both food and a hot drink. Would Franklin's men have devoured each other within a few feet of a supply of chocolate?

It is certainly possible that the Inuit, angered by what they saw as 'theft, or invasion', and armed with plentiful metal-edged weapons, fell upon a severely weakened party of men, completing their ghastly work by hacking at the limbs, joints, and faces to ensure that the vengeful spirits of their victims did not come after them. A most graphic example of this was found in 2013 when a number of scattered bones, originally collected and buried by Schwatka in July 1873, were examined. Among the bones was a right scapula (shoulder bone), the rear of which carried six parallel cuts marks – a clear indication that the victim was lying face-down on the ground as his attacker was standing over him and – at the very minimum – hacking at his right shoulder.

Whilst making his way westwards along the southern coast of King William Island, McClintock came across a skeleton believed by some to be that of Petty Officer Harry Peglar. He recorded that 'The skeleton – now perfectly bleached – was lying upon its face, the limbs and smaller bones either dissevered or gnawed away by animals.' Could it be that the bones were 'dissevered' as the victim lay face down?

Dr Beattie and his allies then returned to the fray. Clutching at an earlier idea (and apparently unaware of Inuit practices preventing pursuing spirits of the dead), they decided that the broken long-bones were cannibalistic attempts to obtain the bone-marrow. This had been done, they claim, by boiling the broken bones. The action of boiling the bones in a container had led to 'pot polishing' – a smoothing or wearing of the bones as they move within the pot during the boiling process. However, a basic rule in osteoarchaeology is that other factors can produce the same 'pot polishing' effect. In the case of bones lying exposed on the Arctic surface, the chief (but not the only) cause of such an effect would be exposure to high winds bearing abrasive sands, grit, or other fragments of shingle or gravel – all of which abound in the region.

Even more absurd is the claim that the fractured end of one of the boiled bones was so splintered that it was used to scrape the remnants of the meal from out of the pot. Although such an idea is nothing more than a guess, it does highlight an interesting aspect that has not been investigated (or if it has, it has been kept well in the background). Different long-bone fractures indicate different causes and circumstances. When a living person breaks one of the leg or arm bones, their natural reaction is to stagger, fall, sway, stumble, or roll. By doing so, the fracture twists, producing a jagged, irregular edge to the break (known as a 'spiral' fracture). Such an action could have easily produced a bone with the fracture point in a shape useful for scraping the inside of a cooking pot. However, unless the victim was being eaten alive, the dead victim of cannibalism would have suffered a 'transverse fracture' where there is no post-breakage movement, and the bone is broken directly – and cleanly – across its length. Consequently, if Dr Beattie's claim is correct, in all probability the victim would have been living when the bone was fractured – a victim of an attack, perhaps?

Furthermore, a cooking pot would have required a fireplace to support it over a fire – none have been found. Fuel would have been required. There were plenty of flammable items in the boat, and the boat itself was made of wood, but nothing was used. So where did the fuel come from? The boat also contained several items that could have been used to construct a spit on which to roast flesh – rowlocks and iron stanchions, for example – but none were so employed. Finally, of all the bones discovered, not one showed any sign of burning.

It remains possible (even likely), that groups of men, debilitated beyond hope with scurvy, could fall victim to a band of Inuit intent on murder and plunder. The subsequent ritualistic mutilation of the bodies following the attack accounting for much of the 'cut marks', smashed skulls and limb bones. Those men who survived and were able to continue their march south, were left alone to bury their dead 'facing upwards & and the hands ... folded in a very precise manner across the breasts'. Hardly the actions of cannibals.

Although relegated to the absolute periphery of the Franklin story, there are two items of possible information that could contain a kernel of truth. Referring to the Austin-Penny search expedition of 1850/51, an article (probably of American origin) in *Fraser's Magazine for Town and Country* of February 1851 told the reader:

A detention took place off Cape York, in consequence of a terrible story having been communicated to Captain Austin by the Esquimaux interpreter (Adam Beck) on board Sir John Ross's ship, to the effect that, in the winter of 1846, two ships had been broken up by the ice forty miles to the northward, and burned by a fierce and numerous tribe of natives; and that the crews, being in a weak and exhausted condition, had been murdered.

The only modern-day supporter of the Adam Beck story of the Inuit massacring Franklin's men was Wayne Davidson, who lived amongst the Inuit. Beck had placed his story of the attack on Wolstenholme Island – just off the northern-most part of the west coast of Greenland. In 1850, the island was investigated, but no evidence of the outrage could be found. Consequently, the version provided by Beck was widely dismissed. Davidson however, placed the assault on Prince of Wales' Island's Back Bay, where the remains of a 27-foot whaler had been found. But there was no evidence of a massacre. There are no bones, nor any other material suggesting a mass slaughter had taken place such as was found at Erebus or Terror Bays.

It needs to be understood that it was in Beck's interest to keep the Franklin search ships off the coast of his part of Greenland for as long as possible. The ships were stuffed with valuable items that could be traded with the seamen, eager to find souvenirs to take home. What could be easier than to take a tale – well-known amongst the Inuit – and simply move the site to a nearby island? And where was the original island? No-one knows, but the most obvious candidate can only be King William Island, a location close by the heart of the Netsilik expansion that drove neighbouring tribes to all points of the compass – including towards Greenland.

Did the Inuit really enact self-protective mutilation of their fallen enemies? *The Fraser's Magazine* report continued:

Letters, however, from the American ships (*Advance* and *Rescue*) mention a circumstance in connexion with Cape York which seems to have escaped the notice of our English friends, and may possibly have had some influence in giving rise to the above report. They state that near that Cape more than twenty corpses of Esquimaux were found: ice-preserved, entire except their eyes and lips, and lying down, lifeless (sledge) dog by lifeless master. The cause of this passing away of life was a mystery. There was food around them, and where food and fuel are nearly convertible terms they could hardly have been without fire or light.

There was, of course, no accusation of cannibalism. Despite the weakness and high improbability of the cannibalism case, in 2014, the *Canadian*

Geographic magazine noted that 'Cut marks on skeletal remains of crew members have since corroborated those gruesome reports.' And in September 2016, *Maclean's*, a highly respected Canadian news magazine, still reported that 'some' of Franklin's men 'resorted to cannibalism in their most desperate hour'.

The final rational, and contemporary, word on the subject should go to Vice-Admiral Sir George Richards who had served as the captain of HMS *Assistance* on Sir Edward Belcher's search expedition of 1852-1854, carrying out long sledge journeys before becoming the Hydrographer of the Navy in 1864. He wrote in the Introduction to Gilder's *Schwatka's Search* of the

> ...strong reasons why the Esquimaux should not be believed... They are said to give as their reasons, that some of the limbs were removed as if by a saw. If this is correct, they were, probably, the operators themselves. We learn from the narrative that they were able to saw off the handles of pickaxes and shovels. At all events the intercourse between the natives and such of Franklin's crews as they met is surrounded by circumstances of grave suspicion, as learned from themselves, and this suspicion gathers strength from various circumstances related on Schwatka's journey. Be this as it may, I take my stand on far higher ground. Of course such things have happened. Strong, shipwrecked mariners, suddenly cast adrift on the ocean, have endeavoured to extend life in this way when they were in hourly expectation of being rescued. But how different the case in point! The crews of the *Erebus* and *Terror*, when they abandoned their ship, were, doubtless, for the most part, suffering from exhaustion and scurvy; death had been staring them in the face for months. The greater part of them probably died from exhaustion and disease long before they got a hundred miles from their ships, and found their graves beneath the ice when it melted in summer, or on the beach of King William Land. It is possible that no more than half a dozen out of the whole crew ever reached the entrance to the Great Fish River. We need not call in starvation to our aid. I fully believe that by far the greater portion perished long before their provisions were consumed. The only thing that would have restored men to convalescence in their condition would have been nursing and the comforts of hospital treatment, not a resort to human flesh.

And the 'lead poisoning'?

The results of Dr Beattie's work on Beechey Island, which led him to the lead poisoning speculation, has been questioned by such eminent Franklin researchers as A. G. E. Jones and S. Trafton, and leading food scientist Dr K. T. H. Farrer (whose work provides the bulk of the following technical detail). One thing has to be stated right at the beginning. Despite the wealth of experience and expertise that has been brought to bear on the matter there remains not a single shred of evidence that lead poisoning occurred at any stage in the Franklin Expedition (the men buried on Beechey Island probably died from pneumonia or tuberculosis, or a combination of the two). Certainly, the levels of lead in the bones and hair of the bodies disinterred on Beechey Island were high when compared to those found in caribou and Eskimo bones, but how high they were, when compared to other mid-nineteenth seamen, remained an unknown (and unpursued) factor.

The accumulation of lead in bones begins at birth, and the amount collected depends upon the environment in which the individual is raised. Humans and animals living in the virtually lead-free environment of the Arctic could be expected to have minimal amounts. Mid-nineteenth century seamen, on the other hand, brought up breathing industrial pollution (Leading Stoker John Torrington was raised in Manchester) or rural adulteration (Royal Marine Private William Braine came from the cider-drinking county of Somerset, and would have been subjected to the 'sugar of lead' used to counteract the acidity of that drink). All were likely to have spent their life drinking water delivered through lead pipes, eating off lead-glazed pottery whilst using pewter-based cutlery, consuming perishables wrapped in lead foil, and being medically treated with lead-based drugs.

The lead found in the hair need not all have been ingested but could have come from the immediate environment. This would account for higher levels being noted in Leading Stoker Torrington's hair. The time spent in the lead-enriched atmosphere of the *Terror*'s boiler room could have raised the levels significantly. Even the wood shavings on which the dead man's head rested could have increased the lead level in his hair. Whatever the case, the use of hair as a measurement of lead ingestion, and as a predictor of lead poisoning, is seriously flawed when it is realised that not all the hair from the same head will necessarily give the same reading, and variations can be found even within a single hair.

According to Dr Farrer, the only sure way to test for recent exposure to lead is to look for it in the blood, the risk to health level being that reached (in adults) by the ingestion of 2.3 milligrams of lead per day for four years, or 3.3 milligrams per day for eight months.

Could, as suggested by Dr Beattie, the canned food be implicated?

The canning of food was not a new technique. The Royal Navy had adopted the principle in 1813 and by 1819 canned food was being issued to the ships of Parry's Arctic Expedition, 26 years before

Franklin sailed. The initial practice of using such supplies for the sick had given way to using canned food as a significant portion of the ship's supplies. Unfortunately, in the case of one manufacturer at least, the standards of the preserved meat began to slip, and such supplies became known as 'canned carrion'. Seamen would throw away the contents keeping the cans as additional mess 'traps' (metal containers are still known in the Royal Navy as 'fannies' from the Victorian legend that the murdered and dismembered body of little Fanny Adams was used to fill the cans).

Franklin's people had more than enough cans for their purposes. On Beechey Island piles of empty cans were found stacked neatly in rows. It was estimated that well over six hundred had been taken ashore. The reason why they were on shore in the first place itself something of a mystery. There was, of course, no reason why they should have been simply thrown away. Even if the seamen had no further use for them, they would have been very useful as gifts to – or to trade with – the natives. Or as a smithy is known to have been set up on shore, they could have been recycled to recover the sheet metal and lead. If that was the case, it is possible that the six to seven hundred cans that were left on the island were still waiting to be dealt with by the blacksmith when the ships sailed, and were simply abandoned. Under such circumstances, it would be reasonable to assume that many more cans had been opened, their contents disposed of, and the scrap metal recovered by the blacksmith. It must be remembered, however, that the contents of the cans, both meats and soups, were supposed to be *preserved* provisions. Other meats were provided in the form of salt beef and pork (almost double the amount of canned), pemmican (for sledging parties), 10 live bullocks, and the huge numbers of birds likely to have been killed by the ship's companies during hunting expeditions and consumed immediately or dried by hanging in the ship's rigging and salted down. In addition to the birds, meat would have been available from the animals once the ice had been reached. When this supply was no longer available the next to be consumed should have been the salt beef and pork.

The custom of the Royal Navy at that time was to have alternate meat and non-meat ('banyan') days. The reason behind this pattern was not simply a question of making the meat last longer, but to cut down on the large amount of fresh water required to soak the salt meat before it could be rendered edible. Such a consideration did not apply in the Arctic where fresh water was almost always immediately at hand. The general issue of preserved meats would have been deferred until later in the expedition. And yet, of the more than eight thousand cans of meat, and the cans of concentrated soup meant to provide over twenty thousand pints, less than seven hundred have ever been found. The bulk remained on Beechey

Island – fewer than a dozen have been found where the voyage ended (one adapted as a pannikin).

Does this suggest that early on in the expedition something was found to be wrong with the meat and soup? The renowned expert on the Franklin Expedition, Dr Cyriax, thought not. He pointed out that the producer of the canned food, Stephen Goldner, had been awarded a medal at the 1851 Great Exhibition for his preserved meats, that there were no complaints during the early years of his supply to the Royal Navy, that Franklin's officers had dined off the contents of cans and declared them to be 'very good indeed', and that a can brought back from the Arctic was found to be in good condition in 1926. On the other hand, no-one would submit a poor quality product to a prestigious exhibition, and it was not unheard of for manufacturers to present the officers with a gift package of their products in the hope of a recommendation. Furthermore, the officers might not have been referring to Goldner's products. Dr Cyriax appears to have assumed that all the canned meats came from Goldner.

Certainly, Goldner was contracted by the Admiralty (the first such contract with that manufacturer) to supply the canned meats for the expedition, but Captain Erasmus Ommanney, searching the coast of Cape Riley in August 1850, 'found some empty preserved meat tins, with Donkin's name'. Perhaps the officers had purchased their own supplies of Donkin's canned meats (founded 1813, and based in South London), and it was these that were considered to be 'very good indeed'.

As for lack of other complaints, Dr Cyriax appears not to have heard of the 'general murmur' prevailing in the troopship HMS *Hercules* during her 1850 voyage when the Goldner's canned meat was issued: 'Some of them were continually bad, and thrown overboard.' Nor of the nine thousand Goldner's cans that were returned to the Royal Clarence Victualling Yard by ships returning from the Mediterranean and East Indies, their contents 'putrid'. Nor of the 'many' cans seen 'weeping' as they were loaded onboard HMS *Plover* (a Franklin search ship) despite being 'passed' by Victualling Yard officers.

In August 1851, civic officials responsible for the area close to the Royal Clarence Dockyard requested Portsmouth Magistrates to take action over a foul smell emanating from the dockyard. The stench was found to have 'proceeded from gases exuding from some thousands of tin canisters of what is called patent preserved meats, polluting the atmosphere of the place so as to occasion fainting and sickness.' *The Times* noted, 'It is a painfully important fact, that this description of preserved meat formed a large proportion of the provisions supplied to the exploring ships under Sir John Franklin, and other vessels which have been sent at various times on distant services.' Two doctors were sent to examine Golder's canned meat held by the Victualling Yard; their resultant report was not for weak stomachs: '...very disagreeable odour ... nausea ... insufferably noxious ... putrefactive fermentation'. The contents of the cans 'dripped a thin, black foetid fluid', whilst some were 'reduced to a state of putrescence'. So bad was the condition of the preserved meats that one of the men assisting the doctors to open the cans 'became ill, and so remained during the night and a portion of the next day'. In all six thousand cans were examined, and 'less than one in fifteen was found fit for human consumption.'

Goldner's agent had sold on some of the cans to the Merchant Marine and others ended up being bought by a soap manufacturer at £2 a ton. He had hoped to extract the fat from the meat 'to use in my trade'. However, 'The stench was so dreadful that I was obliged to order it away.'

Goldner had fled to Galatz, Moldavia, where his factory was still turning out canned meat for the 'principal Italian shops at the west of the town – the canister emblazoned with the royal arms, painted in all colours but the true.' He was said to have 'some powerful influence at the Admiralty', which allowed him to continue supplying the Royal Navy despite the evidence and complaints from competitors that he was selling 'dirty goods' at 'dirty prices'. The rumours were strengthened by an Admiralty ruling which made it permissible for seamen to reject the salt meat ('salt junk') in lieu of fourpence per ration to be spent on alternative food. In the case of rejected canned meat, however, the seamen received nothing. Even more astounding, when the Select Committee on Preserved Meat met in 1852 to enquire into the problem of bad meat, they summoned the Comptroller of the Victualling Department of the Navy, Thomas Grant. When asked by Admiral Bowles if it was accepted that a certain number of cans supplied from any contractor would be bad, Grant replied, 'Yes. We find from an account we drew out, of the quantity supplied to the Arctic ships, the *Erebus* and *Terror*, that 15,422 lbs. were returned into store, and condemned as unfit for service.'

'Whose contract was that?'

'The preserved meats supplied to those vessels were sent in by Cooper, by Gamble, by Cope, by Nichols, by Wells, and by Simkin.'

The Comptroller of the Victualling Department of the Navy listed just about every supplier except the one who had actually supplied Franklin's ships – Stephen Goldner.

Unfortunately, the Select Committee did not have the wit or the knowledge to respond with the fact that the cans found two years earlier on Beechey Island were all Goldner's. Instead, Admiral Bowles' next question was neatly sidestepped. 'When those canisters which were supplied to the *Erebus* and *Terror* were opened, and the meat found bad, was there any objectionable substance found amongst it?'

'In the returns from the *Resolute* and the *Assistance* the other day there were two canisters found which contained improper substances.'

The original question was not repeated, and the matter was dropped never to be raised by the Committee again.

In its final report the Select Committee noted:

It appears that the provisions so specially complained of were supplied by running contract, dated 28th December 1844 (to be determined by three months' notice), between the Government and Mr Goldner, the manufacturing in London, at Houndsditch, as well as at Galatz in Moldavia... Your Committee find, that the complaints were responded to by the Admiralty immediately and a survey ordered, under which there were great condemnations, and the contract of 1844 was (re) determined in 1849. That the same parties then entered into a new and more stringent contract in 1850, and again in 1851; but that the increasing complaints and consequent condemnations and rejections under surveys, ordinary and special, on delivery to the store, and on subsequent extraordinary occasions, induced the Admiralty not only to put an end to and cancel the contract deed of 1851, but to prohibit any deliveries under that contract at the Government Yards.

So, despite a catalogue of complaints leading to the dropping of Goldner's contract, the Comptroller of the Victualling Department of the Navy was still trying to protect Goldner's name. He even allowed Goldner to continue supplying pickled goods to the Royal Navy. Perhaps the 'powerful influence at the Admiralty' was very powerful indeed.

Sir John Richardson, who had been responsible for the production of the expedition's pemmican at the Royal Clarence Victualling Yard, believed that Franklin's canned meat supply had been bad. Captain Sir John Ross, after studying the cans on Beechey Island supported Richardson, as did Captain Ommanney on his return from the 1851 search expedition.

Decades after Franklin's disappearance, the story of the poor quality of Goldner's was still being aired, and yet was ignored by Dr Beattie and his team.

The Marlborough College Classics scholar and biographer for the *Dictionary of National Biography*, A. H. Beesly, wrote in 1881:

The preserved meats supplied had been those of Goldner's patent, and when it was known in England that seven hundred or more of the tins had been found on Beechey Island, while at Portsmouth a large quantity of the meat had been condemned as putrid, it was plain that this vile stuff had been thrown away as worthless.

This was not mentioned in the subsequent book co-authored by Beattie and a journalist and author, John Geiger (of whom, more later).

If then, that was the case, and Franklin had found that most of his canned meat and soups were bad and, consequently disposed of them overboard, would enough of them have remained to poison the ship's companies with the lead used to solder the can's seams? Even if there had been, could the lead actually contaminate the meat?

Dr Farrer maintains that, despite the food being in contact with large areas of lead, electrolysis would have ensured that particles of lead would have migrated to the tin and iron of the can rather than into the food. He was supported by other scientists including Dr Magnus Pyke:

Lead, which is a component of the solder by which the side seam of a can is made, does not contaminate the can's contents. The electrolytic relation between lead and tin is such that, whereas tin migrates into the food, lead tends to be deposited on to the can. Such electrolytic action can be speeded up if the contents of the can had a high acid content but, as meat is low in acid, the subsequent electrolytic action would be extremely slow in any case. Tests carried out on 19th-century canned food support these findings.

Dr Beattie continued to champion his lead poisoning theory by declaring that the isotope ratio value of the lead found in the samples taken from the bodies on Beechey Island was almost exactly the same as that in the lead used to solder the preserved meat cans. At first sight this would appear conclusive in the same way that in a case of suicide from cyanide it would seem reasonable to seek the culprit in the empty bottle of cyanide clutched in the victim's hand. But, just as there is more than one source of cyanide in the world, there is more than one source of lead with the same isotope ratio value. Dr Farrer points to the similar lead isotope ratio value in Italian petrol, Roman bones and lead coffins found in England, and the Victorian plumbing of Edinburgh houses. Whereas different isotope ratio values cannot come from the same source, *similar* values do not have to come from the same source, nor do they provide an unchallengeable link between the two values.

Surprisingly, over a quarter of a century after Dr Beattie's lead poisoning theory, researchers at Canada's McMaster's University carried out an investigation aimed at keeping the lead poisoning theory afloat. A can of ox-cheek soup found in the Arctic and part of the collection of the Royal Ontario Museum was opened up and tested for lead. 'High levels' of the metal were found in the soup, so, using an 'x-ray fluorescence capability' the soldered lid of the can was examined to see if there was any evidence of lead. Much to the surprise of the researchers, 'The numbers showed us lead levels that were pretty much off the scale.' Clearly, no-one knew that solder – particularly that used during the Victorian times – was an amalgam of tin and lead, with lead sometimes accounting for over 50% of the solder content. Furthermore, the can had never been near the Franklin expedition, and was found on Dealy Island, an island not visited (at least, by non-natives) until 1851. In 1961, a Canadian Wildlife Service biologist, C. R. ('Dick') Harington, visiting the island during research on musk oxen and caribou, opened two 108-year-old cans – one of beef, the other of mixed vegetables. He recorded that 'The beef came apart in shreds being light-coloured, dry and fibrous – but absolutely fresh. Mixed vegetables, consisting of carrots, leeks, etc, were preserved in a dark jelly. The resulting stew was very palatable, although the broth was rather fatty.'

No-one explained to him that he was risking being poisoned by the 'off the scale' lead in the can, Doctor Harington survived to receive the Royal Canadian Geographical Society's Massey Medal, the Meritorious Service Award from the Yukon Government, and be appointed an Officer of the Order of Canada.

Incidentally, in 1994, McMaster University analysed the bones of Franklin's men that had been found on King William Island and announced that they 'had found lead levels that were higher than researchers had detected in any living human'. The objectivity and relevance of *direct comparison* remained unappreciated.

The lead poisoning theory was undermined in 1993 when Dr Keith Farrer, a professional research chemist, Fellow of both the Australian Academy of Technological Sciences and Engineering and the International Academy of Food Sciences and Technology, and a highly respected food scientist, studied the question of lead poisoning on the Franklin Expedition. After detailed research, Dr Farrer concluded that 'the contribution of canned foods to body loads of lead or to any incipient ill health in Franklin's crews *was trivial* (author's italics).'

Other specialists then picked up the baton.

In 2013, R. Martin, S. Naftel, S. Macfie, K. Jones, and A. Nelson carried out research into the distribution of lead in bones from the Franklin expedition. The results proved to be '*inconsistent* with the hypothesis that faulty solder seals in tinned meat were the principal source of Pb in the remains of the expedition personnel' (author's italics).

Research carried out by K. Millar, A. W. Bowman, W. Battersby, and R. R. Welbury, and published in the *Polar Record* in 2016, stated, 'There was no clear evidence of lead poisoning despite the relatively high level of lead exposure that was inevitable on ships at that time.'

In 2016, research at Saskatchewan University on a thumb nail removed from the body of one of the seamen exhumed by Dr Beattie on Beechey Island revealed that a severe zinc deficiency may have led to his death from tuberculosis. There was no evidence that lead was involved in his death.

Also in 2016 came the hammer-blow to the lead poisoning theory. A team of researchers from Lakehead University, Thunder Bay, Ontario, carried out the single test that should have been carried out years earlier. They compared *like with like*. With ample information and detail from early nintenth-century Royal Naval personnel who had died and were buried in the Royal Naval Hospital in Antigua, they carried out the same tests on remains of Franklin's men from Beechey and King William Islands. The Antigua men had served in sailing ships and had been subjected to the same sources of lead to those of Franklin's men. The lead measurement in each group were so closely related that little meaningful difference existed. Despite living closely with their shipmates, and latterly, under the supervision of hospital staff, there are no reports of half-crazed men desperate to devour their messmates, and no outbreaks of ill-discipline on the lower decks. Just the sad loss of men carrying out their duty and serving their country and humanity.

The research carried at Lakehead University was finally underlined by a team of scientists at Glasgow University under the direction of Professor Keith Millar of the College of Medical, Veterinary and Life Sciences. The team studied all the forensic data that had been collected from the bodies of Franklin's men buried on Beechey Island, and from skeletal remains found on King William Island. From the fact that the first two years in the Arctic showed a degree of activity which could not have been achieved with men suffering from neurological and psychological problems, and that, in the original research, little (if any) account was taken of the fact that the men came from a country in which lead plumbing was almost universally used, the probable cause of the disaster was a combination of 'bad luck' and a continuation of several years when the temperature was never high enough to melt the ice.

If the canned food proved to be so bad that much or even most of it was rejected, there would not have been enough lead ingested to have poisoned anyone. Furthermore, the chances of the food becoming tainted by the lead from the soldered seams of the cans is remote in the extreme due to any electrolytic action transferring lead particles to the can itself rather than to the contents. Any similarities between the lead isotope ratio values of the samples taken from the bodies and the lead found on the cans has every chance of being merely coincidental. And, finally, tests on the remains of men from similar backgrounds show that effects of the lead ingested throughout was 'trivial'.

There is another frequently forgotten aspect to the illusory lead poisoning. John Smart Peddie was carried on board the *Terror*, and Stephen Samuel Stanley in the *Erebus* as Surgeons. They were assisted by Alexander McDonald and Harry D. S. Goodsir as Assistant Surgeons. All had long experience at sea with the Royal Navy and – if it existed – all would have recognised lead poisoning and known the treatment needed to alleviate it. A. G. E. Jones searched through every surviving medical journal kept by the surgeons on board the Arctic search ships. Over 1,500 sick cases were studied and, in a letter to the author, he said there was 'not one of them that looks like lead poisoning'. This finding was confirmed subsequently following a similar examination of the available records by a team from the University of Glasgow College of Medical, Veterinary and Life Sciences and the College of Science and Engineering.

Consequently, and taken together, despite the books, the television programmes, and the academic promotions, the lead poisoning theory is now as dead as a door-knob. So, with no cannibalism or lead poisoning, what might have happened?

In late May 1859, Lieutenant W. R. Hobson came across a boat on a snow-covered shingle beach on the southern shore of Erebus Bay, King William Island, in the central Canadian Arctic. The vessel was almost entirely covered with snow and would have escaped detection but for a projecting stanchion and part of her gunwale that broke the white surface. On clearing the snow from the inside of the boat, Hobson found, not merely a mass of expeditionary and personal items, but also parts of two skeletons, one more complete than the other, both lacking skulls. Some days later, on 30 May, the site was visited by Captain F. L. McClintock RN who made a more detailed search of the boat and the immediate area.

There was no question that the boat was part of the missing 1845 Franklin Expedition, but it was to give no indication regarding the fate of that expedition – the only clue to that had earlier been found in the document discovered by Hobson to the north. Apart from the bones, the contents of the boat could be classified under three broad headings: expedition goods (axes, saw, guns, ammunition, goggles, knives, etc), personal items (toilet requisites, books, purse, cutlery, watches, etc), and a random collection of items that could represent native trade goods or, simply, 'odds and ends'. The personal items would probably have been 'bundled' together in the number of handkerchiefs found in the boat (and almost certainly undone by Hobson in his initial search). The boat itself (likely to have been a ship's whaler) had been cut down where possible to reduce weight, and a weather-cloth was secured to twenty-four thole pins for added protection. The thole pins would have originally supported oars, but the oars had been cut down to provide paddles as being more suitable for the passage of a river.

The boat was displaced from a solid oak sledge which had been used to transport it to the site. The direction in which the sledge was pointing led McClintock to assume that it was being towed back to the ships trapped in the ice off the northwest corner of King William Island, and that the two men whose bones remained had been left by the return party as they went to collect supplies. When Hobson had cleared the boat of snow, he also found two double-barrelled shotguns, both with one chamber loaded and both cocked ready for use.

Thus began a trail of guesswork and intimation that was to lead to the construction of flimsy theories based on cannibalism and lead poisoning. Theories too often presented as confirmed fact. Franklin's men, the detractors have decided, tried to make their way down the west coast of King William Island, reached Erebus Bay where two boats were abandoned with a number of men – probably suffering from scurvy – who then resorted to cannibalism. Those who continued southwards also took up cannibalism but, despite a constant and easily available supply of fresh meat and fish, still managed to expire on the coast of Northern America.

Natives visited the more inland of the two boats at Erebus Bay, looting, and breaking it up. A native gave Hall a piece of the boat, and Ranford noted the existence of small fragments of wood at the site he visited. The bodies themselves were probably left to nature; their bones being scattered by scavenging wild animals. The canvas awning would not have taken long to collapse and decay, leaving very little to catch the eye of Hobson, McClintock or Schwatka.

The boat by the beach, guarded by its human remains, might have been invested by native spiritual beliefs with a quality absent from the nearby camp (this was the chief reason given by the local Inuit for the survival of the boat remains at Back Bay on Prince of Wales Island, despite the obvious value of its wood). Whatever the reason, historians have every cause to be grateful for its remaining unmolested. Without it there would be too little evidence to ward off the distracting theories, however plausible, that seek to denigrate an exceptional enterprise that ended in tragedy.

When I began my searches in the north, my prime objective had been that Holy Grail of Arctic scholars and historians – the grave of Sir John Franklin. That still remains the highest objective of many serious Arctic historians and researchers. There remains, however, one other major mystery – what happened to the Franklin Expedition ships?

The *Erebus* and *Terror* are not the only ships to have disappeared in the northern mists. Commander McClure's *Investigator* sank at Mercy Bay, on the northern coast of Bank's Island. Of the five ships abandoned during Captain Belcher's 1852-54 Franklin search expedition four have long been lost in the Arctic Archipelago, whilst the fifth, HMS *Resolute*, responding to her name, drifted out into the Davis Strait. The *Breadalbane* was crushed by the ice off Beechey Island but with the incident taking place in front of many witnesses her position was recorded, allowing a visit by Canadian divers in 1981.

The disappearance of Franklin's ships, on the other hand, had been witnessed by no-one – at least no-one that survived to tell the tale to a comprehending ear. Consequently, their fate had been the subject of considerable speculation based almost entirely upon hearsay and two fleeting and unconnected sightings.

The first 'sighting' took place off Newfoundland in 1851 when the crew of a whaling ship spotted what they claimed to be two vessels, complete with masts and bowsprits, trapped on an ice floe. Despite this singular scene, the captain of the whaling ship did not close with the ice floe to investigate. During the same month, off the same coast, a German ship spotted two waterlogged hulks, but was unable to identify the wrecks.

In 1854 Dr John Rae reported that he had met natives who had heard from other natives that a party of white men had used sign language to make the Eskimos 'understand that their ship, or ships, had been crushed by ice'. The following year, according to a second-hand story told many years after the event, an Ojibway Indian, Paulette Papanakies, was a member of a small party detached from the Anderson Expedition searching for Franklin at Back's Great Fish River estuary in the summer of 1855. Whilst separated from the rest of the party, Papanakies claimed to have seen two ship's masts far away across the ice at the eastern end of Simpson Strait. When he eventually revealed this to his colleagues, he claimed that they decided to say nothing to Anderson on the grounds that the delay in following up the sighting would place the entire expedition at risk. More hearsay evidence was reported by McClintock's dependable interpreter, Carl Petersen, in 1859. He was told:

> A large ship which had three masts had come driving with the ice from the north, but had been twisted into pieces between heavy ice floes northwest of King William's Island and thereupon sank, so that they (the natives) had saved nothing from the wreck. All the men, however, had long previously gone to the land with their boats and had travelled southwards.

On another occasion, the same natives told Petersen that the King William Island Inuit '...had seen two large ships, one of which was smashed by ice, and to their great regret, sank, so that they had been quite unable to obtain anything of the wreck; the second one, on the other hand, was driven ashore later in the year.'

Another group of natives (this time actually on King William Island) told him that they had seen a ship 'but the masts were gone'. He was unable to find out how the masts had been removed, whether by burning, or by being 'snapped off' by the elements (the local Inuit, however, had ample access to metal, and could have cut down such masts).

In March 1859, a curious incident happened when McClintock, returning to the *Fox*, reached the west coast of the Boothia Peninsula at Cape Victoria, a headland almost on the same parallel as Cape Felix. Among the Inuit he met there, was an elderly man named Oonalee. The Inuit 'made a rough sketch of the coastline with his spear upon the snow, and said it was eight journeys to where the ship sank, pointing in the direction of Cape Felix. I can make nothing out of his rude chart.'

Nevertheless, McClintock copied the sketch along with notes indicating the spot where Oonalee had claimed the ship sank. For some reason, McClintock oriented the chart in the customary direction – with north at the top. In doing this, the site of the sinking is shown at the southeast corner of the island – the very same spot where the Anderson Expedition's Paulet Papanakies had claimed to see masts over the ice at the eastern end of the Simpson Strait (approximately 108 miles south of Cape Victoria, or 'eight journeys' of 13.5 miles). And yet, students of the Franklin Expedition still continue to insist that Oonalee's 'pointing in the direction of Cape Felix' can only refer to the site where the ships were deserted – on the northwest coast of the island, a position where the depths of the waters are unlikely to have allowed the masts to be showing above the surface.

According to his biographer, the explorer Charles Hall heard in 1869 a hearsay account of natives seeing a ship with three masts and four boats 'hanging at the davits and another above the quarterdeck'. They also told him that the ship had sunk 'very near O'Reilly Island, a little eastward of the north end of said island'. This suggested that at least one of Franklin's ships had drifted south past the mouth of the Simpson Strait and had sunk off the west coast of the Adelaide Peninsula. Ten years after Hall, his fellow American Lieutenant Frederick Schwatka was told by the natives that a ship had sunk a few miles to the west of Grant Point (off the north-west corner of the Adelaide Peninsula).

In 1905 Roald Amundsen, whilst making the first ever voyage through the North-West Passage, heard a story of natives finding a ship off Cape Crozier and, even as late as 1923, Knud Rasmussen was told a tale of Inuit, many years earlier, entering a ship shortly before it sank off the northwest coast of King William Island. The latter account echoing one told to Hall about the vessel found by the natives off O'Reilly Island.

A Canadian soldier, Major Burwash, visiting the area in 1926, learned that the natives believed that a ship had sunk off Matty Island in the James Ross Strait. He examined a number of wood and metal items claimed to have been picked up on the shore and decided that they were, indeed, likely to have come from a ship. An attempt by Burwash to locate the wreck failed to find anything.

Lieutenant-Commander R. T. Gould resurrected the story of the ships on the ice in his popular 1928 book *Oddities* and in 1959 by Rear Admiral Noel Wright. Admiral Wright's book *New Light on Franklin* not only claimed that the ships on the ice were the *Erebus* and *Terror*, but that the ship supposedly seen by the natives off the Adelaide Peninsula was none other than HMS *Investigator*, which had found its way out of Mercy Bay, passed down the McClintock and Victoria straits to end up south of King William Island. This theory was convincingly and comprehensively destroyed by R. J. Cyriax, whose work was further confirmed by aircraft flying over the site and, in July 2010, by a Canadian expedition which, by using side-scan sonar, found the ship in just ten minutes – exactly where the survivor's records said it would be. Nevertheless, the find was used to inflate the stories of Inuit word-of-mouth communication by reporting that Inuit oral history was 'validated for knowing where the intact ship sank'.

Did the natives, on the other hand, actually find any of Franklin's ships at all? Or were they merely substituting actual experiences and reminiscences that had arisen from the abandonment of the steam paddle-ship *Victory* and her escort *Krusenstern* by Ross on the east coast of the Boothia Peninsula in 1832?

Among the items found in the native's possession, very few, if any, could be confirmed as coming from a ship. All could have come from the boats being dragged down the island by the retreating parties. Even the use of sheet copper by the natives to make fishing hooks and cooking implements – used to support the claim that the copper had come from the ship's bottom sheathing – could have come from the copper cooking vessels used by Franklin's people. One such example was discovered as late as 1992 at the Erebus Bay boat site. What then, happened to all the canvas, cordage, metal fixings, and massive wooden fixtures?

Prior to 2014, only two items had ever been discovered that, on balance, seem to have come from the ships. One, discovered on the western shore of Victoria Strait by Captain Collinson in 1853, was a piece of door frame later identified by the foreman who had worked on the ships as being likely to have come from one of them. The second find was a length of wood – variously described as a 'bunk board' or 'bunk head' – obtained from the Inuit by Lieutenant Schwatka in April 1879. What particularly distinguished this item were the initials 'L.F.' made out in copper tacks. None of Franklin's ship's companies had these initials, nor were they to be found on the *Terror* when she was earlier commanded by James Ross and George Back. My view, however, is that the letters could signify one of the 'parts of ship'. In a naval tradition that still exists today, a warship is divided into 'parts of ship' such as 'Forecastle', 'Quarterdeck', 'Engine-room', 'Starboard Flats' etc. These, in turn, frequently identify a 'cleaning station' with a number of men responsible for keeping the area clean. Their cleaning gear would be kept in a locker and marked appropriately – thus 'Fx' for Forecastle, 'QD' for Quarterdeck (although 'Ax' for 'Aftercastle' was, and is, sometimes used), 'ER' for Engine-room etc. Other lockers would have been used to store mess 'traps' marked in the same manner as the cleaning gear. On larger ships, messes were usually numbered, but this may have been unnecessary on ships that were unlikely to have more than four ratings' messes (including the Royal Marines' 'Barracks') and was, in any case, usually linked to a gun position, a situation that did not apply to Franklin's ships (a pannikin was found at Victory Point marked 'No. 15 Mess' – an unlikely number for a ship with only twenty able seamen). McClintock found a knife marked 'W.R.' (Wardroom) in the Erebus Bay boat. Accordingly, 'L.F.' could have meant 'Larboard Forward' (other messes being 'Larboard Aft', 'Starboard Forward' and 'Starboard Aft'). The use of the word 'Port' for the left-hand side of a ship did not come into use until the mid-1840s. It possible, therefore, that the board marked with the letters L.F. *could* have been part of a ship's locker from the *Erebus* or *Terror*. Consequently, it would seem that there is no actual evidence of the natives having visited, or even seen, Franklin's ships.

The vessels could have drifted northwards and reached the waters off Newfoundland. This was not impossible. It is well known that HMS *Resolute*, abandoned by Belcher, passed through Barrow Strait and Lancaster Sound and was found drifting in the Davis Strait. James Ross's ships *Enterprise* and *Investigator* were similarly forced to follow the easterly flow of the ice, as were the American Franklin search ships *Advance* and *Rescue*. There is even an unsubstantiated account of a British merchantman, the *Octavius*, being trapped in the ice off Alaska in November 1762. Thirteen years later, she was found drifting off the coast of Greenland, her crew long dead, having achieved the first ever (but unconfirmed) North-West Passage.

On the other hand, there could have been a simple, straightforward reason for the ships' disappearance. There is one first-hand account of the natives meeting the retreating Franklin party. In May 1869, Charles Hall met two natives (Tuk-ke-ta and Ow-wer) who claimed to have met a party of white men on their way south down the coast of King William Island. In his notes, Hall records them as saying:

> He ('Aglooka' – the leader of the white men) then made a motion to the northward & spoke the word oo-me-en, making them to understand there were two ships in that direction; which had, as they supposed, been crushed in the ice. As Aglooka pointed to the N., drawing his hand & arm from that direction he slowly moved his body in a falling direction and all at once dropped his head sideways into his hand, at the same time making a kind of combination of whirring, buzzing & wind blowing noise. This, the pantomimic representation of ships being crushed in the ice.

R. J. Cyriax suggested, 'The Franklin record (i.e. the document discovered by Lieutenant Hobson at Crozier's Landing) proves that both the *Erebus* and *Terror* were still in existence when the retreat began.' He concluded that 'The officers commanding the main body are most unlikely to have known what had happened to the ships since their departure from them.' But the note does not prove any such thing. What the note (written by Captain Fitzjames, second-in-command to Crozier) actually says is 'H.M. Ship(s) *Terror* and *Erebus* were deserted on the 22nd April, 5 leagues NNW of this, (hav)ing been beset since 12 Septr. 1846.'

'Deserted' (often replaced even by Cyriax with 'abandoned') is the language deliberately used by someone knowing he would have to face a court-martial for the loss of his ship. No captain would be prepared to leave such a hostage to fortune that might later be used against him. However, as proof that the ships were not in a functioning or serviceable condition – why would the ship's companies leave them in the bitter cold of an Arctic April? Why did they have time to build large oak sledges on which to carry the ships' boats – boats that were also adapted for river use? And why were the boats' contents clearly a random selection, hurriedly grabbed and thrown into them?

Could the ships have been crushed in the ice fifteen miles to the north-by-northwest of Crozier's Landing? If they had been, it could account for the choice of landing place by Crozier, and the random chaos of stores and personal effects found at the site.

The ships had found themselves trapped in the ice of northern Victoria Strait, directly in the path of the millions of tons of ice pouring down the McClintock Strait. This immense weight of ice does not stop until it hits the northwest corner of King William Island where it piles up in great ridges and heaps until the vast pressure forces it southwards along the Victoria Strait. Franklin's ships had been 'beset' in the ice since September 1846. They had been subjected to the crushing force of the ice for at least eighteen months. It does not require a great leap of the imagination to conclude that, at last, the ships' ribs began to cave in and bulkheads began to crack in April 1848. This was probably not unexpected, and warnings may have prompted the construction of the 23-foot oak sledges for the carriage of the ship's boats. They and the boats would have been placed on the ice ready for an emergency escape if the ships were, indeed, suddenly crushed.

The *Erebus* and *Terror* were stoutly built bomb-vessels with further strengthening for use in ice-bound waters. In their holds they carried a massive iron steam engine. Once their bulkheads had been breached, unless successful damage control had been carried out, they would have sunk like stones. The *Breadalbane*, a much lighter vessel, was crushed by the ice off Beechey Island and sank in less than fifteen minutes.

Consequently, with Inuit tales of sinkings that placed the lost ships at widely differing sites around King William Island, and the only subsequent, considered, approach being taken by Lieutenant-Commander Rupert T. Gould during his time at the Admiralty Hydrographic Office, the position of the ships according to the details in the Back Bay message deposited by Crozier remained the dominant target for searchers. Gould's chart shows the 'probable drift' of the *Erebus* and *Terror* south through Victoria strait to the Simpson Strait. He then indicated the possible links with Inuit tales of ships in Wilmot & Crampton Bay (at the eastern side of Queen Maude Gulf), with one off Grant Point, the other off O'Reilly Island. Gould also included the possible sighting of ship's masts by Anderson's men at the eastern end of the Simpson Strait.

Then came the welcome news that the *Erebus* had been found. The discovery was all the more thought-provoking as the site was at the eastern end of Queen Maude Gulf – at one of the sites indicated by the Inuit, and at a site located by Gould's work at the Hydrographic Office.

The search had begun in 2008 as a national government enterprise with support from the Nunavut Government (since 1999, the north-eastern sector of the old North-Western Territories). Under the direction of the Ministry of Environment and Climate Change, Parks Canada's Underwater Archaeology Team (UAT) worked in co-operation with other agencies, the Canadian Royal Navy, the Canadian Coast Guard, and the Canadian Hydrographic Service.

As the start of a multi-year project, 2008 was essentially a time for planning. However, an interesting, even absorbing, event was the Canadian Hydrographic Service's survey of 40 miles of water leading towards Wilmot and Crampton Bay at the eastern side of Queen Maude Gulf. Two of the several Inuit local legends placed a ship in that area that could have been one of Franklin's. With 2009 proving to be a year of heavy ice accumulation, combined with some of the project partners being committed to use their vessels for other research projects during the International Polar Year, the planning continued.

The following year, with nothing to show for two season's work, it was decided that the projects should head north to locate one of the Royal Navy's Franklin search ships, HMS *Investigator*. In 1851, Captain McClure, an Irishman who intended to set his own rules for his part in the search for Franklin, had found himself trapped on the north coast of Banks Island in what he named the 'Bay of God's Mercy' (a name soon modified – even by his own ship's company – to the modern 'Mercy Bay'). After three years with no break in the immediate ice, McClure decided to break up his ship's company into small parties, one to stay onboard in the hope of a break-out, others attempting to reach differing possible places of refuge. Fortunately, the ship was found by another Franklin search party. The *Investigator*'s survivors were led eastwards and taken home in another ship. McClure was knighted and awarded a large money prize for being the first expedition leader to pass through the North-West Passage (much of it on foot).

Using the Canadian Coast Guard Vessel *Sir Wilfred Laurier*, the UAT employing side-scan sonar had no difficulty in finding the sunken ship. The position was well-known to the Arctic 'Bush' pilots who regularly flew over the site, and its position was recorded by McClure and others when the ship was deserted. It took the UAT somewhere between ten and fifteen minutes to find McClure's ship. So via the press, the Canadian public could see some reward for the government's spending on the project.

The remainder of the year was spent in passing through the southern waters surveyed by the Hydrographic Service in 2008. In all, using the *Sir Wilfred Laurier*, over 60 square miles of the sea bottom was mapped over six days without result. However, as a familiarisation exercise with the eastern side of the Queen Maude Gulf, it could provide later benefits.

In 2011, ice conditions permitted a search in the north of Victoria Strait. The message left at Back Bay by Captain Crozier of HMS *Terror* had identified the area as the position were the ships had been deserted – a position supported by Inuit folklore which told of a sinking ship. Nothing, however, was found. Further to the south, searches were carried out of the Royal Geographical Society Islands and the Alexandra Strait, separating the islands from the south-west coast of King William Island. A University of Victoria team joined the search using the lasers of LiDAR ('Light Detection and Ranging') and side-scan sonar.

2012 saw the arrival of the Arctic Research Foundation (ARF) vessel RV *Martin Bergmann* – a 64-foot former Newfoundland trawler, named after a distinguished Arctic scientist, and modified to serve as a research vessel. She was to be based at the Cambridge Bay settlement on the southeast coast of Victoria Island. ARF was the brainchild of Jim Balsillie (a philanthropist founder of the Blackberry software and communications firm) who would serve as ARF's leader during the expedition. The Coast Guard's *Sir William Laurier* was made available and provided a base for the Parks Canada team. The search took up from where it had left off the previous year and continued south-by-south-east to the area in the eastern Queen Maude Gulf around O'Reilly Island. The University of Victoria team carried out a trial with the AUV fitted with side-scan sonar. In addition, archaeologists were delivered by helicopter to examine small islands in the area, which were referred to by Parks Canada as 'key islands'. As an additional justification of public expenditure, the survey by bathymetric (sea depth and contours) equipment was done so as to 'make the area safe for the navigation of modern ships'. At the same time, however, a dark shadow appeared on the scene. Whilst under the threat of job cuts from the Canadian Government, all Parks Staff – *including* scientists – were forbidden to express any comments, especially criticisms, regarding the organisation.

During the 2013 season, the *Sir Wilfred Laurier* in company with a small diving and research boat named the *Investigator* spent six weeks continuing to survey the sea bottom in Victoria and Alexandra Straits combined with forays around O'Reilly Island, situated in the south-east corner of Queen Maude Gulf. An archaeological party under the direction of the Government of Nunavut searched the shores of Erebus Bay.

2014 proved to be a year of triumph marred by national politics, local politics, and disputes within the search teams. Those from Parks Canada involved in the search were not only still prevented from expressing any criticisms but, from 2014, they were forbidden any contact with the media without approval, and all media requests had to be submitted via the civil servants at the Parks Canada national office with the entire exasperating process being dealt with by e-mail. This constraint probably came into force as a result of interviews such as that given by the project's senior underwater archaeologist, Ryan Harris, the previous year. Clearly an intelligent man, Harris, being interviewed for television on the aims of the approaching season, has to repeatedly turn away from his interlocutor in order to read from a script. The result was a stilted conversation that failed to reflect well upon the archaeologist, or his employers, and was dismissed by journalists and others simply as 'scripted'.

However, the search was joined by an organisation on whom the Parks Canada restrictions had no bearing. The *Akademik Sergey Vavilov* (still flying the 1993 Russian Federation flag, and bearing the name АКАДЕМИК СЕРГЕЫ ВАВИЛОВ – but, for some reason, re-named *One Ocean Voyager*) arrived on the scene having been offered to the Royal Canadian Geographical Society (RCGS) by One Ocean Expeditions. Onboard the vessel was the RCGS Chief Executive Officer and former President John Geiger.

John Geiger was no shrinking violet. He had a reputation for thrusting himself to the forefront of any situation that won his attention. When Dr Beattie completed his examination of the bodies on Beechey Island, it was Geiger – who had not been present at the disinterment – who co-wrote the book *Frozen in Time*, which foisted the notion of lead poisoning onto a gullible world.

In February 2006, he took part in a documentary *Arctic Passage* aired on American PBS, along with several others including an anthropologist, Anne Keenleyside. Both John Geiger and Dr Keenleyside had an agenda. Dr Keenleyside was quickly on the attack:

Narrator: Could it be that the men actually survived the ice only to turn on each other?

Keenleyside: I think this evidence is strongly suggestive of cannibalism among these Franklin crew members. These look like very definite cut marks as if they were made by some kind of a knife or metal blade. A lot of the cuts were located in the vicinity of the joints.

Geiger then joined in as if had been present at the exhumation:

It was such a profoundly moving experience... You could see their eyelashes; you could see their eye colour... It's as if they had stepped forward in time.

At this stage, the narrator added a common-sense comment:

At the gravesite, the forensic team performed autopsies and discovered that the men had died of tuberculosis.

The point was lost on the lead poisoning theorists who continued to propagate their narrative along with their allies, the cannibalism zealots.

The programme was probably well meant but suffered from an over-eagerness to sacrifice Franklin and his men on the altar of poor academic research. Dr John Rae, for example, was described as a

'Canadian explorer' when he was actually an Orcadian revered by the local people, and an anchor was described as being 'hoisted' (as if it was a flag on a halyard) instead of being 'raised'.

In 2014, Geiger was eager to join in the search for Franklin's ships as leader of the RCGS, after five years of work by Park Canada and others. On 27 July *The Record* published several of his comments in a feature on the approaching expedition. He was introduced as 'John Geiger, CEO of the Royal Canadian Geographical Society, who has personally been hunting Franklin for decades.' This is arguably an overblown description of a few minor activities and was later expanded to include the exhumation of bodies during 'another Geiger-led expedition in the 1980s' (an activity which history has failed to record). Explaining the outcome of the 1845 Expedition, Geiger informs the reader that 'the ships had been abandoned and the remaining crew was setting out on foot. *This is the origin of the terrible death march which led to starvation, cannibalism and poisoning...* We are inheritors of the British Arctic exploration legacy. Franklin's disappearance in the central Arctic, and the subsequent searches are really the reason Canada can claim sovereignty over the Arctic Archipelago.'

The RCGS and its One Ocean partner did not come alone. In a Memorandum of Understanding between the RCGS and Parks Canada, the RCGS was allowed to bring along The W. Garfield Western Foundation (a long-established and well-respected Canadian philanthropic organisation), Shell Canada, and – as a last-minute addition – a British film company, Lion Television.

The 'One Ocean Expeditions' ship, *Akademik Sergey Vavilov* (hereinafter referred to as the *Vavilov)* carried one of the project's remote operated vehicles (ROV) and their autonomous underwater vehicle (AUV). Employing the latest side-scan sonar technology, the AUV was capable of searching the seabed independently, whilst the ROV was controlled by a ship-borne operator. LiDAR – effectively a system that uses lasers as sea-floor mapping radar – was also employed. The enrolment of Canada Shell was entirely a practical proposition. The mapping of the sea floor could provide a significant advantage to any future oil exploration.

Geiger had good reason to become involved – there was something in the air. The Royal Canadian Navy was sending one of its ships, HMCS *Kingston* – a sure sign that the government wanted, and expected, a result with a General Election due the following year. The Minister of the Environment and Chair of the international Arctic Council, Leona Aglukkaq, issued a Press release in which she stated that 'Our Government has made the North a priority. Through exploration and research in Canada's Arctic, we can understand our past and secure our future.'

Furthermore, although the overall costs of the search expeditions had fallen since they began, any slack was taken up by the purchase of the autonomous underwater vehicle (AUV) and an upgraded remotely operated vehicle (ROV) at a combined cost of $475,000. The AUV would continue the sonar work, whilst the ROV now came equipped with a high-definition camera.

With the ships in place, the search teams ready, and the search restricted to the southern area as the ice held its position in the north, the exploration began beneath blue August skies.

On the 23rd of the same month, almost 600 miles northwest of the southern search area, an aircraft landed at the tiny airport of Pond Inlet on the north coast of Baffin Island. Several people disembarked from the aircraft and were taken through the little hamlet to the Sealift Staging Area. From there they were ferried out to HMCS *Kingston* where they were greeted by Stephen Harper, Canada's Prime Minister. This elite band were representatives of the Park Canada search teams, the Royal Canadian Geographical Society, the Arctic Research Council, the Royal Canadian Navy, and other partners.

The previous day, Harper had called into Fort Smith, a Northwest Territory settlement that had been the Territory's capital until Yellowknife took over. A reporter from the National Press asked Harper if increased Russian activity had caused an increase in demonstrations of Canadian Arctic sovereignty. Being en route to observe Canadian forces in action during Operation Nanook – a demonstration of Canada's response to any threats and emergencies – Harper replied that the purposes of such demonstrations was 'search and rescue'. There was no mention of the search for Franklin's ships as a confirmation of sovereignty.

The 'Russian activity' centred around a titanium plaque representing a Russian flag being deposited on the Arctic seafloor by a Russian AUV in 2007 close to the North Pole. A far better demonstration of Canadian concern over sovereignty would have been for the Canadians to have collected the plaque and announced to the world that a wholly alien lump of titanium had been found in Arctic waters. Such unwanted junk could have been publicly auctioned off to a scrap-metal merchant with the money raised being given to a charity for something vital, such as the preservation of snowy owls.

On the day following the Fort Smith visit, Harper, now on board the *Kingston* anchored off Pond Inlet, addressed the group of about 20 individuals packed onto the ship's bridge. With the press out of the way, he told the search teams, partners, and sponsors of the Franklin search that 'It ultimately isn't just a story of discovery and mystery and all those things, but it's also really laying the basis for what's in the longer-term Canadian sovereignty.'

He then expressed his confidence in the search for Franklin's ships by adding, with a mixture of levity and a certain lack of regard, 'One day we're just going to come around the bend and there's going to be the ship and Franklin's skeleton slumped over the helm, and we're going to find it.' The time had come to pass glasses of whisky around, and John Geiger proclaimed a toast to Franklin, and then to his fellow searchers

The muster of those involved on the bridge of the *Kingston* had not only interrupted a search that was already under way, it also produced great hilarity throughout Canada with a series of photographs of the Prime Minister in heroic mode – especially one of him standing on the forecastle of the ship in the manner of a 21st-century Derick the Viking.

Once everyone was back on site, an air of competition seeped into the event. Geiger, convinced that Franklin's ship (or ships) would be found in the northern search area, set off in that direction on board the very comfortably fitted-out *Vavilov* (still flagged as a Russian vessel, but described as the RCGS's 'floating headquarters'). He was accompanied by a group of well-heeled tourists eager to be part of the great discovery. His choice of search area, and his reasons behind such a decision was explained to his Society's magazine *Canadian Geographic*: 'A note left behind by Lieutenant Graham Gore in a stone cairn on King William Island was critical to learning what we know about the expedition. It is because of that note that we know the ships were last seen in the Victoria Strait, the site of this year's search.'

He then added – in the manner of a 'virtue signalling' 21st-century Dr John Rae – further justification:

> Cut marks on skeletal remains gave us physical evidence to support historic Inuit testimony that the expedition explorers eventually resorted to cannibalism. Recent years have seen advanced technologies and forensics add to what we know. University of Alberta anthropology professor Owen Beattie's work on the crews' remains has helped us understand their health and given us insight into what happened. There's clear evidence of scurvy, cannibalism and lead poisoning.

Such 'clear evidence' did not exist.

Geiger also had on board the very latest, up-to-date, AUV – the *Arctic Explorer* – for searching the sea bottom. Unfortunately, no-one seems to have checked if the AUV was configured for searching under ice – it was not.

In the meantime, the ARF, using the *Martin Bergmann*, began the vital task of exploring the sea floor at the northern end of the southern search area. The work carried the promise of two prizes – firstly, the greatest prize; they could be lucky and stumble upon either of Franklin's ships; or, secondly, if the search produced nothing, it would considerably reduce the search area for subsequent explorations. The vessel carried members of the UAT who would, if the conditions permitted, operate the 'towfish' ROV as it passed over the sea floor. They had an unfortunate start to the search when some of their sonar equipment needed repair, whilst other kit had been delayed before arriving at the sheltered settlement of Cambridge Bay. The weather proved also to be a hindrance, particularly the combination of a rolling swell – which made ROV work difficult – and a breeze too light to break up the ice.

On 30 August the *Vavilov* sailed from the Inuit settlement at Resolute Bay. Following the wake of two Canadian Coast Guard ships, she came down Prince Regent Inlet, through Bellot Strait, and down Victoria Strait to a point off Jenny Lind Island, where a rendezvous was made with the *Sir Wilfred Laurier*. At that point, a successful transfer of the diving tender *Investigator* and two Canadian Hydrographic Service launches, the *Gannet* and the *Kinglett*, to the icebreaker was completed. The CHS launches were carrying high resolution multi-beam sonar – equipment that could produce a three-dimensional image of anything found on the seafloor. The *Vavilov* was then free to return northwards to the top of Victoria Strait – the area where Franklin's ships were known to have been deserted. At most, the ice was fragmentary 'brash', or widespread, thin surface ice that parted easily under the pressure from the *Vavilov*'s bows. Although there were several areas of ice-free water, and the ship's helicopter easily located other areas within reach, there was a marked reluctance to put the 'state of the art' (and hugely expensive) AUV into the water if there was the slightest risk of the equipment being damaged, or even lost (this reluctance was borne out in 2019 when a similar piece of technology, searching for Shackleton's ship, the *Endurance*, was lost under the ice of the Antarctic's Weddell Sea).

A few very restricted trial runs were carried out in open areas – but that was it. Accordingly, Geiger sent messages to the southern component of the expedition, heavily hinting that he wanted an icebreaker. Unfortunately for him. the effect was to drive the *Sir Wilfred Laurier*, along with ARF's *Martin Bergmann* to the south-east – into the eastern part of the Queen Maude Gulf.

Monday, 1 September provided a providential – if rather strange – combination of circumstances. In order for the CHS launches to carry out an accurate survey of the seafloor, global positioning apparatus had to be set up nearby. Such equipment assists in keeping the long sweeps by the launches in a tight pattern of searching and reduces the chances of inaccuracy caused by drifting off track. Consequently, it was decided that such equipment would be placed on the small Saunitalik Island about twenty miles north of O'Reilly Island, and about ten miles west of the Adelaide Peninsula. It would be set up by the hydrologist-in-charge of the CHS group (however, it should be noted that the crew of the *Sir Wilfred Laurier*, including the Captain, were led to believe that the selected destination was to be Hat Island – some miles further to the west).

The hydrologist was to be set down on the island by the *Sir Wilfred Laurier*'s helicopter, whose pilot, Andrew Stirling, was the son of a Royal Naval Chief Petty Officer. This left the aircraft with two empty seats which went to two very senior archaeologists; one was Professor Robert Park, who specialised in the Inuit of Arctic Canada and the other Douglas Stenton, the Director of Heritage of the Nunavut Government who was later to claim that some of Franklin's men were actually *women*. Both men had been associates for a long time and shared the authorship of a number of books. The reason given for the presence of the two archaeologists was so they could carry out ground-based LiDAR surveys of the island and investigate Inuit stone tent rings – a surprisingly low-level activity for two very highly qualified archaeologists.

After landing on the island, the helicopter pilot, whilst keeping an eye out for marauding polar bears, was walking along the sea edge when he spotted an incongruous, rusty iron object lying on the shingle. Nearby was a wooden disc, broken into two semi-circular parts. Calling over the two archaeologists, he showed them his finds.

The iron object was approximately seventeen inches long and shaped like the business end of a huge tuning-fork. But it was not the size or shape that caught the archaeologist's eye – it was the pair of 'broad arrows' stamped either side of the number '12'. Clearly, the broad arrows indicated the item was Royal Naval property, and the number '12' may have been the equipment inspector's personal number. It did not take long to identify the item as part of a ship's davit, used for raising and lowering the ship's boats.

The wooden object, on the other hand, carried no markings and was declared to be a 'hawse plug'. This would have been intended to cover the hawse-pipe which led the anchor cable down into the cable-locker. However, the object looked somewhat fragile, had no horizontal battens to keep it rigid, nor did it have the sort of holes that would have allowed the anchor cable to be fed through.

No one seemed to question how the two widely differing objects were found in the same place. The wooden object was small enough, and light enough, to have floated to the surface. Once there, it would have been subjected to the random tidal effects and winds that could have carried or blown it to any point of the compass. On the other hand, the solid iron davit fitting (weighing about 12 pounds) must have detached itself from its parent vessel, rolled along the upwardly sloping seabed and thrown itself on land just a few feet from where the wooden object lay. What were the chances? Furthermore, the fact that the archaeologists were studying native tent rings strongly suggests that the local Inuit stayed on the island. In a region where both wood and metal were normally extremely scarce (apart from the abandoned remains of John Ross's ship), how did they miss such an opportunity to add the items to their property?

The two objects were collected and returned to the *Sir Wilfred Laurier* before being flown promptly to Ottawa. In the meantime, Geiger continued his demands that the icebreaker should proceed north and carve a path for the *Vavilov* towards the northwest corner of King William Island. Being very well connected, he raised his sights above the collection of ships in Victoria Strait and the Queen Maude Gulf, and aimed at the Vice President of Parks Canada, a government department. He was, however, floored when an e-mailed reply informed him that 'the operation has been managed effectively by the *professional teams*' and his part of the expedition was not in possession of the required Nunavut 'permit authorisations'. Consequently, the icebreaker would not be on its way north, and would concentrate on the activities in Queen Maude Gulf.

Geiger's case was not improved when, at the same time, a Canadian Ice Service research scientist on the expedition (one of the 'professional teams') reported that 'most of the ice in the area is first-year ice, and it's rotten, meaning that it's close to disintegrating.'

The day following the discovery of the two artefacts, the *Investigator* left her parent ship to have a closer look at the area. Three days later, a radio message arrived on board the *Vavilov* – it was for the Parks Canada chief underwater archaeologist, the highly respected and vastly experienced Marc-Andre Bernier. He was in the cruise ship to give enlightening talks to the dignitaries and wealthy tourists who had bought passage in the ship in the hope that they would be present at a moment of great discovery. The message, however, required him to return urgently to the *Sir Wilfred Laurier* as there were staff and finance problems that need his immediate attention. Once he had returned to the icebreaker, he was spirited off in the *Investigator*, the small diving launch carrying an ROV armed with a battery of high-resolution cameras.

Just a short time later, the *Investigator* returned and tied-up alongside the much larger vessel. Expecting the divers to make their way to the bridge to report to him, the captain was shocked when he heard himself being requested, on the ship's broadcasting system, to return immediately to his own cabin. Clearly, such a peremptory request could only mean bad news – a death, perhaps, or mutiny? The captain stiffened his back and made his way to his cabin.

Waiting outside the door, he found the two underwater archaeologists, Bernier and Harris, holding a lap-top computer. The three entered the cabin and, with little conversation, the laptop was opened up and switched on. To his amazement and utter joy, the captain saw an image of a rather battered-looking wreck on a rubble-strewn sea floor. The sunken vessel had actually been found on the first day of the search, in the area prompted by the discovery of the two artefacts. For the next two days, the dive team – and everyone else on the launch – had to bite their tongues and never mention the discovery. Now, with the arrival of the chief underwater archaeologist, the wreck was confirmed to be one of Franklin's ships – although whether the *Erebus* of the *Terror*, it was not possible at this early stage to say. Sat upright on the bottom, the dismasted vessel was just 36 feet below the surface (or, as Franklin, Crozier, Fitzjames, and the others would have approved – '6 fathoms').

With commendable foresight, a pre-existing protocol snapped into operation. Parks Canada's chief archaeologist used the ship's satellite 'phone to inform the Department's Vice-President of the news. The following evening, Bernier was in Ottawa. Even before he had left the ship, the captain oversaw the strict security requirements insisted upon by Parks Canada. Both the ship's internet links and the satellite 'phone were put out of action, and everyone on board was required to sign a nondisclosure agreement. It was imperative that no mention of the find made its way to the media before an announcement could be made by the Prime Minister. However, the stately procedure received a severe jolt on the morning of 8 September when a press release was issued by the Government of Nunavut. The release revealed the discovery of the two artefacts found in the island beach by the *Sir Wilfred Laurier*'s helicopter pilot. Inevitably, the news was picked up for the first time on board the *Vavilov* – still somewhere north of the Royal Geographical Islands. Last-minute invitations were rushed out, inviting representatives of the partner organisations to pack their bags as flights were being organised to enable them to reach Ottawa in time for the great announcement. There were, however, certain restrictions. The occasion was to be nothing more than a photo opportunity accompanied by a statement by Stephen Harper. No journalists were to be invited, nor would the Prime Minister accept any questions.

The event, held at the Parks Canada laboratories in Ottawa, opened at 10am with Harper and the invited guests sat around three sides of a long, rectangular table. The top edge had the Prime Minister in the centre with the Parks Canada senior underwater archaeologist, Ryan Harris, on his right. To Harper's left sat the Minister of the Environment, Leona Aglukkaq, who, as part of her portfolio, had responsibility for Parks Canada. On the Minister's left, next around the corner of the table, sat John Geiger. The Prime Minister began with a rather surprising lack of confidence. Smiling weakly, he began, 'This is a, er, day of, er, some very good news, and that is that, er, we have found one of the two Franklin ships.'

He need not have worried. At that moment, with the precision of a North Korean Guard of Honour, the table burst forth with a round of applause that would have done credit to a Politburo being told by Stalin who was next to be executed – everyone wanted to be first to start applauding, no-one wanted to be the first to stop.

Over the next few days, journalists of every medium wanted to hear the story. The government employees, however, forbidden from direct contact with the press, could not help. There was, nevertheless, one person who could – John Geiger.

Seizing the opportunity, Geiger spoke at press conferences and appeared on television shows. His general demeanour may be judged from an interview he attended on the morning of the 11th. Whilst appearing on the *The Morning Show* of the Global News channel, he was asked 'What's the moment of discovery like? How do you know you have found it?' Geiger replied:

> I wasn't in the room when the hit was made, but after looking for, actually years, for something up there, and suddenly seeing, what was obviously a ship across the screen, you can imagine the excitement and sense of euphoria, that spread throughout the expedition.

Not only was he not in 'the room' – he was 75 miles to the north of it and was not to find out about 'the hit' until six days afterwards. Nevertheless, the tale was worthy of expansion. When interviewed later, he was asked 'Describe that moment when you realized, "That's it!"'

> I was euphoric, obviously. I was extremely excited. Very happy. There was a toast proposed very shortly thereafter. But I was also haunted by it, a little bit, as I have always been by the expedition, by the fact that 129 men died. And, y'know, when we were there, I had the Anglican *Book of Common Prayer* with me, and I cited a prayer for those lost at sea as well. And I reflected on the fact that there may well be human remains on that ship.

Toes curled across the Victoria Strait and Queen Maude Gulf as the dropped jaws of men and women on board the expedition ships hit the deck. One of those most affected on the *Sir Wilfred Laurier* was a Pulitzer Prize-winning journalist working for a leading Canadian newspaper. Concerned by Geiger's interpretation of the events around the discovery, he sent off a more factual version. To his consternation and disappointment, not only was his work at first ignored, but it was soon heavily suggested that he drop the matter completely. Consequently, he resigned, and spread the details on the internet.

Then came the film, made on behalf of the RCGS, and starring John Geiger. In a peculiar manner, the film was made in two versions. The first, made for Canadians only (and not to be shown outside the country) the second for the world in general.

The first was aired on 9 April 2015 on the well-respected CBC's *The Nature of Things* programme under the title of *Franklin's Lost Ships*. It did not take long to detect the tenor of the film's thrust. Before long the commentary informed the viewer that 'The British Navy would be the first to navigate the North-West Passage – and then they would have a stranglehold on the trade to the east.'

This was nonsense. The British had no more intention to bar the North-West Passage to other nations than they had to do the same in the English Channel. The only objection to the achievement of such a passage came from the Hudson Bay Company and its employees (including Dr John Rae). The last thing the HBC wanted was a threat to its monopoly across Canada – a monopoly which would vanish once the passage was achieved.

Then one of the leading 'usual suspects' popped up; someone I had last seen leaving the Admiralty Board Room when I challenged one of his assumptions. Ken McGoogan, the author of a hagiographic history of John Rae, informed the viewer that the Admiralty felt, 'Well look, it's going to be a piece of cake. We're going to send Franklin in – all he has to do is join this northern channel with the southern channel, and he's going to emerge into the Pacific trailing clouds of glory.' He later added, 'There was a lot of evidence that they'd started to eat each other, er, the dead bodies.'

He did not expand on his 'lot of evidence', but the viewer should be grateful that he clarified matters by admitting it was 'the dead bodies' that they were supposed to be eating – rather than biting lumps out of each other as they dragged the boats behind them.

The 'outside Canada' version – entitled *Arctic Ghost Ship* – had a better (but still inaccurate) commentary. Geiger – described in the commentary as 'One of the driving forces behind this year's search', told the viewers

'Ships don't just disappear. If there is a Franklin expedition ship, we will find it.' However, there are centuries of evidence that ships *do* 'just disappear', there *was* a 'Franklin expedition ship', and he did *not* find it.

The talking heads did their usual business. A lugubrious British academic, talking about the three graves on Beechey Island, pronounced with an air of bafflement, 'This shouldn't happen, three men should not die in the first winter of an Arctic expedition. They've only been out of Britain six months – what's killing them?'

There is no evidence that there was a 'Queen's Regulation and Admiralty Instruction' forbidding seamen to die at a particular time. There is, however, overwhelming medical evidence that suggests the cause of the deaths was tuberculosis or pneumonia. Such illnesses were commonplace in ships up until the late 1950s when the hanging of wet clothing on the messdecks was banned and 'drying rooms' were introduced. In 1880, when Frederick Schwatka arrived at Marble Island off the northwest coast of Hudson Bay, he took passage in the whaler *George and Mary*. The whaling ship had suffered two casualties during the whaling season; one was an officer who froze to death on the ice, the second was recorded as having died from 'consumption' – tuberculosis.

Geiger was frequently filmed on the upper deck of the *Vavilov* to a background of rotten ice-floes. Other shots showed the ice covering the sea, only to be easily parted by a gentle nudge from the *Vavilov*'s bows. This did not stop Geiger addressing the camera and saying, 'As hard as this may be to believe, this is summer in the Arctic. This part of the Arctic is as good as it is going to get this year.'

The commentary continued, 'There's more ice here this summer than there's been for several years – *it's thought that Franklin faced exactly the same conditions 170 years ago*.'

Thought by whom? The ship's companies of *Erebus* and *Terror* would have danced the hornpipe and spliced the mainbrace at the sight of so much open water. Even if, by some fluke of nature, the ships were still fast in the ice, the ship's boats – based on the design of 27-foot whalers – could have sailed south though the floes and established supply dumps all along the western shore of King William Island. There would have been no dragging boats on sledges – in fact, nobody would have been able to trek across the ice from where the ships were trapped to Crozier's Landing.

Interestingly enough, a rather different American attitude to broken ice was demonstrated in 1850 by the USS *Advance* under the command of Lieutenant Edwin J. De Haven. A visitor from the recently encountered British ship, the Lady Franklin-sponsored *Prince Albert* was William Parker Snow, who later earned a reputation for rather erratic and unfortunate behaviour in Arctic exploration. Snow wrote of the American approach towards ice-floes streaming past the vessel (the italics are Snow's):

> The way was before them – the stream of ice had to be gone through boldly, or a long *detour* made; and despite the heaviness of the stream, *they pushed the vessel through in her proper course.* Two or three shocks as she came in contact with some large pieces were unheeded; and the moment the last block was past the bow, the officer sang out, 'So: steady as she goes on her course,' and came aft as if nothing more than ordinary sailing had been going on. I observed our own little bark (the *Prince Albert*) following nobly in the American's wake; and, as I afterwards learned, she got through it pretty well, though not without much doubt of the propriety of keeping on in such a procedure after the 'mad Yankee' as he was called by our mate.

Inevitably, much of the film's screen-time centred on the alleged 'cannibalism'. The commentary accompanying film of dead-eyed men chewing on raw bones informed the viewer that cut-marks found on the bones 'were made by metal blades. Flesh was stripped from these bones by knives forged from British steel in a last, desperate bid for survival.'

Does 'stripping' flesh from bones produce cut-marks? Unlikely. Does hacking with an edged weapon produced cut-marks? Probably. Would shipmates of several years stand by and watch as one of their number is hacked down to provide food? Highly unlikely. Were the retreating seamen the only source of metal knives? Certainly not. For years, the Netsilik people had a huge source of metal from Captain John Ross's abandoned ships. There was, of course, no mention of the 40 pounds of 'navy chocolate' remaining in the boat found by the McClintock expedition. Geiger rounded off the film. Addressing the camera he said:

> This is a great moment for exploration. We've been searching for a hundred and sixty years for answers to what happened to the Franklin Expedition. To finally have something significant, to finally have a ship is incredible. (Switch to film of the *Sir Wilfred Laurier*, over seventy miles to the south.) I've spent most of my adult life dreaming of this day, and it's here.

The entire film was a project too far for the philanthropist and founder of the Arctic Research Foundation, Jim Balsillie, whose vessel, *Martin Bergmann*, had done valuable work in seafloor searches that led to the narrowing down of the eventual target area. He wrote a letter to the

Environment Minister (who was responsible for Parks Canada), with a copy to the Prime Minister's office. In it, he stated:

> I am concerned that the documentary contains information that runs contrary to the planning meeting that was held in your office on June 9th 2014 and filmed for the Prime Minister's online news channel... The narrative, as currently presented, attempts to minimize the role of the government and its respective agencies and private partners. It also creates new and exaggerated narratives for the exclusive benefit of the Royal Canadian Geographic Society and its own partners.

After giving a detailed explanation of the actual achievements of the expedition, Balsillie underlined his objections:

> While I don't want to speculate about the motivation of RCGS and its partners in creating an alternative narrative for themselves and their role in the Victoria Strait partnership, I am concerned that official communication outputs, such as this documentary, contain versions of the search that are misleading to the Canadian public.

There was no reply. However, Geiger appeared in a photograph showing just himself triumphantly holding up the iron davit pintle. And then came the medals. The RCGS had just instituted the Lawrence J. Burpee Medal:

> Awarded to recognize an outstanding contribution or other achievement that greatly enhances the ability of the Society to fulfil its mission of making Canada better known on a national or international level, and/ or contributes to the general advancement of geography.

The first Lawrence J. Burpee medals were awarded to the Parks Canada dive team – Marc-André Bernier, Ryan Harris, and Jonathan Moore. John Geiger did not qualify. Then came the hurriedly introduced Erebus Medal:

> Established by The Royal Canadian Geographical Society in 2015, the Erebus Medal recognizes participants in the 2014 Victoria Strait Expedition and their contributions to the discovery of HMS *Erebus*, the exploration ship commanded by Sir John Franklin and lost during his ill-fated 1845-48 expedition. All participants in the discovery, including those in the field and those who worked behind the scenes are recognized with the Erebus Medal. Together, they have rewritten the history books and underscored the importance of the geography of Canada's Arctic.

This time John Geiger's name was on the list. One was given to the Prime Minister, and another to his wife, Laureen Harper, and to Dr Owen Beattie.

In 2015, the Canadian Government introduced its own 'Polar Medal' based on the design of the white-ribboned British Polar Medal. The very first presentation had four recipients: Louie Kamookak, an Inuit historian; Ryan Harris, the senior underwater archaeologist at Parks Canada; Douglas Stenton, the Government of Nunavut's Director of Heritage – all of whom had their contribution highlighted in the medal's citation. They had 'played essential roles in the success of the 2014 Victoria Strait Expedition. Their passion, dedication and perseverance contributed directly to the discovery of the wreck of Sir John Franklin's HMS *Erebus* in September 2014 – and to the resolution of one of polar exploration's greatest enigmas.' All their achievements and the prestige of the award, however, were in my opinion somewhat undermined by the presence of the first name on the list – John Geiger.

As he left the building after the presentation ceremony, Geiger was approached by a reporter who asked him what he and the RCGS had done towards the discovery of the *Erebus*. Geiger ignored the question and marched on. When the reporter tried to take a photograph, a woman at Geiger's side jabbed her hand at the camera lens. No prayer was said for those lost at sea.

The actual discovery of the *Erebus* was, under any circumstances, a great achievement. There remained, however, a number of unanswered questions. Firstly, how did the ship get to where she was found?

A practical reason for the ship's final resting place could be found in a long-forgotten incident that took place on a dark Christmas Eve 1789, 2,000 miles south of the Cape of Good Hope. Commanded by Lieutenant Edward Riou, the 45-gun HMS *Guardian,* carrying the ship's company, passengers, and convicts bound for Australia, collided with a large iceberg. Within minutes, the lower hull had been ripped open, the rudder had been wrenched away, and the stern frame shattered. After an exhausting night manning the pumps, and with attempts to stop the flow of water by lowering sails over the side having failed, Riou allowed those who wished to try and escape to take to the five ship's boats. Two hundred and fifty-nine people took to the boats, leaving sixty-two in the ship (including one young woman).

By mid-morning, there was 16 feet of water in the hold, and the ship's ballast had fallen through the bottom of the fractured hull, leaving the ship lurching wildly. Riou went down to the lower gun-deck to see the situation – and what he saw, dictated his subsequent actions.

It was the usual practice when wooden casks had been emptied, to re-seal them, and store them until they could be returned to the dockyard.

Peering into the hold, Riou saw that the empty casks had broken free and were floating on the top of the water, which was within inches of the hold deck-head. Acting quickly, Riou gathered all available hands and had the lower-gun deck hatches closed and sealed with pitch and oakum (the strands of shredded rope). Any other gaps in the deck were caulked with the same mixture. In effect, Riou had converted the lower-gun deck into the ship's bottom, with the casks acting as flotation aids.

Nine weeks later, the *Guardian* ran ashore near to Table Bay. There had been heavy losses amongst those who had taken to the ship's boats, but all those who had stay with Riou survived – including a midshipman who was not only closely related to the First Lord of the Admiralty, but also to the Prime Minister. 'The Gallant Riou' died nine years later under Nelson's command at the Battle of Copenhagen, and Parliament voted him a large and distinguished memorial in St Paul's Cathedral (the midshipman proved to be a particularly obnoxious specimen who was to sail under the command of Captain George Vancouver. He later gained infamy when, meeting his former captain in a London street, he thrashed him with his walking cane). The lesson of Riou's survival would not have been lost on the subsequent generations of Royal Naval officers – particularly those entering the hazards of icy regions.

It would not be at all fanciful to suggest that the *Erebus* was 'nipped' by the ice when trapped off northwest King William Island. With the lower hull beyond repair, the loss of ballast – and, possibly, the fall of the railway engine through the ship's bottom, or perhaps it was thrown overboard – Crozier was faced with a single option; temporarily supported in the grip of the ice, all he could do was to try and keep the ship afloat by sealing off and caulking the main deck (and possibly employing empty casks to add to their flotation aids – the 'hawse plug' found at 'Davit camp' on Saunitalik Island, may well have been a cask lid).

Before the work of sealing the ship was completed, and in accordance with naval tradition (and to keep the local natives from ransacking the ship if they came on board, or if the ship fell into the hands of any fellow mariners), it is highly likely that abandoned clothing would have been stuffed into outer clothing and the crude mannequin propped up at the Wardroom dining table. A naval cap would have been placed on the head, and the face decorated with large staring eyes and huge grimacing teeth. Little would the jokers have thought that when the Inuit reported the sight to Franklin searchers, their 'jack-acting' effort would have been interpreted as everything from the grimacing grin of a corpse, to a stuffed walrus. An article in *The Guardian* (12th September 2016) mentioned 'startled Inuit stumbling upon a large dead man in a dark room on the vessel, with a big smile. Experts have suggested that may have been a rictus smile, or evidence that the man had suffered from scurvy.' Hall heard the account from the Netsilik themselves: 'To get into the igloo (cabin), they knocked a hole through because it was locked. They found there a dead man, whose body was very large and heavy, his teeth very long. He was left where they found him.'

Professor Potter came close to a similar idea based upon an Inuit tale of visiting one of the ships. According to Inuit folklore, when they went on board they were terrified when three 'black men' appeared from below decks. The Professor thought that the ship's company were celebrating 'Mayday Festivities' where 'Blackamoors' may have taken part, or Guy Fawkes's Night, when there would, of course, be a Guy. Had such an incident actually occurred, it would, however, be more likely to have been the popular entertainment of a 'Minstrel' troupe taking part in an evening's 'Smoking Concert'. An example of self-organised entertainment was recorded by Lieutenant Sherrard Osborn in his *Stray Leaves from an Arctic Journal*:

> Vocalists and musicians practised and persevered until an instrumental band and glee-club were formed to our general delight; officers and men sung who never sang before, and maybe, except under similar circumstances, will never sing again, maskers had to construct their own masks, and sew their own dresses, the signal flags serving in lieu of a supply from the milliner's; and, with wonderful ingenuity, a fancy dress ball was got up, which in variety and tastefulness of costume, would have borne comparison with any one in Europe.

With regard to the idea of the steam engine being thrown overboard, *Fraser's Magazine* of 17 February noted that:

> Availing ourselves of the official documents relating to the arctic expeditions, which have recently been published by the admiralty, and of information derived from authentic sources... The fitting up of the (*Erebus* and *Terror*) differed in one respect from that of all ships previously sent out on Arctic expeditions. This consisted in their being furnished with a small steam-engine and Archimedean screw. But in the experimental trip made by the *Erebus*, to test the power of the screw, the utmost speed that could be attained scarcely amounted to three knots an hour, although every means, as we ourselves can attest, were taken to increase this rate. The ships were supplied with fuel for twelve days – a quantity manifestly insufficient for their probable wants, but still as much as they could afford to stow away, having to carry provisions for three years. When the very small speed producible by the steam-power

is considered, coupled with the great inconvenience likely to result from the most valuable part of the ship being occupied by the boiler and machinery, not to mention the great probability of the screw being nipped by the ice, we shall be quite prepared to hear that this part of the equipment has turned out a failure, and that the steam machinery has been thrown overboard, as happened in the case of the *Victory*, commanded by Sir John Ross, in his expedition undertaken in 1818.

With the onset of scurvy, and no prospect of the ice releasing them, the isolating of the main-deck would have given Crozier and his men time to construct stout oak sledges to carry the cut-down ship's boats that gave them their only chance escaping to the south.

If the *Terror* had not, as according to Inuit tales, been crushed by the ice, or been driven ashore, it is entirely reasonable to assume that she would have been prepared to face the rigours of yet more isolated Arctic winters. Still watertight (although leakage in wooden ships was considered normal), she would have had her upper-deck hatches secured and between-decks doors and hatches closed and fastened.

In the end, whatever the condition of the *Terror*, it made sense that the journey southwards of the ship's companies should be a unified effort. There seems little doubt that in late April 1848, one hundred and five men left Crozier's Landing to head in the direction of Back's Fish River. They were dragging at least three ship's boats on sledges. Each would have had room, under normal circumstances, for about twenty men when placed on the water – twelve rowers 'double banked' (McClintock's twenty-four 'rowing thowells'), one steersman, the remainder at the head and stern sheets. They would not – as is usually suggested – have attempted to go around the coast on the ice. Firstly, the ice off the northwest coast of King William Island, even in the summer, tends to be very uneven and roughly piled up. The ship's companies would have had enough experience of the effort involved when transporting their stores over the broken ice from the ships to the shore. Secondly, from Crozier's Landing (on Back Bay), the view to the south consists of the land beyond Collinson Inlet sweeping far to the west until it reaches Franklin Point. To have gone in that direction would have added many unnecessary miles to a route which led directly to the south, and which was already known. And, thirdly, sledge travel, even over bare tundra and shingle, is possible with a sledge of sufficient length (over 23 feet in the case of the Franklin survivors' sledge found by Hobson). Add to that a surface of snow, and thirty or more men of average fitness should have been able to get their loads under way overland without too much difficulty (it is also likely that Franklin's men had experience of summer overland sledging during their stay at Beechey Island).

However, those men who were suffering from scurvy at the outset of their journey would soon find that the effort involved in hauling the sledges would cause their condition to worsen with great rapidity. Significantly, among the symptoms of scurvy are swollen ankles and pain in the leg joints. Some of the men would soon have difficulty in simply walking, much less sledge hauling. Such men would have become useless to the task in hand, requiring the support of their comrades or riding on the sledge-borne boats. Then the inevitable deaths would begin – two men were granted release as the procession made its way along the shingle ridges by Point Le Vesconte and were given a formal, simple burial.

At a place on the south coast of Erebus Bay, between De La Roquette River and Little Point, the senior officer called a halt. By now enough of the men were quite incapable of further progress and the original plan to strike out straight for Back's River was out of the question for all but the fittest.

There was a good reason why the survivors had halted at this particular spot. In May 1847 a party consisting of two officers, Graham Gore and Charles Des Voeux, left the ice-bound ships with six men and travelled south along the west coast of the island. Thanks to the Gore expedition, the hundred or so men who had stopped on the southern shore of Erebus Bay would have known two things: that they could continue overland directly south to Terror Bay, thus avoiding a long haul around the Graham Gore Peninsula; and that there was a very good chance that the ice at this point would break up for the summer.

Suddenly, a new series of options would have presented themselves. The fittest of the men, taking one boat, would be able to press on to Back's River as originally planned, taking the bulk of the victualling supplies with them. Those unable to make that journey would camp at the spot where they had found themselves until the ice broke up. To support them, a group of moderately fit men, who were to remain with the sick at Erebus Bay until the breakup of the ice, were first sent back to the ships (or the landing at Back Bay) to obtain more supplies (assuming the ships were still afloat). The supplies would see the sick through the weeks before the boats could have been launched. They could then launch their boats, and either sail south and eastwards to re-join the advance party or sail south and then west through Coronation Gulf. This latter course would, in turn, open up a number of further alternatives. They could continue on towards the Bering Straits in the hope of meeting rescue ships which were bound to be attempting to find them along that route, or they could ascend one of the rivers which were known to flow northwards to the edge of the North American continent along that coast.

The parties then divided. Between thirty and forty men took up the tow-ropes of their boat-topped sledge and headed south, directly across the neck of the Graham Gore Peninsula. The return party – of about ten to fifteen men – headed back northwards, dragging one of the (possibly cut-down) boat sledges behind them. A similar number, suffering from various degrees of scurvy, established a camp at the Erebus Bay site.

Left to their own devices, the men at Erebus Bay who were capable would have begun to organise themselves. Using the sails of the two boats, they constructed a shelter alongside the boat whose sledge had been taken north by a return party. All the fuel and most of the food was taken from the other boat and brought over to the shelter base and replaced with trade goods and unnecessary personal items ('bundled' naval-fashion in cotton handkerchiefs). With the second boat now much lightened and still on its sledge, it was dragged further along the coast and down to the beach ready to launch (this would have been done over several days and provided a distraction and opportunity for competition that would have appealed to officers and men alike). Despite their desperately weak condition, all they then needed to do would have been to tend to the most seriously ill and keep their morale high. Summer was on its way, supplies from the ships would be plentiful, and – once the ice had broken up – they had the means of escape over the element they knew best.

After an absence of about two weeks the return party arrived – and found a scene of horror.

Sometime after the lead party had left, a group of Netsilik Inuit including women and boys appeared approaching the site. At first, just for a very short time, hearts leaped at the thought that help was at hand. But their hopes were instantly dashed as the strangers raced in amongst them slashing, stabbing, and spearing. Few could defend themselves, but two men grabbed shotguns and ran to one of the boats, fending off the Inuit as they approached. Before long, the two men by the boat were the only survivors as the attackers pillaged the bodies before hacking at the limbs of the fallen. The slaughter, which echoed, and exceeded, the massacre of the Inuit by the Copper Indians at 'Bloody Falls' almost eighty years earlier was only the beginning. After the women had looted what they could carry from the site, the Inuit could not help but notice the sledge tracks heading towards the south-east – and set off to follow, already elated by their success.

Behind them, the two men by the boat stayed at their post. Gradually weakening over the next few days, they propped their loaded weapons against the boat's side, climbed on board and snuggled down beneath the furs they had found. They were never to wake again. Their bones were to be the 'sad relics' found by McClintock a decade later.

As the return party dropped the drag ropes of their sledge and stumbled towards the first boat they found ample evidence of a massacre. Broken and mutilated bodies lay scattered on the snow-dusted tundra. Several were piled together at the entrance to a canvas shelter and others lay huddled inside. More bodies were sprawled in the direction of the beach as if they had been killed during a desperate attempt to flee. A further search found more damning evidence of the carnage when the second boat was reached. No-one remained alive of the party left at Erebus Bay.

The return party was left with a single choice – to go south and try to catch up with the advance party, or to return to the north. There would have been no possibility of them hauling a sledge-mounted boat with so few numbers.

In all probability the small group chose to return northwards. From there, fortified with the ample supplies remaining onboard or at the landing, they would have slowly hauled one of the remaining ship's boat northwards across the ice until open water was provided by the summer's thaw. Taking advantage of the northerly current through Franklin Strait, they entered Peel Inlet and sailed northwards in the hope of meeting rescue ships that had entered the Arctic through Lancaster Sound, or even meeting up with whaling vessels that hunted their prey in the open waters south of Devon Island. Upon reaching Back Bay on the northeast coast of Prince of Wales Island, they were forced ashore by the freezing of the water. With the boat hauled up the beach, and using their sail as a tent, they waited for the sea to freeze enough to take their weight whilst they broke up the boat to make a sledge. The boat nails were removed and either hammered through the soles of their boots to gain a grip on the ice or kept as trade items for the natives. When the time came for them to take to the ice, they dragged their sledge out of Back Bay and into Peel Sound. Their bones now rest on the shores of Prince of Wales or Somerset Islands, or beneath the cold waters of the Sound.

The lead party set up a similar camp to the one at Erebus Bay on the northeast shore of a deep, island-scattered, bay (officially 'Terror Bay' after 1910). The exhausted, the dying, the despondent, were left in tents around a fireplace as the remainder of the party – now probably reduced to no more than thirty men (Inuit reports of 'thirty to forty men' are probably exaggerated) – set off along the southern shore of the island in search of a place where they could cross over to the mainland. Not long after their departure, the Netsilik descended upon the camp, killing all who remained, with the bodies piled up in the tents.

The hauling and pushing of the last of the ship's boats eastwards continued to take a heavy toll of the remaining men. Some simply dropped where they were and were given a rudimentary burial beneath a pile of rocks, others, wracked with pain, disoriented, and disheartened, wandered away from the main group and lay down, their bodies left to wild animals and the elements.

After a journey of about seventy-five miles, a small island (Todd Island) was seen off the shore. Crossing the ice, the party rested on the rocky islet. Yet several of the men, perhaps as many as eight or nine, succumbed to fatigue. They had given their all, and the time had come to accept their fate. Again, the bodies were covered with rocks before the survivors – probably no more than fifteen men – headed south towards the beckoning thin brown line some twelve miles away on the horizon.

Fortunately, the ice at the crossing point was generally smooth, greatly easing the passage of the boat-laden sledge. Consequently, instead of heading towards the first point of land (Point Richardson), they decided to enter a narrow inlet which took them a mile or so further south. There they found a low beach and dragged the boat clear of the ice (the site later became known as 'Starvation Cove'.

The leader walked up the gently sloping shingle only to be appalled by what he saw. To the east, south, and west, the land lay flat, barren, and cheerless. Behind him, the party on the beach had been reduced to such a low number of men capable of continuing with the hauling that the idea was completely beyond consideration. At least a third of the men had reached the latter stages of scurvy. There was no alternative other than to leave those incapable of further travel behind, whilst those who could still walk would attempt to reach Back's Great Fish River. The boat was overturned to provide a shelter. Guns, ammunition, and fishing lines were left behind as the temperature was beginning to rise, and foxes, hares, birds, and even caribou might appear within range.

After walking about six miles towards the east, they experienced a moment of hopeful anticipation when they came across a frozen passage leading to the south. Taking the ice to be Chantry Inlet, they continued along its western edge until they reached its southern rim where their expectancy reached another height on the discovery of a river. It was, however, little more than a steam, and clearly not the mighty river descended by Commander George Back in 1834 (the inlet was later 'Barrow Inlet', and the river 'Squirrel River').

After a rest, they continued on in a south-easterly direction until, after about seventeen miles, they came to Chantry Inlet's wide channel. Already, clear water could be seen on its eastern side, but the western half remained frozen enough for them to cross over on to Montreal Island. It was widely known that Back had left a cache of supplies on the island and, with the snow clearing off much of the open ground, they probably hoped that some of the supplies still remained. However, the site had been visited in 1839 by a Hudson Bay Company expedition led by Peter Dease and Thomas Simpson. As in the later case of Dr John Rae, neither they, nor their employers, had any interest in encouraging others outside the company to explore the area. The cache was removed and not replaced. This action may have been reflected in the 1855 Anderson and Stewart expedition when the cache was neither searched for, nor acknowledged. The Anderson and Stewart visit, it needs to be remembered, was the Hudson Bay Company expedition requested by the Admiralty that took no Inuit interpreter, had poorly constructed canoes, had some men who saw ship's masts appearing out of the water but did not tell their leaders, and had leaders who detested each other. Their claim to have found a piece of wood with the name *Erebus* inscribed on it, along with another piece marked with the name 'Mr Stanley' (who was a surgeon in the *Erebus*), should be viewed with great caution.

Fourteen years later, the explorer Charles Francis Hall wrote to his chief sponsor, Henry Grinnell: 'The result of my sledge journey to King William's Land may be summed up thus: None of Sir John Franklin's companions ever reached or died on Montreal Island.'

Franklin's men probably did reach Montreal Island – but finding that the cache had been obliterated, and with the small chance that there was early season Hudson Bay Company traffic on the river to the south, they set off immediately – only to walk into oblivion.

And what of the *Erebus*? The idea that some of Franklin's men would have returned to the ship and sailed her south into Queen Maude Gulf is absurd. Both the *Erebus* and the *Terror* had ship's companies of over sixty men – because they *needed* over sixty men. When 'All hands on deck!' was sounded on the Boatswain's pipe, everyone from the captain to the cook had duties to perform. The basic requirement of re-siting the ship's massive rudder was an enormous 'evolution' directed by men of great experience. Even if such a task had been achieved, there would still have been the want of a Captain of the Foretop, a Captain of the Maintop, Upper Yardmen, Quartermasters, leadsmen, and a navigator just to take the ship through uncharted waters.

When the ice of Victoria Strait eventually began to break up, there was every possibility that the ship, locked into an ice floe, began to drift south. It need not have taken a long time for the drifting ice to reach the southern end of the strait. Urged on by wind-blown ice passing down Peel Sound to be joined by that from the McClintock Channel entering Victoria Strait, ice floes would have been crowded into streams of broken and gradually rotting ice jostling to break out into the waters of Queen Maude Gulf. Some – particularly those passing to the west of The Royal Geographical Society Island – would have been forced to the west by currents emerging from the Simpson Strait, whilst others could be influenced by counter-currents driving to the east. Some, however, after passing through the narrows of Alexandra Strait, continued south, passing to the west of the scatter of small islands lining the western approaches of Wilmot and Crampton Bay. Trapped once again by the ice of approaching winter, the *Erebus* spent one – and possibly more – years before the support of the ice disappeared, the flimsy hatches of the main deck gave way, and the ship slowly descended the six fathoms to the floor of the bay, where she sat upright just as the *Investigator* and the *Breadalbane* did in their later searches for Franklin's ships.

By whatever means the ship arrived in the Queen Maude Gulf, whether carried on an ice-flow, or drifting in open waters, there remains a large number of questions that have been ignored or allowed to fade beyond the scope of enquiry. The following are raised in no particular order.

1. Sonar – including 'side scan sonar' – works on the principle of transmitting a sound signal which strikes a solid object, and then returns to a receiver. It will not work, however, if the object is not solid, in which case the signal is absorbed rather than reflected and no return signal is received. A good example of a non-solid sound target is seaweed – in particular large kelp forests. From the subsequent diving photographs taken of the *Erebus* it is clear that the ship – particularly the upper deck – is covered in seaweed, including kelp. Yet the discovery of the vessel is shown on a laptop screen to be entirely clear of any indistinct or hazy areas covered by the kelp.

2. The *Erebus* was fitted with iron bow-plates to add additional strength to the bows. There is, however, no sign of this modification. Furthermore, although *The Times* reported that the ship was not fitted out with copper sheathing on her hull, both Richard Cyriax and A. G. E. Jones concluded that the reason for the non-provision of the copper plating – no threat from hull-damaging organisms in the cold Arctic waters – was wholly invalid. Not only do barnacles exist in the Arctic but if the expedition was to be successful, it would have to return through the Pacific where the main threat, the teredo worm (or 'shipworm') was prolific. It was, therefore, considered that during the two weeks between *The Times* report and the ships sailing, both the *Erebus* and the *Terror* were 'copper-bottomed'. Such a decision seems to be confirmed by the discovery of copper sheets (or 'strips') and nails discovered at the Erebus Bay boat site and the 'piece of sheet copper' on Montreal Island – all noted by McClintock. Furthermore, Schwatka's second-in-command,

William Gilda, mistakenly assuming that the boats dragged across King William Island were copper-bottomed, in discussing the boat found at Starvation Cove, noted that 'most of the kettles that we saw in use among the Netchilliks were made of sheet copper that they said came from this and the other boats in Erebus Bay. The sheet copper was reserve supplies intended for the ships, and just being carried inside the boats.

3. According to the senior underwater archaeologist from Parks Canada, as he conducted a filmed underwater tour of the *Erebus*, the presence of cannons in the ship was as a result of a homeward voyage where there might be enemies of Great Britain to be encountered. This, despite Franklin's orders saying that 'Should Great Britain become involved in hostilities with any other power before the conclusion of the voyage, he is to maintain a strict neutrality.' There was certainly no role for cannon when traversing the North-West Passage, other than as saluting guns, or as signalling guns – neither duty being suitable for a 6-pdr cannon (or, according to the underwater archaeologist, the 12-pdr cannon also onboard). Franklin's ships' first port of call on emerging into the Pacific would have been the southern tip of Vancouver Island (soon to be the Hudson Bay Company's 'Fort Victoria'). Naval activity in the area was increasing with Royal Naval ships expected to be on station, a certainty which led to the establishment of a large naval dockyard at Esquimalt just a few years later. Under such circumstances, it would not have been difficult to obtain cannon, ammunition, and gunpowder – even from the Hudson Bay Company. Neither of Franklin's ships carried a Gunner to train gun-crews, or to maintain the equipment associated with operating the guns (the Armourers carried on board were borne on the ship's books for the maintenance of small arms, and as blacksmiths). It is worthy of note that Commander Fitzjames in HMS *Erebus* had served in the gunnery training ship, HMS *Excellent*. Nevertheless, if the hull of the *Erebus* had been badly damaged before being deserted, any cannon would have been thrown overboard in an effort to lighten the ship. There was also the question of the heavy 'truck' gun carriages needed to operate the 6-pdr cannon. The guns and their carriages would not have been easily assembled in a hurry, and yet there is no sign of the carriages in the immediate vicinity of the guns. Finally, for a vessel that had been brought to the highest design and technology level for her mission, why had she been given two outdated guns that had not only been manufactured during the reign of George III, but were made of cast bronze (not 'brass' as the archaeologist's

report says)? The Royal Navy had not employed bronze cannon for decades since the price of cast-iron had dropped dramatically with new manufacturing processes. Furthermore, in time, bronze cannon not only sagged over the length of the barrel and, with the only shot available being iron cannon balls, the damage done to the inside of the barrel soon rendered the guns inefficient. Even if they were intended for use, there is no evidence from the photographs of the guns that they were fitted with the lanyard-operated flintlock 'fire-locks' that had been in use for well over fifty years with the Royal Navy. There might, however, be two possible reasons why bronze cannons were supplied to Franklin's ships. Firstly, they would not interfere with any magnetic observations which were being carried out (the mass of iron of the steam engine could be compensated for by 'swinging' the ship – the vessel was swung around its anchor for 360°, both with the ship on an even keel and with the ship heeling to either side, to allow the compass to be adjusted), and, secondly, that they were provided to Franklin as high status gifts to reward any help from the natives (whether Indian or Inuit), or to be presented to senior Hudson Bay Company officials on Vancouver Island, where the probability of establishing a naval base (Esquimalt) was already being actively considered. The possibility of cannons was first brought to the attention of Franklin scholars following notice of an expedition discovered by Professor Potter in an old newspaper at the end of 2016. In 1935, the 28-year-old Francis Kennedy Pease was searching for Franklin relics. Leaving Fort Churchill (in company with his Irish terrier), he linked up with Derek Graham, an English trapper, and an Inuk named Kubloo. The Inuk told Pease that Inuit talked of having seen 'red-bearded men near Chesterfield Inlet about 50 miles north'. As they crossed the Tha-anne River and trekked up the eastern side of South Henik Lake, Kubloo also told them that 'his grandfather had seen part of Franklin's party in that region and that there was an old camp nearby.' Two days later, on the shore of Yathkyed Lake, they came across a cairn containing relics. The next part of the story is not clear from the newspaper account, but it seems that the party travelled to the shore of Hudson's Bay (nearest point about 152 miles away), built an igloo on the sea ice, only to have the ice break away and float clear of the coast. Three days later, the giant floe returned to the shore, and, with much of their supplies and equipment damaged or lost, they returned to Fort Churchill. During his time in the area, Pease claimed he saw evidence of gold and silver ores and declared that he would return to mine the precious metals and use the income to find the grave of Sir John Franklin. There is, however, no evidence that he achieved this noble ambition. And the relics? Pease told the newspaper that he had found 'The grave of a white man believed to be the last survivor of the Franklin expedition, dated 1851, about 250 miles north of Fort Churchill; and a cairn beneath which was buried the remains of a sea chest, other wood which appears to be from a boat, patches of blue Navy cloth, canvas, old cannon shot and nails'.

The thought of a Franklin expedition member having reached 400 hundred miles south of any other supposed Franklin remains is intriguing, even encouraging. However, to have brought 'old cannon shot' (six-pound minimum, and possibly twelve-pound) south from King William Island without a single rational purpose would not only have been heroic – it would have been preposterous. Professor Potter encapsulates the doubt surrounding much of this story with the considered question – 'Might he have made the whole thing up?'

4. In a sailing ship of a substantial size, the ship's wheel needs to be of such proportions that would allow at least two men to operate the ship's steerage (on many ships, the wheel is doubled so that at least four men could man it). In uncharted waters, and in conditions of hazardous navigation, orders such as 'Hard-a-starboard!' need a rapid response. This would be best achieved by having one man reaching up and heaving down on the wheel, whilst on the other side, another man pushing up – the extended wheel spokes acting as handles having been specifically designed for such a purpose. The photographs of the wrecked *Erebus*'s wheel appear to show it as having approximately the same size as one on an early 20th-century large sailing yacht, such as that onboard King George V's yacht *Britannia*. A ship's wheel of such dimensions on board a vessel such as *Erebus* would be wholly inadequate. Fortunately, history has left an actual photograph of *Erebus*'s ship's wheel behind the ship's Second Lieutenant, Lieutenant Le Vesconte. Not only is it much larger than the one on the wreck, it is also doubled for operation by four men.

5. Once the *Erebus* had been located, an aircraft flew over the site and took a photograph of the ship clearly visible under the waters of the eastern Queen Maude Gulf. As the region experiences large numbers of low altitude flying, why was the ship not seen before?

6. During his filmed tour of the *Erebus*, the Park's Canada underwater archaeologist took the viewer to the stern of the ship. There he pointed out the gap between the stern post and the rudder post. This interesting feature had been designed so that the ship's screw could be raised and lowered as required when bringing the steam-engine into use.

However, the image of the ship obtained by the Remote Operated Vehicle (ROV) and the Autonomous Underwater Vehicle (AUV), also the superb graphics produced by the Hydrographic Service, clearly show that the ship's stern had been completely destroyed – including the stern post and the rudder post.

7. The biggest mystery of all was the finding (and removal) of a bell found on the upper deck of the *Erebus*. The bell – judged by the length of the diver's hand – was somewhere between eight and nine inches in height. There was no immediate evidence nearby of a wooden frame for it to hang on. Subsequent photographs of the inside of the bell revealed that it had no clapper, nor was there any evidence of a means of attaching a clapper. On the outside of the bell were the numerals '1845' and an Admiralty broad arrow (known in the Royal Navy as a 'Pusser's arrow' – from the old naval Civil Branch rank of 'Purser'). The numbers on the bell were later announced to be the date on which the *Erebus* sailed under Franklin – thereby proving the ship to be unquestionably the *Erebus*. However, for centuries, ship's bells – especially Royal Naval ship's bells – have always borne the date of their *launching* – even when having to be replaced through (for example) action damage. In addition, the size of the bell is clearly inappropriate for a vessel of *Erebus*'s dimensions. The bells were not merely used to mark the passage of time and the changing of the watches. They also were used to sound an alarm (fire, for example), and to warn off other ships when enveloped in fog. The sound of the bell had to be able to reach every corner of the ship from the mastheads to the bilges, from the forepeak to the captain's cabin. An interesting aspect of this find is that, in almost every photograph, the bell is placed much closer to the camera than other measurable features (including the underwater archaeologists themselves) – thus making the bell appear to be much bigger than it actually is. From its size, markings, and lack of a clapper, the bell is probably a Royal Naval dockyard bell, sounded by an external striker (usually controlled by a length of rope), and used to mark the start and end of the working day.

Two other issues, connected with, but not directly part of, the discovery of the *Erebus* need to be mentioned. Firstly, there seems to be an obligation in almost every article written or interview given on the subject to praise Inuit folklore. Certainly, local legends and place-names are passed down the generations of Inuits, but in just the same manner as they are in all cultures. An Inuit woman told the viewers of Geiger's film *The Arctic Ghost Ship* that 'Oral history is our science, it's the science of Inuit.

That's how we learn about where to go and get the food, or you may know about the ice conditions in the springtime.'

Describing Inuit oral history as a 'science', does not, however, make it unique, or even rare. Bedouin Arab history also depended strongly upon oral folklore, but it existed within an unyielding framework of poetry that was recited and sung for thousands of years until the beginning of the twentieth century. Not subject to the personal variation of the Inuit storyteller, the Bedouin poems – despite a tendency to glorify individual tribal legends – are a reliable history of that people. Franklin's own English county of Lincolnshire is bursting with tales of 'Black Shuck', a monstrous dog that roams the county; in Lincoln, a visitor could be sent flying by the head of St Hugh rolling down some stone steps; a horse named 'Byard' made a record jump when a witch dug her fingernails into its rump; and an imp, sent by the Devil to cause chaos in Lincoln Cathedral, was turned into stone. All such tales could be classed as 'indigenous folklore' – but no-one demands that they be believed. Franklin's birthplace – 'Spilsby' – means the 'village belonging to Spila', after a Viking chieftain. Much of Europe follows the same pattern in naming places. The use of possession, or indicator names, is not unique to Inuits. It is merely an ancient system that is still evolving today and has been found to be useful around the world.

Most astonishing of all is the Inuit's folk-memory regarding Franklin sites. Their claims include a ship being crushed and lying on its side in Erebus Bay, Hall was told of a ship that 'was crushed in the ice – not very far from Neitch-il-le', an Inuit from near Bellot Strait also told Hall that a ship 'was seen to sink in deep water', whilst another added that the ship 'had been crushed by the ice out in the sea west of King William's Island' (which had already been told – verbatim – to McClintock). McClintock recorded that he learned:

> Two ships had been seen by the natives of King William Island; one of them was seen to sink in deep water, and nothing was obtained from her, a circumstance at which they expressed much regret; but the other was forced on shore by the ice, where they supposed she still remains, but is much broken.

Nevertheless, McClintock had to resort to guesswork to calculate where '…the one was crushed and sunk, and the other driven on shore. But as the natives had not visited the northwest shore since the landing there of the lost crews, it seems tolerably certain that it was off the south-western shore of King William's Island that the abandoned ships were destroyed.'

Incongruously, two sites in Queen Maude Gulf were claimed by the Inuit to be the site of the sinking of one of Franklin's ships. One was indicated to Hall in 1866, the other to Schwatka in 1879, and both appeared on Gould's chart of 1927, and on a later adaptation by William Gibson. Accordingly, much time and effort was spent searching off Kirkwall Island and forty-six miles away to the south, off O'Reilly Island. With cruel unpredictability, the *Erebus* was found almost in the centre of a line between the two points – effectively, nowhere near to where the Inuit's folklore had suggested.

Irony is piled upon irony when it is pointed out that the first suggestion that the area might be the site of one of Franklin's ship came from over three thousand miles away, and 164 years before the discovery of the *Erebus*. On 6 February 1850, Lieutenant Sherard Osborn RN, wrote to Lady Franklin suggesting that two parties should be dispatched to Felix Harbour on the eastern Boothia Peninsula. From there, they would cross the peninsula to the eastern end of the Simpson Strait. One party would then go northwards searching towards the James Ross Strait, the other '...to examine the estuary of the Great Fish River, and thence to proceed westwards along the coast of Simpson's Strait, and, if possible, examine the broad bay formed between it and Dease's Strait.' The 'broad bay', of course, being the Queen Maude Gulf.

Despite the numerous and well scattered locations claimed by the Inuit to be the site of the sinking of Franklin's ships, John Geiger, the CEO of the Royal Canadian Geographical Society, told the *New York Times* magazine in March 2016, that 'The Inuit had always known where one of the boats sank — their oral history is incredibly accurate.'

To counterbalance the inaccuracies in Inuit folklore, it became necessary to claim that Franklin's failure was due to the lack of Inuit consultation by the Royal Navy. Clearly (it was claimed), if only Franklin had sought advice from the local indigenous population, his expedition would have been a resounding success.

The second issue is there is an almost invisible thread that has linked every stage of the discovery, a thread that appears only now and again to remind those involved of the real reason for the highly connected and ambitious search – Canadian national sovereignty. Geiger insisted in an interview with CBC a month before the discovery of the *Erebus* that to find one of Franklin's ships would be 'the foundation of Canada's claims of sovereignty over the Arctic'. But the government was pushing at an open door.

Canadian sovereignty was taken firmly in hand in 1903 by the then Prime Minister, Sir Wilfred Laurier. By 1930, American and Norwegian claims had been quietly dealt with and the entire Arctic Archipelago in Canada's north was generally accepted as being Canadian. The Russians claimed the North Pole and the surrounding ice in 1937. However, the claim was easily dismissed as ice, consisting of water, was treated simply as part of the sea – and no-one can claim water outside coastal limits. In 1969, the Americans sent the super-tanker *Manhattan* through the North-West Passage from east to west with Canadian assistance. After picking up a ceremonial barrel of oil from Prudhoe Bay (named by Franklin after a naval friend), they returned and safely reached New York claiming the first commercial voyage through the Passage. Although the Canadian Government had agreed to the voyage, an outraged Canadian public forced the government to put a stop to foreign vessels using the waterway without Canadian approval.

In 1985, the US informed the Canadian government that it intended to send one of its Coastguard vessels, *Polar Star*, through the Passage – without seeking prior permission. Again, the Canadian government reluctantly agreed against a background of increased public and political outcry. This time, seeing an opportunity to cause friction between the US and Canada, the Soviet Union joined in. As they would not allow open transit through the North-East Passage they would support Canada in its determination to keep the Americans out of the waterway.

The discovery of the *Erebus* sitting on the bottom of the eastern Queen Maude Gulf went a long way to proving Canada's claim of sovereignty over the Passage – especially when supported by 'incredibly accurate' Inuit stories of men having been seen on board, and footprints leading away from the ship.

And yet, it is well known that there is no requirement for the discoverers of the Passage to have done it on board a ship. In fact, McClure, the winner of the prize for being the first to pass through the Passage completed much of the final part as a passenger on board a sledge. All that would have been required would have been for some of Franklin's men to have reached Simpson's Strait for the Passage to have been achieved. There is, of course, plenty of evidence that they did so in 1848. There is also the strong likelihood that the Passage had been achieved even earlier.

The note found at Crozier's Landing states that Lieutenant Graham Gore, Charles Frederick Des Voeux, a Mate (equivalent to the later Sub-Lieutenant) and six men were landed close to Back Bay on the northwest coast of King William Island. What their man objective was is not known, but there would have been a single, overriding requirement that Franklin needed that could only have been achieved by a shore party. By May 1847, the rate of drift of the ships would have been calculated, the condition of the ship's companies would have been known, and the stocks of supplies would have been listed. All that was needed was to learn how far to the south the ice extended; and was the passage unobstructed all the way to the western point of Simpson's Strait? McClintock had no doubts about their objective – in referring to a second note that had been discovered, he wrote:

It was a duplicate of the Point Victory record, and shows that Gore and Des Voeux merely left them under cairns, without adding further particulars at the time of depositing: their attention was probably directed to a more important matter, the completion of their discovery of the North-West Passage.

Dr Richard Cyriax clarified a particular question regarding the completion of the Passage:

To navigate the ships from the Atlantic to the Pacific Ocean was, of course, highly desirable, but not indispensable, because the two extremities of the known regions were connected by continuous channels; whether this proof was obtained in a ship, in a boat, or on foot, was a matter of indifference.

As was later demonstrated by McClure and the ship's company of the *Investigator.*

Both Gore and Des Voeux had wide experience. Gore's grandfather had served with Captain Cook and, when both Cook and his second-in-command died, Lieutenant John Gore brought the expedition ships safely back to England and was promoted to Captain on his return. Graham Gore was signed onto the books of his father's ship at the age of eleven (a legal fiction which assured the boy seniority and status) and entered the Royal Naval College at Portsmouth when he was fourteen years old. As a Midshipman he served in HMS *Albion* at the Battle of Navarino (20 October 1827) and was promoted Lieutenant the following year. In 1836, he was appointed to HMS *Terror* and served in the Arctic under Captain George Back, narrowly escaping disaster when the *Terror* was badly damaged. After serving off the coast of China in the First Opium War and surveying the coasts of Australia, Graham Gore was appointed as First Lieutenant in HMS *Erebus* on 8 March 1845.

Although still in his early twenties, Des Voeux, the son of a clergyman, had served in the Second Syrian War in 1840 and the First Opium War in 1842. During the second conflict he was temporarily attached to the British Army as Naval aide-de-camp to General Sir Hugh Gough. He passed his Lieutenant's examination in May 1844 but would have to wait for two-and-a-half years before he could be promoted. A popular young officer, he had been appointed to the ship through the influence of Commander Fitzjames, with whom he had served on several occasions. McClintock (who knew him only by reputation) was impressed by his 'intelligence, gallantry, and zeal'. Fitzjames considered him to be 'a most unexceptionable, clever, agreeable, light-hearted, obliging young fellow'. That Gore and Des Voeux worked well together may be seen in a note that Fitzjames made off an island on the southwest coast of Greenland as they searched for new aquatic life-forms. He wrote 'Gore and Des Voeux are over the side, poking with nets and long poles, with cigars in their mouths.'

No record exists of their journey, but an account exists of how a similar sledge journey would have been undertaken. George Brown was a seaman in the *Investigator*, firmly trapped in the ice of Prince of Wales Strait under Captain McClure. Sent as part of a sledging party under two officers in search of any evidence of the Franklin Expedition, Brown wrote:

> We were a party of eight with one sledge, and we were away some fourteen days. We six men seized the sledge-ropes, and harnessed, we set off to drag some eight miles a day what contained all our then worldly goods. This consisted of a tent, bedding, cooking materials and food. Night – from 6 p.m. to 6 a.m. was the time chosen for the march, the order of which was the Commander's, some way ahead with his gun over his shoulder, the sledge following in his track. As the time for the halt drew near, the leader would seek some soft snow, and sticking his musket into it, would walk about to keep himself warm, whilst the men pitched the tent. This done, the officer in command, passed to the further end and had one cloth of canvass all to himself; the men two and two, head to feet, the cook at the entrance. All but he, as soon as the tent was pitched, putting themselves into their woollen bags, lay down to rest and sleep till roused to the sound of 'Dinner Oh'. Pemmican and bread dust, or salt pork and ½ gill of rum. This repast was eaten sitting up with our legs in the bags. Commander and men fared alike. To him only was allotted the cold comfort of greater space, and the duty of taking an observation or of making a note of the day's proceedings whilst the men slept.

Starting out on 24 May, Gore's journey to the nearest point on Simpson's Strait on the southwest coast of the island was about 65-70 miles. The only potential barrier was the large 'Collinson's Inlet', but that would have been completely frozen over making the passage very easy. Even a leisurely trek to the Simpson Strait would only have taken a week. Given a day of rest before returning, Gore and Des Voeux would have had ample time to inform Captain Sir John Franklin that his expedition had achieved what it set out to do before his death on 11 June – he had indeed 'forged the last link'. The most westerly point of King William Island, the point overlooking the Simpson Strait, was given the name 'Graham Gore Peninsula'.

Whilst it remains a great achievement to have found HMS *Erebus* beneath the waters of the Queen Maude Gulf, the sovereignty of Canada had been established over the North-West Passage one hundred and sixty-seven years before and needed no political augmentation or ornament. However, there was more to come – an equally impressive achievement lay just over the horizon. On 11 October 2016, CBC News quoted the Franklin scholar Professor Russell Potter: 'It's like it just got more complicated. Somebody dropped a huge bag of evidence that you have to wait to open.'

What had become 'more complicated' was the Franklin story – and the 'huge bag of evidence' was the discovery of HMS *Terror*.

Almost in accordance with an unwritten set of rules, the discovery of Franklin's second ship was not straightforward. On 3 September 2016, the *Martin Bergmann* left Gjoa Haven and set off westwards along the south coast of King William Island. She was on her way to join a Franklin search at the northern end of the Victoria Strait. There she would join the Coastguard's icebreaker *Sir Wilfred Laurier*, and one of the Royal Canadian Navy's coastal defence vessels, HMCS *Shawinigan*.

Whilst passing through the Simpson Strait, one of the vessel's crew, Sammy Kogvik, an Inuk from Gjoa Haven, said that six or seven years earlier he had been hunting with a friend near Terror Bay on the southwest coast of King William Island. Looking out into the bay, they saw what appeared to be the top six feet of a ship's mast. They walked out onto the ice to investigate and took a photograph of each other by the object. On their return to Gjoa Haven however, Kogvik found that the camera had disappeared. Assuming it to have simply fallen from his pocket, an idea supported by a suspicion that 'bad spirits' emanating from Franklin's men were involved in the loss, he told no-one about the incident (his accompanying friend died a year after the discovery). By now, any event occurring in the Franklin search had to be accompanied by accusations of searchers disregarding the inevitable Inuit story that they knew about it all in the first place. The skipper of the *Martin Bergmann* decided to steer towards Terror Bay.

The vessel arrived at about 4am by the light of the early morning sun. At first opportunity the boat's aluminium skiff was put over the side and began to tow a side-scan radar. The elements, however, combined against the search and by midday, winds and choppy waters led to the decision to continue westwards and join the other search vessels. Then, with the skiff recovered and as the vessel pointed her bows to the bay's entrance, a recognisable shape scrolled across the sonar screen. The image took the form of a large ship – complete with projecting bowsprit. A remotely operated camera was sent down, only to become snagged and lost somewhere near the bowsprit (a bizarre echo of Kogvik's lost camera).

Undeterred, the captain and the Operations Director, Adrian Schimnowski (a man Jim Balsillie described as a 'northern Swiss Army knife') decided to press on with the investigation. There was, however, a degree of caution to be observed. After the events of 2014, with the laurels being grabbed by the Government and the Royal Canadian Geographical Society, the facts of the discovery were to be firmly established before the news was spread. Eventually, on Monday, 12 September, an article – coincidentally written by the same journalist who resigned from another national newspaper over the RCGS's manoeuvrings – appeared in *The Guardian*. It said not only had the *Terror* been found but a remotely operated vehicle had entered the ship through an open hatch and photographed plates, wine bottles, and furniture. The ship appeared to be in a remarkable state of preservation with some hatches open and a number of the stern windows broken. Schimnowski explained, 'It seems like everything was battened down for the winter, everything was shut down as fast as possible, and everything seems to be in place.'

The Parks Canada reaction was muted. Their press release described the find as a 'potential' discovery and remarked that 'The discovery of HMS *Terror* would be important for Inuit communities and Canada.'

For the *Terror* to have arrived at Terror Bay a hundred miles from the last point where she was locked in the ice, and subjected to the whims of wind and tide, there is no requirement for her to have drifted in company with the *Erebus*. What is required, however, is that the ice should have broken up. If the *Terror* proved to be generally sound in her hull, she could have returned to the water without damage and drifted south in the general flow of the broken and rotting ice. On the other hand, if like the *Erebus* she remained in the grip of an ice flow, there still remains only one direction in which she was likely to have travelled – south.

However, when multi-year ice breaks up, especially when the ice comes from two massive sources such as Peel and McClintock Sounds, the floes would generally crowd together driven by the wind and currents. As the *Terror* survived as far as the southern end of Alexandra Strait, she would have found herself in an even worse confusion of ice with the southern-bound ice colliding with the west-bound floes passing through

Simpson Strait, and the surface turbulence caused by the collision of the two prevailing currents. The resultant counter-currents and pressure from jostling ice-floes could have forced the *Terror* the dozen or so miles to reach the entrance to Terror Bay where the predominant easterly current edged her out of the main stream and into the quieter waters of the bay itself. The movement of ice on water does not have the stability of a fleet at sea. During our visit in 1993, we had seen the crowding and buffeting of ice floes that had edged large numbers of them out of Peel Sound and into Back Bay on Prince of Wales Island.

It is highly improbable that after such a buffeting, whether as a freely floating vessel or – more likely – one still in the grip of an ice-floe, her watertight integrity would have remained intact. Once free of any supporting ice, she would have begun to settle and slowly sink to the seabed seventy-eight feet (thirteen fathoms) below. As she did so, the air remaining trapped below decks would have applied pressure to the upper deck hatches, which would eventually have been forced open. At the same time, and for the same reason, several of the stern windows would have blown out.

Such an unmanned Arctic voyage would not have been unique. There are two well-known examples. In 1931, the SS *Baychimo* was abandoned in the Beaufort Sea and eventually drifted into oblivion. The entire distance of her drift is unknown, but she was seen off Point Barrow and not again until she was located five months later near Herschel Island of the coast of Yukon – apparently driven there by strong westerly winds. During that time she had drifted over four hundred miles against the prevailing current. Most famous of all, however, was the Franklin search ship, HMS *Resolute*, whose captain was ordered by Captain Belcher – the senior officer of the search expedition – to abandon his ship, ice-locked in Viscount Melville Sound, in May 1854. To the eventual embarrassment of the Royal Navy, the *Resolute* appeared in Baffin's Bay sixteen months later – 1,200 miles from where she was abandoned. The embarrassment was increased when the Americans restored the ship, sailed her across the Atlantic and presented her to Queen Victoria. When the ship was eventually decommissioned, the Queen ordered that three items of furniture should be made from her timbers – a desk for the President of the US, which has seen much service over the years in the Oval Office of the White House; a smaller version to be presented to the widow of the American philanthropist, Henry Grinnell, who had sponsored several of the Franklin searches, and an even smaller desk for the Queen herself.

Within days of the announcement of the discovery, Parks Canada underwater archaeologists were at work. Much of their observations agreed with the high probability that the ship had no-one on board during her southwards drift. Neither the rudder nor the propeller was in position (the archaeologists assumed that the rudder 'could be on the seafloor covered with silt'), and all four anchors were secured for passage. However, the capstan had been displaced, still carrying a heavy rope which appeared to trail over the ship's side. This had probably been employed for an ice-anchor before she was deserted, and, at some stage during the drift the capstan sustained damage as the ice broke away. The ship's bell was found but not removed (nor its inscription recorded), and a single, unmounted, undescribed cannon was seen.

The discovery was a great achievement and deserving of every accolade. Nevertheless, as is common in Franklin-related events, there remain a few mysteries. Both Hobson and McClintock's combined journeys along the south and west coasts of King William Island failed to spot any sign of a vessel drifting southwards. It must, therefore, be assumed that the *Terror*'s voyage took place before their visit. McClintock and other subsequent searchers passed by Terror Bay and failed to spot the masts – or mast – of the ship rising above the waters. Would this have been possible? On the other hand, Kogvik seems to have been spectacularly fortunate. On his arrival at Terror Bay, the *Terror*'s mast had been standing upright for about a hundred and sixty years, yet within six years of his taking a photograph, it had vanished. In the underwater photographs of the wreck, no fallen masts can be seen.

In fact, very little flotsam appears to have been recovered from around the shores of Terror Bay. This may, in part, be the result of the ship having been well secured when she was deserted by her ship's company. On the other hand, the key reasons given why there has been no evidence found of the supposed cannibal's camp site at Terror Bay is that either the sea has encroached onto the shore and has washed all the proof away, or that it was destroyed by the movement of ice. There remains, nevertheless, a singular oddity that has a link with the *Erebus*. Just as a heavy iron part of the ship's davit managed to detach itself from that ship, and made its way unaided along the sea floor and up a sloping beach, so in 1931 the battered remains of what might have been been a metal ship's water tank was found on the shore of Terror Bay. The artefact was promptly declared by the press to be 'A relic of the Franklin expedition'.

Finally, inevitably, the discovery of the *Terror* was made not with sonar, not by accident, but by the Inuit. The praise and admiration for the Inuit was unstinting. On 12 September Schimnowski told *The Star* 'It's important the community is involved. After all, Inuit have been telling stories of the ship's final resting place for generations.' Parks Canada's first news release (14 September) although rather muted regarding the actual find, said 'This extraordinary find underscores the importance of Inuit knowledge and would make a significant contribution to completing the fascinating story of the lost Franklin Expedition.'

On the same day, *MacLean's Magazine* reported that Schimnowski had muddied the waters by admitting that – regarding Terror Bay – he had been told by the Inuit '...during the spring when the ice recedes and the sun is setting, you can see the silhouette of a masted ship in the water.'

Later, the same article carried a comment by a well-known Inuit historian and Franklin searcher who claimed that Parks Canada had been told earlier where the *Terror* would be found: 'I don't think they really took it seriously. There's a lot of modern information of a ship being seen there, under the water, from hunters and also from airplanes.' The following day, Schimnowski told CBC News that the Inuit

> ...had seen something there many years ago, and over time, and for some reason, the stories were not told or not listened to. I heard a story four years ago [about] Terror Bay and Parks Canada was there. I thought that was amazing and we should go there, but Parks Canada didn't seem interested at the time.

Surprisingly, he was supported by none other than Kogvik who (also on 15 September) forgetting his story about a lost camera, told the *Nunatsiaq News*, 'I heard a lot of stories about *Terror*, the ship, but I guess Parks Canada don't listen to people. They just ignore Inuit stories about the *Terror* ship.'

This was expanded in the *Nunatsiaq News* on 23 September, when Schimnowski was reported as saying that the Inuit '...had seen something there many years ago, and over time, and for some reason, the stories were not told or not listened to'. Why then did Kogvik not return to take another photograph?

Three days later, with the find confirmed, the Parks Canada issued a news release designed to raise the tempo:

> The essential role played by Inuit in the search for HMS *Erebus* and HMS *Terror* underscores the importance of Inuit knowledge that led to these amazing discoveries... The discovery of the long-lost Franklin ships has generated a lot of positive attention on Canada's northern legacy and will provide a unique opportunity for economic development, increased tourism and other long-term benefits for Nunavut.

The big guns were brought up in the same release. The Minister of Environment and Climate Change and Minister responsible for Parks

Canada joined in with 'The essential role played by Inuit in the search for HMS *Erebus* and HMS *Terror* underscores the importance of Inuit knowledge that led to these amazing discoveries.' The Government of Nunavut's Minister of Culture and Heritage said 'I would like to congratulate our partners and the community of Gjoa Haven and the Kitikmeot region as they played a vital role in sharing Inuit oral history, making the discoveries possible.' The President of Nunavut Tunngavik Incorporated (an organisation involved in many aspects of Inuit culture and society), added 'I am also very pleased that Inuit traditional knowledge is receiving the attention it deserves for the role it played in leading to the discovery of both ships from the Franklin expedition.'

The underlying cause of all the hyperbole may exist in the programme of the Nunavut Tunngavik Incorporated. One of their key strategies that had begun in 2014 and was resuming in 2016 was to establish devolution for Nunavut. The successful devolution of the Yukon and Northwest Territories had led to the demand that Nunavut's devolution prospects should 'go beyond' the expectations achieved by the other Territories. Clearly, with the imminent possibilities provided by the strong likelihood of oil, mineral, and other resources in the Territory, it was in the interest of the Canadian Government to keep the people of Nunavut on board. Professor Potter posted the following on his blog site, 'Visions of the North':

> All the Inuit I know on King William Island have hoped, over many years, for a find like this, not simply because it would vindicate their ancestors' stories or bring media attention – but because it would bring economic growth, which is so sorely lacking in the North.

If that is the case, why did the Inuit fail to tell those who were searching for the ship? Had they for a long time known where she sank, and were just waiting for the right opportunity? It is known that Terror Bay officially received its name on 30 June 1910, and yet Hall wrote to his sponsor, Henry Grinnell, from Resolute Bay on 20 June 1869 that a tent was found 'a little way inland from the head of Terror Bay', and further mentioned 'a large camping place at the head of Terror Bay'. Even earlier, McClintock names 'Terror Bay' on the map included in his 1859 *Voyage of the Fox* (Fourth edition). So, who gave the place its name – and why?

Whilst recognising that the Inuit should have all the advantages and opportunities of every other Canadian, the Federal Government's perpetual applause sometimes looks patronising.

Possibly the best view of these hardy people was given on 3 February 2015 at a presentation held at the Centre of International Governance Innovation (more usually known as 'CIGI' – pronounced 'See-Gee'). This highly respected organisation was another of Jim Balsillie's numerous foundations, along with the attached Balsillie School of International Affairs. The event was held to present an account of the discovery of the *Erebus* with speakers including Park Canada's chief underwater archaeologist, Marc-Andre Bernier; the Arctic Research Foundation's Director of Operations, Adrian Schimnowski; and – speaking from Nunavut – the Nunavut Government's Director of Heritage, Douglas Stenton. The President of CIGI, Rohinton P. Medhora, opened the event and, during a part of his speech addressed to 'Culture', he told the audience:

> In 1845, there was a deep, sometimes misguided, fascination with how northern people could survive in their harsh environment. Today, we endeavour to maintain the highest degree of respect for Inuit culture, and a shared desire among all Canadians, including our aboriginal peoples, to preserve and celebrate its culture as an integral part of the Canadian identity.

It was a simple statement of the truth; no-one could ask for more.

On the question of sovereignty, there would be very few people in the whole world who would not support the fact that the Canadian archipelago belongs to Canada in its entirety, along with all the natural resources. The North-West Passage, on the other hand, whilst unquestionably Canadian, has provided the people of that country with a golden opportunity to show the rest of the world that they can stand tall in the company of nations. By 2019, the Russians had recognised the opportunities afforded by the opening up of the North-East Passage. New icebreakers were to be launched to keep the Passageway open, a new seaport was being built to support the production of natural gas and would be connected to the railway system. Whilst the North-West Passage would continue to be recognised as a Canadian inland waterway, Canada should show generosity and wisdom by allowing the channel to be open to international shipping on payment of a levy or toll. Not only would such a scheme provide a useful income, it would also provide employment for the people of northern Canada. It would also respect the efforts of many brave men who, over centuries, gallantly sought a route through the archipelago. Both they, and the rest of the world, would be grateful.

There still remains the greatest mystery of all – the grave of Sir John Franklin. When ships were trapped in the ice, early Arctic expeditions occasionally resorted to burials in the ice – only to find that with the summer melt the body tended to float to the surface. Burials ashore were difficult as they required intensive labour to work through the iron-hard permafrost encountered just a few inches below the surface.

Clearly, there were enough fit men available to dig the deep graves of the three Franklin expedition men found on Beechey Island, and little reason to think that there were not enough men who were fit enough to bury Franklin on the nearby northwest coast of King William Island. There is also the question of the '9 officers and 15 men' who had lost their lives 'in the expedition'. The deaths of Franklin, 'the late Commander Gore' and the casualties buried on Beechey Island, reduce the unknown deaths to 7 officers and 12 men. The different figures often prompt comment – usually to no satisfactory conclusion.

The officers were not just seamen with skills limited to seamanship. Several had been trained to carry out scientific observations which were intended to play an important part in the expedition. Franklin's Sailing Instructions told him that he would be carrying

> ...instruments of the latest improvements for making a series of observations on terrestrial magnetism, which are at this time peculiarly desirable, and strongly recommended by the President and the Council of the Royal Society, that the important advantage be derived from observations taken in the North Polar Sea, in co-operation with the observers who are at present carrying on a uniform system at the magnetic observatories established by England in her distant territories, and, through her influence, in other parts of the world...
>
> The only magnetical observations that have been obtained very partially in the Arctic Regions are now a quarter of a century old, and it is known that the phenomena are subject to considerable secular changes ... the passage through the Polar Sea would afford the most important service that now remains to be performed towards the completion of the magnetic survey of the globe...
>
> We direct you, therefore, to place this important branch of science under the immediate charge of Commander Fitzjames; and as several other officers have also received similar instruction at Woolwich, you will therefore cause observations to be made daily on board each of the ships whilst at sea (and when not prevented by weather, and other circumstances) on the magnetic variation, dip and intensity, noting at the time the temperature of the air, and of the sea at the surface, and at different depths; and you will be careful that in harbour and on other favourable occasions those observations shall be attended to, by means of which the influence of the ship's iron on the result obtained to sea may be computed and allowed for.

Weather permitting, these observations would have been carried out on the quarterdeck of the ships whilst at sea, with shore observation sites being established as the opportunities arose. As events turned out, the ill-fortune of being trapped in the ice was also a golden opportunity to carry out research in the vicinity of the Magnetic North Pole some 100 miles to the east. On 1 June 1831, James Ross had been the first to discover the Magnetic North Pole whilst his ship was stuck in the ice on the far side of the Boothian Peninsula. Ross was well aware that the Pole moved. Now Franklin's expedition was not only in the perfect place to measure the rate and direction of the magnetic movement (by 2019, the Pole's position had left Canadian territory and was moving towards Siberia at the rate of 34 miles a year). They also had on board the *Terror* as Ice Master, Thomas Blanky, who had been with Ross at the time of the discovery. It would have taken a relatively short time in the area to calculate the magnetic drift, but dangers from the ice, polar bears, aggressive natives, accidents and illnesses would have always been present. And those in the greatest danger would have been the officers trained in magnetic observations required to stay on site. In the case of a death, it would have been expected that the body would be brought back to the ship, sewn into canvas, and deposited in the lowest part of the vessel to freeze solid whilst decisions were made regarding its later – almost certainly summertime – disposal.

After Gore's breezy note of 'All's Well', it seems unlikely that the deaths would have occurred before the death of Franklin. Crozier would have wanted his ships' companies to avoid a prolonged period of despair and despondency after the death of their popular leader. Ordering Fitzjames to set in motion the unique opportunity of increasing the magnetic observations could have led to such work being the main reason behind the succeeding deaths.

What to do with the body of Franklin? The earliest detailed description of Franklin's grave is claimed to have been given to Hall in 1866 by an Inuk named Su-pung-er. According to Hall, the Inuk described and drew a sketch of a 'vault' in the vicinity of Victory Point which may have been the site of an important burial, or even a place for the storage of Franklin's expedition records.

Richard Cyriax would not entertain the idea that the expedition records were buried. In his 1959 work on 'The Unsolved Problem of the Franklin Expedition records supposedly Buried on King William Island', he dismissed the idea completely. Quoting the words of the explorer, Major Burwash, he wrote 'an army of men might work without exhausting the possibilities of finding relics or records.' Crozier 'was under no obligation of any kind to bury the records he could not take with him. Franklin's (orders) contained no reference whatever to such a possibility, and Crozier had no official precedent to guide him.'

The weakness in Su-pung-er's story (which is always ignored by Franklin researchers) is the word 'vault'. Hall noted in his account, 'Su-pung-er who has been to King Williams land knows just where the sealed record vault (*as I think it is*) really lies.' (Author's italics.)

The comment suggests that Hall had already made his mind up and imposed a domestic American practice onto Su-pung-er's tale. Beginning earlier in the nineteenth century, the Americans began lining graves with a wooden 'rough box' into which the coffin was placed. The extra lining became known as a 'vault' and stone or concrete slabs eventually replaced the wood. Somehow, the Inuk appears to be describing a type of burial practice that would have been wholly unfamiliar to a nineteenth-century Englishman (and to himself), but commonplace to an American who 'thought' it was a grave or a depository for records.

Needless to say, many searches have been made for the 'vault' (including by the author), but nothing has been found. Interestingly enough, however, the rough sketch made by Su-pung-er has been frequently reproduced in the manner of a modern draughtsman replicating the Inuk's work as an American version of a 'vault' – whereas the original (actually, a possibly second attempt by Se-pung-er as the first had been 'defaced') suggests something quite different. The sketch looks like a plan view of an extended rectangle with its corners rounded off. Inside the outline is another representation of a long rectangle with one end drawn straight across, and the other ending as a slightly deflected point. A depiction of a ship, perhaps, or a mound, or a ship inside a mound?

Conjecture as to the whereabouts of Franklin's grave did not loom large in the apparent objectives of Franklin searchers. McClintock did not mention it – possibly out of respect for Lady Franklin. Hall did not survive to offer a possible site of Franklin's burial. Schwatka offered nothing beyond deciding that Franklin was not buried in the Cape Felix area. Peter Bayne – of the 'Bayne Map' controversy – declared an ambition to return to the Arctic to find the grave in 1913. The *Morning Oregonian* reported that the 69-year-old Bayne had 'purchased the old Arctic schooner *Duxbury* and is outfitting her here (in Seattle) for a cruise to Victoria Land, to again search for Sir John Franklin's body, which is buried in a tomb made by his own men.' Nothing came of it.

Francis Pease declared the same in 1935 but failed to pursue any such ambition. In 1967, Henry Larson, the first man to take a ship through the North-West Passage from west to east, wrote:

The greatest mystery, perhaps, is still the whereabout of Sir John Franklin's grave. It had been a lifelong ambition of mine to find the grave of this great man, but this was not to be. I firmly believe, however, that at some future date the grave will be found, for in my opinion this outstanding explorer must have a resting place somewhere on land in the heart of the North-West Passage. No British Navy explorer of that day would have buried his beloved leader at sea, when land was so near. And what about Franklin's records? Where are they, and what story can they tell?

During a stay in Gjoa Haven, the author was stopped in the street by an Inuk who (with a broad grin) offered to take him to Franklin's grave for $20,000. The offer was courteously rejected.

In September 2014, the Inuit historian, the late Louie Kamookak, told *The Star* that 'One group of Inuit said they saw a burial of a great chief under the ground, under stone. They said he was a great shaman who turned to stone.' He then added that another group of Inuit had come across 'a large wooden structure... They managed to get a cross piece they took for a sled. The man who was telling the story said there was a flat stone and he could tell the stone was hollow.'

With no discernible evidence from Kamookak's account, *The Star* concluded that 'If he's right, Franklin is probably still lying beneath the tundra on King William Island's rocky and windswept *northeast* coast.' (Author's italics.)

In Dorothy Harley Eber's 2008 book, *Encounters on the Passage – Inuit meets the Explorers*, the author has a most enlightening interview with Jimmy Qirqut, a native of Gjoa Haven. From him she learned that his father told him:

They buried their leader, the captain, *on a hill* – on the rocks – somewhere on the northwest of King William Island. They gave him a proper burial. They buried him with respect. There was a proper burial in the north of King William Island. Or somewhere else. Nobody knows exactly where. A lot of people went looking for him, but nobody ever found him. In the same place, they packed up his logbook or some papers. They wrapped them up properly so they wouldn't get wet or damaged by the weather, so that those who found him would be able to understand exactly what had happened. They probably didn't leave them on the ground, so probably they covered them with rocks. We heard they buried the captain on a hill – *a long narrow hill*. So if people were looking in the right place, they would probably find him. If he was a captain, if he was buried properly, there should be signs of something – the ground stirred round, rocks as a marker. We heard they buried the captain carefully. That's the way we heard it. We heard he was buried on *a long narrow hill*. The people who went looking for him probably were not looking in the right places. (Author's italics.)

In July 1879 Schwatka's expedition reached the northwest coast of King William Island north of Wall Bay – an area that was described by William Gilda, the second-in-command and recorder of the expedition, as 'the abomination of desolation'. It had clear evidence of considerable activity by Franklin's men who would have been delighted to take part in a 'run ashore' away from the ships. At the midway point between Wall Bay and Cape Felix (just over three miles from both), Schwatka came across the camp first discovered by Lieutenant Hobson twenty years earlier. Hobson described the site in great detail despite searching in poor weather conditions. After finding nothing of interest in or around a tall cairn, he and his party checked the tented area. There remained an astonishing amount of debris, cast off clothing, and equipment. The three tents themselves had been collapsed, and their poles removed, but inside were blankets, clothes, bear skins, and small items ranging from a packet of needles to parts of telescopes. 'Rubbish' lay strewn everywhere outside – broken crockery, including blue and white china, cooking equipment, canvas, yarn, iron hoops used to cooper caskets, three iron heads of boarding pikes, and even a Royal Marine's shako badge.

The camp had obviously been abandoned in a hurry and had probably occurred during the first weeks of April 1848. A small party of ten to twelve men (Hobson's approximation) had been sent ashore to set up the camp in preparation for the next summer's activities. Hobson thought the camp was 'an observatory or shooting station', McClintock judged it to be a 'shooting, or magnetic station'. On examining the cairn at the site, Hobson thought the original height of the structure 'could not have been less than 8 feet high and was 9 feet in diameter at its base'. As nothing of importance was found within the cairn, it probably means that it was intended as the support for a flagpole from which flew an ensign to help locate the camp from a distance. Such an aid would have been of great use to a magnetic observation party setting out to, or returning from, working on the Boothia Peninsula a few miles to the east.

Little had changed at the site when Schwatka arrived. He did, however, find two more large cairns – one inland from Cape Felix and west of the first camp, which proved to be empty. Schwatka thought the cairn 'had been erected in the pursuit of the scientific work of the expedition, or that it had been used in alignment with some other object to watch the drift of the ships'. Whilst returning south, another cairn was encountered just to the north of Wall Bay. This cairn contained 'a piece of paper with a carefully drawn hand upon it, the index finger pointing at the time in a southerly direction.' This item could have been left by a magnetic observation party acting upon a decision to proceed in that direction.

As for the camp itself, little now remains other than a scattering of rocks assumed to be the base of the large cairn. The size of the camp in the summer of 1847 cannot be used to indicate the size of the camp in April 1848. During the summer following the death of Franklin, the number of men ashore could have been very much larger, having been sent there for a very specific purpose.

There is a very faint chance that a clue to Franklin's grave may have been found in the papers discovered by McClintock on the skeleton he found lying face down on the south coast of King William Island. The remains are usually claimed to be those of Petty Officer Henry Peglar, the Captain of the Foretop in HMS *Terror*, but other researchers have found at least two other possible claimants.

From within the remnants of the clothing, McClintock recovered a wallet containing papers, all of which had suffered from long exposure to the elements. Written on one paper was a series of lines with most of the words written backwards. To add to the difficulty in deciphering the contents, the paper had deteriorated to such a degree that many of the words and letters had been obliterated. Much time and effort has been spent over the years to clarify the script, and a recent combined attempt by Professor Potter and the late William Battersby, an author whose works included a biography of Captain James Fitzjames, is as follows:

O Death wheare is thy sting
The Grave at comfort cove
For who has any douat how
Nelson (?) look
The Dyer was and whare Traffalegar
as .s.. Of him
and ... to .. frends a. Laitor. a. Cors. (?)
Best
and w.. ... addam and eve
a Nother
Death ... right hands
... new (?) grave
I ..ham ... to(?) will be a veray
signed ... me yes and a splended
And
That [m]akes trade Florrish
That way the world
... round
Florrsih

The work of Professor Potter and Battersby has been used as the basis of this author's attempted explanation:

O, Death, where is thy sting? (1)
At the grave at Comfort Cove. (2)
For who has any doubt how
like Nelson the dyer was (3)
when asked about Trafalgar, (4)
and was to all a friend,
and later a corpse at rest?
And, would you Adam and Eve it, (5)
another death.
Might the hands dig a new grave (6)
for Hammond? (7)
It will be very well signed by me, (8)
Yes, and splendid.
That makes tradition flourish. (9)
That's the way the world goes round. (9)
FIN. (11)

1. A rhetorical question. 2. Answer – In this case, 'Comfort Cove' is probably a Victorian naval euphemism for a communal grave. 3, 'Dyer' – the person who has died. 4. Franklin was at the Battle of Trafalgar, 5 'Adam and Eve it' is rhyming slang for 'believe it?'. 6. In this case the 'hands' probably mean the men – as in 'All hands on deck!' (7) Possibly John Hammond, a Royal Marine serving in the *Terror*. 8. 'signed', in this case, means to mark the grave in some way. 9. The original 'trad' – for 'tradition' existed in the eighteenth century but is better known from the twentieth century 'Trad Jazz'. It would, of course, have been expected that formality attended any burial. 10. 'That's the way the world goes round' is idiomatic of death and sacrifice. 11. Possibly a popular contraction of 'ad finem' – 'The end'.

There remains one wholly unrecognised item of information regarding Franklin's grave that stems from an extraordinarily weak (indeed, unlikely) source, yet one that provides an interesting result. About a year after Franklin's ships were deserted, Captain Ker, the Master of the whaler *Chieftain* was approached by an unknown Inuk whilst visiting Pond's Bay (modern Pond Inlet). The stranger presented Ker with a piece of paper but further communication was impossible as neither could speak the other's language and no interpreter was available. The paper contained an image of a central series of vertical lines which bulged to the right and was otherwise unidentifiable. However, as much of the Arctic land below the broad seaway that extended from Lancaster Sound to the Arctic Ocean was generally orientated north-south, it was widely assumed that the central image represented one of those land-masses – in particular, the Boothia Peninsula. Consequently, the paper was orientated with the image vertical. The peculiarity of this assumption is made immediate by the paper also bearing images of four three-masted sailing ships which, in the orientation adopted, appear to have their hulls in a vertical position with the masts pointing to the right. Two of the ships were to the right of the paper with a single line separating them from the central image. The other two, with the same orientation, were positioned in the centre of the central image. The foremost of the right-hand ships was connected by a single line to the aftermost of the central ships.

Many of the original interpretations assumed the ships to be those of the John Ross 1829 expedition. Other thought it to be James Clark Ross in command of an 1848 Franklin search expedition with the ships HMS *Enterprise* and *Investigator*. However, it was soon revealed that neither expedition, or any others, could be made to fit the image, and the paper was soon discounted and forgotten about.

It appears that no-one thought to turn the paper 90° to the left. By doing so, it takes on a completely different character. If the image is intended for a searcher approaching from the east (the bottom of the image), it then shows two ships at the top, off a coast (represented by a single, gently curving line), and with their bows to the south. The second two ships are represented as being inside a mound on the shore with a single line connecting them to the offshore ships. Would it be considered absurdly irrational to suggest that the paper is trying to communicate the fact that something connected with the offshore ships is now inside a mound? Even worse, could the paper have been handed over to an Inuit by a survivor of some dreadful event with the desperate plea that he (or she) makes sure that it reaches white people? Pond Inlet is about 500 geographical miles from the northwest coast of King William Island, and about 650 miles by water. Could the paper have reached the *Chieftain* the following summer, only to be dismissed in the same way as the discovery of two mounds on the same coast in 1992 was dismissed?

What then, is the summary of support for the idea that the two mounds I found on the northwest coast of King William Island are the burial place of Franklin and some of his men? The Inuk Se-pung-er's sketch could be interpreted as a mound; the Inuit historian, Louie Kamookak, was convinced that Franklin was buried somewhere on the island; the camp found by Hobson on the northwest coast and re-visited by McClintock and Schwatka confirms non-native activity at and around the site at the right time; the reverse-writing paper attributed to Petty Officer Henry Peglar suggests burials ashore; the 'Chieftain map' indicates ship-related burials in a mound, and Dorothy Harley Eber's conversation with the Inuk, Jimmy Qirqut, repeatedly highlights 'a long narrow hill'.

Was there any clue in the shape of the mounds? Crozier, who took command on the death of Franklin, was an Irishman who would have been well acquainted with the type of 'barrow' or tumuli that are found in Ireland and throughout the British Isles. In such monuments were placed Celtic, Saxon and Viking heroes, warriors, and chieftains, sometimes laid in their wooden longships. Franklin's own Lincolnshire had many of them. During his time at school at the nearby town of Louth, he would have regularly had to pass by an ancient burial mound 184 feet long, 40 feet wide, and over 6 feet high known as 'Spellow Hills' (or the 'Hills of the Slain').

Were the mounds on the northwest coast of King William Island suggested by such barrows? There would certainly be an additional benefit from such an arrangement. If a normal churchyard row of graves, with headstones or crosses, had been adopted, they could have been the target of native despoliation – the local Inuit desperate to obtain the metal, cloth, and wood the graves contained (a people continually living on the edge of survival has to carry out acts that may seem repugnant to more comfortable societies).

If they had been raised by Crozier, why have two mounds? Firstly, it could be a simple case of there being two ships, both with their own dead. Crozier could have ordered one to be raised for the *Terror* and one for the *Erebus*. Such an idea may be supported by sailing instructions which had one ship keeping station on the starboard quarter of the leading vessel (the smaller mound is to the right rear of the larger). On the other hand, he could have had one built for officers and one for ratings and Royal Marine other ranks – an unremarkable arrangement for those times. Finally, he could have had the smaller mound for Franklin with the remaining dead interred in the larger.

Were there any further clues? The site of the mounds is place conveniently midway between Franklin's summer camp and Cape Felix. The Peter Bayne map said 'the place where the men were buried ... was situated on a flat-topped mound' (Burwash). The Inuk who was terrified by the 'black men' when he visited one of the ships claims he was approached by the ship's captain who pointed to a spot ashore and told him that the 'black men' lived there and 'that neither he nor any of his people must go there'. The captain could have been Crozier pointing to the spot ashore where his men were building mounds as a burial site. It would be important to ensure that the Inuit remained unaware of the real purpose behind the work.

It was my assessment of these incidents that made me think that there may have been more (and remains more) in connection with the two large mounds I found south of Cape Felix. Not only did the experts declare that the features were 'natural' – based upon the fact that they consisted of apparently undisturbed soil and rocks – but also that they were too large to have been man-made. I, on the other hand, would contend that any sign of soil disturbance would have been lost after 146 years of settling. In addition, the mounds – if man-made – had suffered from almost a century and a half of repeated long periods of compaction from the weight of a heavy covering of snow. As for the inability of the ship's companies of the *Erebus* and *Terror* to have constructed the mounds, it is only necessary to look at the feat of the men from the naval squadron based on the Brazilian Station in 1844. The paddle steamer, HMS *Gorgon* (with Acting Mate F. L. McClintock RN onboard) ran ashore off Montevideo. Not only did stores, masts and weapons have to be removed to free the ship, but huge water-restraining embankments had to be constructed and, finally, 19,000 tons of wet sand shovelled clear by hand. Their achievement was considered to be 'a monument bearing silent testimony to the unflinching, unconquerable endurance of British sailors'. Could such words be applied to two isolated mounds on the shores of King William Island? Could they be the work of approximately eighty seamen, unemployed and keen to be ashore during the 1847 Arctic summer, in an effort to mark the passing of their ships – something seamen have done throughout the ages, whether Vikings defacing an English village church with a scratched drawing of their longship, or British sailors painting the name of their ship on the sun-baked cliffs of the Persian Gulf? Two ships; two ship-shaped mounds running in a north-south direction close to where the ships were drifting southwards, locked in the ice of Victoria Strait. Could they have been prepared to receive something in the event of a possible tragedy – or the result of a tragedy itself?

Following the 1993 visit, I heard that – after Peter Wadhams' observation at the Royal Geographical Society – one or two comments had been made in Yellowknife and elsewhere that I was the man who had discovered 'some lemming holes'. Having lived on a stoker's messdeck for long periods, I was well armoured against sarcastic remarks. However, what did annoy me was the word getting around that the mounds were nothing more than ordinary 'drumlins'. It did not take much research to convince me that they were nothing like drumlins.

There is no doubt that drumlins are usually made of 'till' – sediments deposited by glaciers – and come in a variety of shapes, but they are generally known for their egg shape with one end higher than the other. They are also known known for their grouping together in 'swarms' numbering hundreds and even thousands. Single, isolated drumlins are rare, and double drumlins hardly more common. They are formed by retreating glaciers depositing the till, a geological mixture, often containing clay and rocks varying in size from sand to boulders. Till also forms ground moraines, again created from glacier till, but known for their level, or gently rolling areas. Much of the northwest of King William Island is covered by ground moraines.

Falling well within naval practice, Crozier would have had working parties sent ashore accommodated in the prepared camp. From their base they could then scrape off the ground moraine till down to the level of the permafrost to build the mounds ('Comfort Cove'?) intended for their deceased shipmates. Instead of indulging in morale-sapping grief and despondency, they would be celebrating the lives and achievements of their companions and mess-mates. In addition, the mounds would have provided a shelter for those records they would not be later carrying with them to the south. Following the subsequent spring melt, the area surrounding the mounds – the area which had provided the till from which the mounds had been constructed – would have created an encircling, moat-like body of shallow water.

When the specialists Professor Savelle (who had been on the Beechey Island disinterment episode with Dr Beattie) and the geologist Art Dyke started work on the southern mound, it was not long before Dyke decided it was 'marine till'. This meant that the mound had been created by the slow rise of the island from below the waters (therefore not a glacial drumlin?). When the top soil and rocks were cleared off, Dyke later informed us that the exposed slabs were 'megaclasts'. These are sedimentary marine rocks formed by a succession of layers of smaller materials which, over time, harden into layered – or stratified – rock. Megaclasts occur through natural events such as frozen ice splitting the rocks to slab-like layers. What I have never been able to accept, however, is the idea that a rock, having split into separate layers, could arrange itself neatly on the top of a grave-shaped, rock-cleared area on the slope of a mound; and then surround the newly created paved area with smaller slabs standing upright.

Then Savelle joined in, lifting the slabs. He had hardly lifted more than a few when he came to the conclusion that the till beneath the slabs was too 'compacted' to have been disturbed by human activity during the last couple of hundred years. It followed, therefore, that there could not have been any grave dug on that site 146 or 147 years ago. On reflection, however, it seemed to me that about a century and a half of prolonged pressure from deep snow on crumbly till barely more than a couple of inches below the surface might have at least a degree of compaction.

Nevertheless, Art picked up a spade and began to dig, followed by Dan Weinstein who, taking another spade, started digging in the centre of the area. In almost an instant, the 'compacted' till proved to be almost as easy to disintegrate as a currant bun. A flash of hope surged through the onlookers until Dykes' spade hit something hard. This time, it really was the end of the show. At about eighteen inches below the surface, the spade had hit permafrost. According to Savelle, again the completely expected circumstance meant that the site had not been used for a grave. This reasoning disregarded the fact that a decade or so earlier, he had passed through six feet of permafrost over the graves of three of Franklin's men who had been buried only two years earlier than the possible use of the mound site.

Whether they were deposited as glacial or marine drumlins, reared as giant frost heaves, even directly by *Manus Domini Dei*, as far as Crozier would have been concerned they were perfect for the burial of his lost companions – even a late death could be accommodated on the feature's flanks overlooking the North-West Passage.

The mounds' potential was dismissed out of hand in a manner which perhaps suggested that there was no interest in finding Franklin's grave in that part of the Arctic. Such a possibility would play no part in the establishment and confirmation of Canadian sovereignty.

Captain Sir John Franklin, Royal Navy, Knight Commander of the Royal Guelphic Order, Knight of the Greek Order of the Redeemer, Fellow of the Royal Geographical Society, and holder of the first Gold Medal to be awarded by the French Société de Géographie, was a hero to his contemporaries, immensely popular with all those with whom he served, with a kindly, approachable, leadership supported by a firm sense of duty. He had served under Nelson at the Battle of Copenhagen, was in the first ship to circumnavigate Australia, survived a shipwreck, fought off the French at the Battle of Pulo Aura, and arrived back in England just in time to take part, under Nelson once again, at the Battle of Trafalgar – all when he was still under twenty years old. In 1814, under Vice Admiral Sir Alexander Cochrane, Franklin led the boats of HMS *Bedford* against the Americans in the Battle of Lake Borgne where all the enemy ships were captured or destroyed. Four years later, as Captain of HMS *Trent,* he took part in the 1818 expedition towards the North Pole, and, between 1819 and 1827 Franklin led two Arctic expeditions overland. After serving as Captain of the 28-gun HMS *Rainbow* he was appointed as Lieutenant-Governor of Van Diemen's Land, at that time a penal colony. His active interest in the welfare of all on the island – including the indigenous peoples as well as the convicts – led to a corrupt administration having him removed. Two years later, Franklin was appointed to command the North-West Passage expedition that was to lead to his demise. Although his career and death was overshadowed by the well-known details of the circumstances surrounding the gallant death of Captain Robert Falcon Scott RN (which served as a more direct and powerful example to all those Britons serving in the First World War), Franklin's name lived on in statues and many geographical sites, and even in school 'houses' (including the school of the author), most famously, perhaps, at King Edward VI Grammar School at Louth where Franklin was educated.

But then the stifling pall of academia fell upon the story in the mid-1980s when the bodies of three of Franklin's men on Beechey Island were disinterred. This event led to the absurd conclusion that the three men had died of lead poisoning, which eventually led to the ludicrous suggestion that *all* of Franklin's expedition probably suffered from the same condition. As a result, a book was published and a television documentary presented the 'lead poisoning' as an enduring fact.

From then on, the Franklin story became the academic fast track to career success, whether by pen, office, or archaeologist's trowel. More books were rushed off to the publishers, a good example of which was one of the very earliest, *The Arctic Grail* by the late Pierre Berton, a holder of a dozen honorary degrees. Referring to Franklin's first overland expedition, Berton wrote:

He was prepared to leave his bride of 17 months, even though he knew she was dying of tuberculosis. It was this reckless ambition, this hunger for fame and promotion that had been Franklin's undoing in that first expedition when, with little preparation and no experience he had set off blindly across the Barren Ground of British North America. It would be his undoing again a quarter of a century in the future, when he and 129 men vanished forever into an unexplored corner of the Arctic.

The fact that his wife pleaded with Franklin to go and had even sewn a Union Flag for him to fly over unexplored lands, is ignored. Furthermore, is 'ambition, a hunger for fame, and for promotion' a bad thing in a naval officer? In exploration, how does anyone prepare and gain experience for something which is unknown?

Ken McGoogan in the *Smithsonian Magazine* of 6 April 2018 wrote, 'Franklin and his crew became martyrs to science, good Christian men who suffered a cruel fate at the hands of Mother Nature.' However, to that reasonable summation of the story, McGoogan added (from his biography of John Rae) that Rae suffered from 'a smear campaign initiated by Lady Jane Franklin, the explorer's scandalized widow, supported by racist writings from the likes of Charles Dickens.' The author is besmirched with a late twentieth-century label that would have meant as much to Dickens as the concept of astronomical black holes would have meant to the nineteenth-century Inuit.

In the introduction to his book, *The Arctic Journals of John Rae* McGoogan refers to an article published in the May 2012 edition of *Literary Review of Canada* written by Professor Adriana Craciun of Boston University in which she regards the Franklin expedition as 'tragic uselessness', and as 'a failed British expedition whose architects sought to demonstrate the superiority of British science over Inuit knowledge'. Professor Craciun returned to the fray in 2017 with 'In Arctic Modernities: Culture and Literature'. In her paper 'The Disaster of Franklin: Victorian Exploration in the 21st Century Arctic', she described Franklin's expedition as 'a Eurocentric disaster predicated on the superiority of British technology and (mis)understandings (*sic*) of the Arctic'. This sounds very much like her earlier comments, but with a choice of words that used to be known in America as 'fog factor'.

The fog factor features in a 1996 paper by Anne Keenleyside and others. Keenleyside is an Associate Professor at Trent University. 'The Final Days of the Franklin Expedition: New Skeletal Evidence' claims that the question of lead poisoning is settled because the lead found in the bones of Franklin's men was 'significantly higher' than that found in 'modern cadaver samples', a 'nineteenth-century Inuit bone', and a bone from a 'nineteenth-century caribou' – none of whom are likely to have dined off pewter plates; eaten food wrapped in lead foil; drank water from lead pipes supplied from a lead water tank; or supplied with lead-laced medicines, and who were not considered at risk by medical professionals immediately at hand. Keenleyside's conclusion is that:

> Elevated lead levels in the remains are consistent with previous measurements (Beattie and Geiger, 1987; Kowal et al., 1989) and support the conclusions of Beattie and colleagues (Beattie and Geiger, 1987; Kowal et al., 1989, 1991) that lead poisoning had greatly debilitated the men by this point.

All this at a time when the lead poisoning irrationality was already being dismissed by experts in the field.

The case presented by the paper begins to fall apart at two points. Firstly, a tooth was used to suggest that one of the victims was aged 12 to 15 years old who had served in one of the ships as a 'cabin boy' – a style of 'rating' which has never been adopted by the Royal Navy. When it was learned that there was no-one on any of the ships who was of that age, the detail was not raised again – despite the fact that an Inuit boy of that age might well have been involved in an attack (the possibility that women might have been involved was not raised until some time after the paper was written). However, the greatest misdirection in the paper is this:

> The most noteworthy aspect of the analysis was the discovery of cut marks on 92 bones, or approximately one quarter of the total number of bones... Most of the affected elements were recovered from the western end of the site, where the densest concentration of bones and artifacts was found. The cut marks, which ranged in length from 2 to 27 mm, were easily distinguished from animal tooth marks by their sharper borders, narrower width, and wider spacing (Ubelaker, 1989: 105). In contrast to cuts made by stone tools, the observed cuts, examined under a scanning electron microscope, exhibited features characteristic of cuts made by metal blades, namely straight edges ... a V-shaped cross section, and a high depth-to-width ratio (Walker and Long, 1977; Walker, 1990).

Associate Professor Keenleyside is right. The bone cuts are a clear indication of edged metal being used. What is missing is the fact that the Netsilik Inuit not only had access to metal, they also used it to drive the neighbouring tribes out of the area (this is often disguised in some accounts that suggest that the retreat was due to hunting failures with resultant starvation). Dr Birket-Smith noted that metal was so rare in the Arctic that one tribe – the Chugach – had an ancient tradition of celebrating the finding of a piece of iron by a man. The lucky finder would lie on the ground in the doorway of his tent with the fragment of iron at his side as other members of the tribe stepped over him in order to gain some of his luck for themselves. The Netsilik, however, had so much metal from John Ross's *Victory* that some parts (including engine parts), still remain in situ. At the time of writing her paper, Keenlyside seems to have been of the opinion that the Netsilik were still a Stone-Age people. If so, they were a Stone-Age people with metal weapons.

Ignoring this fact clearly supports Dr Rae's, Dr Beattie's, and John Geiger's 'cannibalism' theory. So firmly had the presumption become rooted – even before Keenleyside's paper – that in a 1994 CBC interview the much-respected Canadian writer, Margaret Atwood, albeit one with an unknown expertise in Royal Naval Arctic matters, told her viewers that the tragedy was caused by Franklin's 'stupidity' and that he was was 'a kind of a dope'.

In 2015, an Inuk, Jack Anawak, in seeking political office, had strong opinions on a memorial to Franklin and his men. Anawak told his listeners that 'Honouring somebody who's a failure, I don't think is a good idea...

I mean, he failed at first at finding the North-West Passage, and secondly, failed at surviving in the North when he could have survived by using the expertise from the North.'

A section of the British national press joined in. On 28 October 2009 *The Guardian* noted that Inuit oral accounts yielded non-existent 'eyewitness descriptions of starving, exhausted men staggering through the snow without condescending to ask local people how they survived in such a wilderness'.

Perhaps the most depressing academic paper on the subject of Franklin's expedition was published by the Arctic Foundation and written by two Métis (Canadians of indigenous and French descent), one of whom was a professor, the other a holding a Master's degree. The paper was described as a 'longstanding project writing a scholarly edition of Kennedy's Short Narrative'. William Kennedy's 1851-52 search for Franklin (Kennedy was also a Métis), was the subject matter, but the work was written with the aim of comparing incompetent British sailors with Inuit knowledge and experience. The reader is told, for example, that Franklin's expedition, instead of being a search for the North-West Passage for the benefit of mankind, was a 'British Imperial adventure of colonization'. Writing of Kennedys' mission, 'The reason for this successful search (although Kennedy did not find Franklin or his ships) was Indigenous Knowledge.' They also believed that 'Kennedy knew before he set sail in the *Prince Albert* where the ships of the Franklin expedition were located. He had at his disposal the advantages of Cree spirituality, Indigenous Knowledge, and the lived experiences of the North.'

Ignoring the core purpose of exploration, the reader is told that the Franklin expedition '...did not understand the Place (*sic*) in which they found themselves... Kennedy realized that Franklin's search for the Northwest Passage was severely hampered by a lack of Indigenous knowledge. In short, the islands, the waters, and the ice were not on the Admiralty maps.'

The vast majority of the Arctic's islands and waters were, of course, on Admiralty charts. Even the coast near to where the expedition was held in the ice had been surveyed and charted. The Scottish-born Briton, John Rae, was apparently 'vilified in a highly racist fashion (by) the British press'. Once again, a late twentieth-century adjective being applied to a mid-nineteenth-century situation.

The writers also claimed that in 1851 Kennedy called at the Admiralty, 'literally demanding that his ship have pemmican to assist in avoiding scurvy on board.' There is no record, that pemmican has ever been seen as a means of avoiding scurvy. It contained no Vitamin C – any fruit or vegetable matter would have lost its antiscorbutic qualities during processing. In addition, Kennedy himself noted that '*By the liberality of the Admiralty*, we were supplied with a ton and a half of excellent pemmican, which proved invaluable in the extensive winter journeys we were afterwards called upon to undertake.' (Author's italics.) No suggestion of 'literally demanding', or reluctance to provide.

In summary of these interpretations: by consulting the Inuit, Franklin would have succeeded in his enterprise. This notion ignores the fact that Franklin's 1845 expedition was a *maritime* venture using three-masted former bomb-ketches, with reserve propulsion coming from a former railway steam-engine. What advice could the Inuit have given about such vessels and their deployment in Arctic waters? There is no evidence that Franklin, or any of his men, ever encountered any Inuit until their fatal encounter at Erebus Bay. Tales of the Inuit meeting the ship's companies as they made their way southwards are at best second-hand, wholly unsubstantiated, and probably supplied initially by Dr John Rae.

An adjective which is frequently employed to explain the cause of the deaths of Franklin's men is 'unprepared'. An example of this supposed lack of preparation is the outfitting of the men to face the fiercely cold conditions of the Arctic, a process in which the Inuit of the central Arctic were not consulted. Bernadette Dean, who was an Inuktatuk interpreter and a specialist on Inuit clothing appeared on the credits of the film *The Passage*. As far as I am aware, she was not present during the filming at the Admiralty, but ten years later, I came across her name once again. She was part of a delegation of 'highly skilled caribou and sealskin clothing makers' visiting The Smithsonian Institution's National Museum of the American Indian's Cultural Resources Center. I would have liked to have met her during my time on set, as a report on the delegation's visit stated:

> Bernadette Dean ... (was) well aware that non-Inuit could never have survived in the Arctic, one of the most forbidding environments on earth, without the knowledge (her) ancestors had gained over thousands of years of the land, ocean, ice, and sky, and of animal behaviors.

In fact, the bitter chill of the Arctic was no mystery to seamen on the earlier attempts to penetrate the North-West Passage. By the mid-nineteenth century, with the experience of earlier expeditions and the ever-expanding whaling fleet, the rigours to be combated were widely known.

The Franklin expedition took place before the advent of naval uniform for ratings, and the clothing that the seamen wore had to be purchased by the men themselves (a situation that existed – even with the introduction of a standard uniform – to well beyond the mid-twentieth century, by which time a 'Kit Upkeep Allowance' had been introduced). No captain, especially men with experience like Franklin and Crozier, would have allowed his ship to sail without first inspecting the 'cold weather' clothing of his men. Furthermore, if any shortcomings were discovered after sailing, the trade and bartering that existed at Disko Island in Baffin Bay, just off the west coast of Greenland, would have easily supplied the deficiencies. The main settlement, Godhavn, had been a whaling centre since 1773, and had an indigenous population who would have eagerly accept the opportunity of trading native clothing for western goods

No-one ever told the hundreds of seamen and officers who dragged sledges thousands of miles through the Arctic in search of Franklin that they would never survive because they were in the 'wrong rig'. Subsequent expeditions in the region, such as The British Arctic Expedition under Captain Nares, never wore Inuit clothing. Their forced return after reaching the 'Furthest North' was due to an outbreak of scurvy, not defective clothing. Captain Scott's expeditions eventually reached the South Pole, but the fatal outcome was nothing to do with the clothing they wore, which was manufactured in England by Burberry and Wolsey.

The Inuit did not *choose* to use caribou and seal skins. Their only options were to adapt to the environment, move to warmer climes, or become extinct. In reality, as soon as mass-manufactured clothing became available, most of the Inuit abandoned their traditional attire and wore the same clothing as any other outdoor workers in cold weather. This change was brought about in the early 1950s, when the Canadian Government, with northern resources becoming more evident, and with the changing lifestyle of the Inuit, wished to ensure that the northern people had the same opportunities as any other Canadian. It is an extremely rare sight to see any of the Inuit wearing skins from animals. Whilst it is right and proper to admire the skills evolved by Inuit women over centuries in the making of durable, warm, and waterproof clothing from a very limited source, it is wrong to claim that it remained the only effective option for men who went by sea to explore amongst polar ice.

The final ignominy attached to Franklin is 'Imperialist'. As the Lieutenant Governor of Van Diemen's Land, Franklin was appointed to his post having had nothing to do with its creation. What he and his wife did do, however, was to try and improve the life and conditions of both settlers and convicts. They also reached out to the aboriginal people by bringing a young orphan girl into their own home as a companion for their daughter, Eleanor. The girl had originally been named 'Mary', but was known as 'Mathinna' in the Franklin household. When Franklin returned to England, rather than take her away from her cultural roots, he found her a place at an orphanage. Sadly, despite Franklin's worthy intentions, Mathinna found herself trapped between two disparate cultures – one who would not accept her, the other whom she rejected – and she died aged about 21 years. A Tasmanian gold-mining town was named in her honour.

There was nothing about Franklin's attempt to open the North-West Passage that had a hint of 'Empire building'. Most of what later became the Dominion of Canada, including the Arctic Archipelago – was already part of the Empire. The Passage itself was intended to be of advantage to humanity as a whole, an advantage that was to have been bought at the risk of British lives. In the case of Franklin and his men, accusations of 'Imperialism' are entirely dishonourable.

Facts are never allowed to get in the way of opportunities to disparage a tragic enterprise that was born of hope, in favour of a disingenuous pandering to a section of Canada's people who have demonstrated their significance over many centuries, and have no need for condescending manipulation by politicians and academics looking for scandal amongst the bones of long-dead men. The Franklin Expedition needs no dark myths to underline its tragedy. One hundred and twenty-nine British seamen went out on an adventure that would bring little advantage to their country yet could benefit the world. All volunteered in the knowledge that they might not return. At most they would get a 'mitten allowance' and a story to tell their grandchildren.

What sort of people were those who were the targets of this contrived academic hostility? Little remains of the clues which may have been obtained from their personal possessions. But close by their bones were copies, or portions of copies, of *The Vicar of Wakefield*, *Christian Melodies*, *A Manual of Private Devotions*, a Bible, remnants of a New Testament and a Church of England prayer book. They were not angels of course, nor would they have considered themselves to be heroes, but they were men of moral standing who were raised with a firm concept of self-discipline and duty.

From the people who had sent them out on the voyage, came the words carved onto on the Franklin memorial erected at London's Waterloo Place:

TO THE GREAT NAVIGATOR
AND HIS BRAVE COMPANIONS
WHO SACRIFICED THEIR LIVES IN COMPLETING
THE DISCOVERY OF THE NORTH-WEST PASSAGE.
A.D. 1847-8
ERECTED BY THE UNANIMOUS VOTE
OF PARLIAMENT.

A rare and precious discovery was found among the items obtained from the Inuit by Dr John Rae. It was two pages torn from John Todd's 1835 book, *The Student's Manual*. On the first page was printed the words:

We gaze upon the ocean rolling in its mighty waves, and listen to its hoarse voice responding to the spirit of the storm which hangs over it, and we feel an awe, and the emotion of sublimity rises in the soul. So it is with the desires. There is something inexpressively delightful in having the mind filled with a great and noble purpose.

On the second page is the following dialogue:

'Are you not afraid to die?'

'No.'

'No! Why does the uncertainty of another state give you no concern?'

'Because God has said, "Fear not; when thou passeth through the waters I will be with thee: and through the rivers, they shall not overflow thee."'

As *The Boy's Own Paper* noted. 'The poor victim perhaps treasured the page, read and re-read it and gazed on it until the mists of death crept over him. He was not found, but the page told those who were searching how one, at least, of those brave seamen died.'

REQUIEM AETERNAM, REQUIEM IN PACEM,
REQUIEM IN HONOREM.

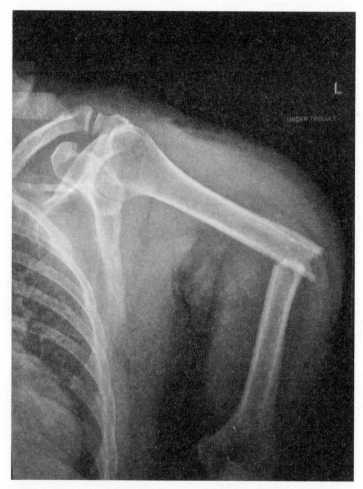

An example of a 'spiral' fracture. (With kind permission of Dean Groves)

Se-pung-er's sketch for Charles Hall. The suggestion of a mound or ship (or a combination of the two) is obvious.

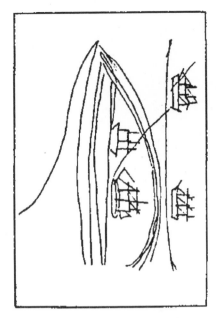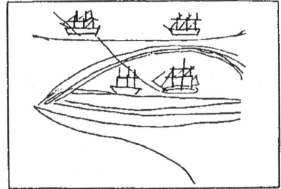

The 'Chieftain map' as originally viewed.   The 'Chieftain map' re-orientated.

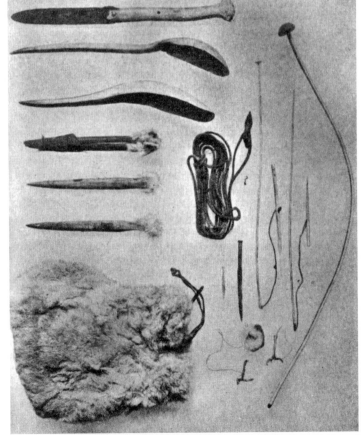

A National Museum, Copenhagen, collection of nineteenth-century Netsilik hunting tools. Almost all are made from stone, bone or antler, the exception being (top left) a snow knife made with a metal blade. Such a tool (or close variations) could easily have been used as a weapon.

Bibliography and Sources

Anonymous *The Great Arctic Mystery*. Chapman and Hall, 1856.

Arctic Rewards and Their Claimants T Hatchard, 1856.

Beattie O. B. and Geiger J. *Frozen in Time: Unlocking the Secrets of the Franklin Expedition*. Saskatoon, Saskatchewan Western Producer Prairie Books. 1978.

Birket-Smith K. *The Eskimos*. Methuen & Co, London, 1959 revised edition.

Canadian Hydrographic Service *Chart 7083 (1980). Cambridge Bay to Shepherd Bay*.

Channel 4 *The Wild West* 1995 Chairman & Committee. *Report of the Select Committee on Preserved Meats (Navy)*.

Cyriax R. J. *Sir John Franklin's Last Arctic Expedition*. Methuen and Co. 1939.

Department of Fisheries and Oceans, Ottawa.

Dickens C. 'The Lost Arctic Voyagers'. *Household Words* November 1854.

Eskimo Evidence and the Franklin Expedition. Unpublished Paper. personal letters to the author.

Farrer K. T. H. 'Whence Came the Lead in Franklin's Crewmen?' *Food Science & Technology Today* 1989.

Hearne S. *A Journey from Prince of Wales's Fort in Hudson's Bay to the Northern Ocean in the Years 1769, 1770, 1771 & 1772*. A Strahan & T Cadell 1795.

House of Commons 3 May 1852 Chairman & Committee. *Report to the Lord Commissioners of the Admiralty on the Cause of the Outbreak of Scurvy in the Recent Arctic Expedition; on the adequacy of the provisions made in the way of Food and Medicine; and on the propriety of the Orders for provisioning the Sledge Parties*. Admiralty Committee on Scurvy 3 March 1877.

Jones A. G. E. *Newly Discovered Human Skeletal Remains From the Last Sir John Franklin Expedition, 1845-48, and the Franklin Graves on Beechey Island*. unpublished paper, May, 1986.

Keenlyside A. 'The Last Resort: Cannibalism in the Arctic.' *The Explorers Journal*, winter 1994/5 King R. *The Franklin Expedition from First to Last*. John Churchill 1855.

Kowal W., Beattie O. B,. Baadsgaard H., Krahn P. M. 'Did Solder Kill Franklin's Men?' *Nature* 25 January 1990 'Arctic Enigma.' *The Lancet* 3

February 1990 'Source Identification of Lead Found in Tissues of Sailors from the Franklin Arctic Expedition of 1845.' *Journal of Archaeological Science* 1991.

Lead and the Last Franklin Expedition. Journal of Archaeological Science 1993 *Lead is Presumed Innocent.* Letter to Food Australia November, 1994.

Letter to Rear-Admiral Richards, Hydrographer to the Admiralty. 25 April 1872.

Lewis-Jones, H. 'Nelsons of Discovery': Notes on the Franklin Monument in Greenwich.

Lloyd-Jones R. 'Ships Companies of HM Ships Erebus and Terror'. Unpublished paper.

Newman P. *Company of Adventurers.*Viking 1985.

Nourse J. E. *Narrative of the Second Arctic Expedition Commanded by Charles Francis Hall.* Washington. Government Printing Office 1879.

Ommanney E. *Diaries kept onboard HMS Assistance 1850-51.* Unpublished.

Parry W. E., Richardson J. 'Letter/Memorandum to Captain Beaufort.' 1854.

Phillips C. 'Preserved Provisions – Food for the Royal Navy in the High Arctic, 1818-1875.' Canadian Collector September/October 1985.

Poulsom Lt Col N. W. and Myres J. A. L. Rear Admiral CB *British Polar Exploration and Research. A Historical and Medallic Record with Biographies 1818–1999.* Savannah Publications, 2000.

Pyke M. *Food Science and Technology.* John Murray 1964.

Rae J. *Voyages and Travels of Dr Rae, in the Arctic Regions.* Letter from Rae to Sir H. Dryden. February 1856.

Ranford B. *In Franklin's Footsteps.* Equinox 1993 'Bones of Contention.' *Equinox* March/April 1994.

Review of Admiral Noel Wright's Books *New Light on Franklin* and *Quest for Franklin* with an Introduction by A. G. E. Jones. Unpublished paper 1990.

Robertson U. A *Mariner's Mealtimes & Other Daily Details of Life On Board a Sailing Warship.* The Unicorn Preservation Society 1979.

Shackleton, Sir Ernest. *South. The Story of Shackleton's 1914-1917 Expedition.* William Heinemann, London, 1922.

Sutherland P D (Editor). *The Franklin Era in Canadian Arctic History 1845-1859.* National Museums of Canada 1985.

Trafton S. J. 'Did Lead Poisoning Contribute to Deaths of Franklin Expedition Members?' *Information North* November 1989.

Wallace H. N. *The Navy, The Company, and Richard King – British Exploration in the Canadian Arctic, 1829-1860.* McGill-Queen's University Press 1980.

White W. *Probable Fate of Sir John Franklin and Crew; or, The Scurvy in the Arctic Seas, and Correspondence of Captain W White with the Lords of the Admiralty, and the Principle Commanding Officers of the Late Arctic Expedition, on its Prevention and Cure.* Piper Brothers and Co. 1852.

Woodman D. C. *Unravelling The Franklin Mystery.* McGill-Queen's University Press.

Index

272, 275, 280, 283, 293, 299, 322, 332, 334

Belcher, Captain Sir Edward   151, 178, 264, 275, 278, 317

Bell, Doctor William   254

Bellot Strait   183, 186, 216, 244, 245, 255, 287, 310

Bellot, Lieutenant Joseph-Rene   162, 165–66, 234

Beluga whales   156, 160, 181

Bernier, Marc-Andre   289–90, 295, 321

Berton, Pierre   73, 78, 87, 334

Bertulli, Margaret   32, 40, 51, 54, 110, 181, 185, 190–92, 196, 202, 220, 222, 227

Birket-Smith, Doctor Kaj   259–60, 336

Blanky, Ice Master Thomas   170, 323

Booth Point   25, 27–33, 219, 252, 254, 259

Boothia Peninsula   16, 49, 89, 113, 183, 186, 203, 244, 255, 276–77, 311, 326, 329

Bouet, Francois   118

Braine, Private William   264

Bray, Donald   36–7

British High Commission   13

Brize Norton, RAF   44

Brown, George   314

Browne Island   154, 156, 164, 172

Browne, Lieutenant W. H. J.   154, 165, 175, 216

Burwash, Major L. T.   41, 60, 93, 277, 323, 330

Calgary   13, 38, 40, 44, 47–8, 50, 54, 106, 108, 124, 188–89, 192, 194, 196, 220, 222

Calgary Light Horse   13, 33, 40, 47, 108, 189

Cambridge Bay   107, 112, 117–18, 120–21, 123–24, 147, 197, 219–20

Canadian Geographic   286

Canadian High Commission   12, 14

Canadian national sovereignty   311

Cape James Ross   30

Cape Jane Franklin   91, 93, 125, 197, 199, 201

Cape Maria Louisa   42–3, 53, 56–7, 70, 82, 99, 123–25, 142–43, 202, 217

Cape Queen Adelaide   16, 18

Cape Riley   267

Cape Walker   162, 166, 170, 179

Cape York   171, 262, 263

Captain Ker, Whaling Master   328

Caribou   18, 23, 25–7, 29–32, 57, 67, 78, 159–60, 163, 214–15, 218–19, 260, 265, 271, 303, 335, 338–39

Carr-Harris, Mike   104, 106, 122

Casarini-Wadhams, Maria Pia   36, 113

Castor and Pollux River   244

Chantry Bay   244

Chantry Inlet   121, 124, 245, 303

Chariot Carriers   192, 194

Chieftain, whaler   328–30, 343

Churchill   117–18

Coleman, Patrick   42

Collinson Inlet   91, 93, 125, 205, 207, 219, 299

Collinson, Captain Richard   14, 55, 93, 120, 221, 243–44, 277

Continental Polar Shelf Project   113

Cooper Lake   198, 204

Coppermine River   14, 221, 236, 253

Coronation Gulf   55, 221, 300

Craciun, Professor Adriana   334–35

Crack, Lieutenant I. D. (Jim)   13, 47

Crozier, Captain Francis R. M.   10–11, 15, 35–6, 77, 80, 170, 189, 193, 216, 232, 249, 279–81, 290, 297, 299, 323, 330, 332–33, 339

Crozier, Captain F. R. M.   10–11, 15, 35–6, 77, 80

Crozier, Rawden   35–36

Crozier's Landing   42, 60–1, 87–97 passim

Culgruff Point   89–90, 94

Curley, Tagak   234–38, 248

Cyriax, Doctor R. J.   37, 170, 233, 248, 260, 267, 277, 279, 305, 313, 323

Dagmar Aaen   152
Daily Telegraph   13, 191
Davidson, Wayne   148, 154, 171, 178, 185, 202, 263
De la Billière, General Sir Peter   226
De La Roquette River   300
Dealy Island   271
Dean, Bernadette   338
Dease, Peter   304
Des Voeux, Mate Charles Frederick   312–34
Diamond, Betty   36–5, 40, 43, 233
Dickens, Charles   235–37, 247-48, 335
Dickens, Gerald   236–37
Dictionary of National Biography   269
Disko Island   339
Drift Calculator   80, 134, 138, 146, 149, 223
Dyke, Arthur ('Art') S.   134, 332–33

Eider ducks   27, 32, 61, 101, 213, 260
Erebus Bay   15, 111, 182, 185, 189–90, 240, 251, 255–57, 261, 273–74, 277–78, 282, 300–02, 305–06, 310, 338
Erebus Medal   295

Farrer, Doctor K. T. H.   265, 270–71
Fisherman's Friend   111
FitzJames, Captain James   91, 249, 258, 279, 290, 306, 314, 322, 323, 327
Fort Crozier   218, 222–24
Fort Simpson   242
Fort Smith   196, 285
Fox   60, 92, 101, 183–84, 187–88, 216, 276
Fram   56
Franklin Point   60, 88, 90, 93, 102, 105, 189, 199–200, 219, 299
Franklin Project   92

Franklin, Jane, Lady   334
Fury Beach   11, 170

Gander, Newfoundland   46
Geiger, John   61, 259, 270, 283–84, 286–87, 289, 291–96, 309, 311, 335–36
Gibson, William   25, 311
Gilder, Colonel W. H.   76, 140, 252
Gilder, William Henry   97, 256, 264
Gjoa   15, 221
Gjoa Haven   15, 18–20, 23–5, 27–9, 32, 40, 121–22, 124, 189, 219–20, 315, 320, 325
Glenbow Museum   48
Godhavn   339
Goldner, Stephen   267–69
Gould, Lieutenant Commander R. T.   80, 139, 146, 277, 280, 311
Graham Gore Peninsula   15, 189, 300–01, 315
Grant Point   276, 280
Great Fish River   30, 35, 124, 169, 184, 244, 245–46, 249, 264, 275, 303, 311
Great Northern Diver (loon/s)   20, 64, 88–9, 94, 101, 133, 213, 260
Great Slave Lake   49
Griffith Island   151–52, 166
Grinnell, Henry   304, 317, 320
Guelphic Order   15, 245, 333
Gulf of Boothia   15, 169
Gutersloh, RAF   44

Hall, Captain C. F.   25, 30, 42, 239–41, 250–51, 255, 261, 274, 276–77, 279, 298, 304, 310–11, 320, 323–24, 342
Hammond, Private John Royal Marines   328
Harley Eber, Dorothy   325, 330
Harper, Stephen   285, 290–91
Harris, Ryan   282, 290–91, 295–96
Hartnell, Able Seaman John   36
Hat Island   288